Engineering Physiology

Karl H. E. Kroemer ·
Hiltrud J. Kroemer ·
Katrin E. Kroemer-Elbert

Engineering Physiology

Bases of Human Factors Engineering/Ergonomics

Fifth Edition

Springer

Karl H. E. Kroemer
Blacksburg, VA, USA

Katrin E. Kroemer-Elbert
Westfield, NJ, USA

Hiltrud J. Kroemer
Blacksburg, VA, USA

ISBN 978-3-030-40629-5   ISBN 978-3-030-40627-1   (eBook)
https://doi.org/10.1007/978-3-030-40627-1

1st and 2nd editions: © Springer Berlin Heidelberg 1986, 1990
3rd edition: © VNR (now Wiley) 1997
4th edition: © Springer Berlin Heidelberg 2010
5th edition: © Springer Nature Switzerland AG 2020

This Springer imprint is published by the registered company Springer Nature Switzerland AG
The registered company address is: Gewerbestrasse 11, 6330 Cham, Switzerland

A Few Words about the 5th Edition

This fifth edition of Engineering Physiology has the same purpose as the earlier prints: To provide physiological information which engineers, designers, managers and many other persons need to make work and equipment "fit the human".

As appropriate, chapters have been revised, figures and tables updated. New material relates especially to:

- Recent experiences with biomechanics and modeling of the body.
- Effects of shift work/sleep loss on body functions, attitude and performance.
- New measurements of body sizes and the resultant changes in applications of that information.

We chose the title *Engineering Physiology* in the 1980s to indicate our treatment of the topic. The book does not take the place of standard (biological-medical-chemical) textbooks on human physiology; instead, it models and describes the human body in terms that provide practical, design-oriented information on essential features and functions.

Even an audacious engineer wouldn't dare to devise a system as complicated as the human body. Yet, what we design, whether simple tools or complex structures, must suit the humans who use them. Therefore, understanding how the human body functions, what it can do with ease (or, worst, barely tolerate) is the basis of human-centered engineering.

This book helps lay the foundations for teamwork among engineers and physiologists, chemists, biologists and physicians. "Bioengineering" topics concern bones and tissues, neural networks, biochemical processes, anthromechanics, biosensors and prosthetics, perception of information and related actions, to mention just a few areas of common interest.

Such understanding provides the underpinnings for devising work tasks, tools, workplaces, vehicles, work-rest schedules, human-machine systems, homes and designed environments so that we humans can work and live safely, efficiently and comfortably. This is the field of *ergonomics* or *human (factors) engineering*, terms often used interchangeably.

About Models

Every chapter starts with a "model":

When developing models we must realize that selecting certain features, drawing distinctions, making classifications usually imposes artificial divisions of our own choosing upon a universe that is, in many ways, all in one piece. We do such modeling because it helps us in our attempted understanding of the intricate system. It breaks down a set of objects and phenomena too complex to be grasped in their entireties into smaller realms that we can deal with one by one. There is nothing objectively "true" about such models; the only proper criterion of their value is their usefulness. (Slightly altered from Isaac Asimov, 1963, The Human Body, p. 13. Signet, New York.)

About References

Basic human physiological characteristics did not change in recent years, though body sizes did. The previous editions of this book (in 2010, 1997, 1990, and 1986) contain listings of earlier publications. In this fifth edition, we are referring mostly to recent publications, along with selected classic references.

Traditional practice was to support statements in the text by listing the names of the authors, and of their co-authors, who wrote previously on that topic. That wordy custom disrupts the flow of reading, especially when there are strings of names and dates. To avoid that problem, we simply place a small marker, * , in the text where references or explanations are desired. These appear, at the end of the chapter, in a separate "Notes" section, which the reader may skip or consult.

We Would Like to Hear From You!

We appreciate your comments, telling us what we did well and what we should do better. You can contact us at kroemer@vt.edu or katrin.ergonomics@gmail.com.

Karl H. E. Kroemer
Hiltrud J. Kroemer
Katrin E. Kroemer-Elbert

Contents

Chapter 1
Skeletal Structures

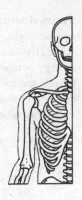

Overview
The skeletal system of the human body contains some 200 skeletal bones, their articulations, and associated connective tissue. These structures all consist of special cells embedded in an extracellular matrix of fibers and a ground substance. Bones provide the structural framework for the body. Ligaments connect the bones at their articulations, while tendons connect muscle with bone. Mobility is provided by the action of muscles across the articulations of bones. In particular, the spinal column is often of great concern to the ergonomist because it is the locus of many overexertion injuries.

The Model
The bones provide the basic solid framework for the body. They serve also as lever arms at which muscles exert torque about the articulations of the bones. These joints permit the linked bones to move through various kinds and amounts of motions. The resulting mobility depends on basic joint design, on support by cartilage, and on constraints by ligaments and muscles.

Introduction
Connective tissues, including bones, cartilage, tendons and ligaments, are composed of cells (mostly fibroblasts) embedded in fibers and ground substance. The extracellular matrix encloses the cells and contains collagen fibers and elastic fibers. Collagen fibers (subdivided into several types) have high tensile strength and resist deformation, particularly stretch; whereas elastic fibers can elongate. The ground substance has large sugar molecules with a protein core (proteoglycans), lipids and water and, in bone, calcium and other minerals. The actual composition of the

© Springer Nature Switzerland AG 2020
K. H. E. Kroemer et al., *Engineering Physiology*,
https://doi.org/10.1007/978-3-030-40627-1_1

tissues determines their physical properties, especially regarding their deformation and strength, which obviously differ considerably between bone and the other connective tissues. Some connective tissues are loose, irregular, even fatty to form spongy wraps for nerves, blood vessels and organs. Dense aligned connective tissues are called *fascia* when they wrap muscles or other organs, *cartilage* at the ends of bones in their joints, *tendons* when they connect muscle with bone, and *ligaments* when they connect bone with bone.

1.1   Bones

One main function of human skeletal bone is to provide the internal framework for the body, see Fig. 1.1; without its support, the entire body would collapse into a heap of soft tissue. Bone also serves as mechanical lever arms at which muscles, attached via tendons, articulate body parts against each other in skeletal joints. (More about muscle pulls at bone levers in Chaps. 2 and 4.) Some bones act as shells to protect body organs: the elastic rib cage contains lungs and heart; the skull, firm in adults, encloses the brain. Long hollow bones also provide room in their cores for bone marrow, which serves as a blood cell factory.

The 200-some bones in the human skeletal system appear in two forms: flat or long. Examples of flat bones, also called axial or appendicular, include the skull, sternum, ribs, and the pelvis. Long bones are essentially cylindrical such as in the arms and legs. The long bones consist of a shaft (diaphysis) which, toward each end, broadens (metaphysis) and develops into a bulbous form (epiphysis) that contains the actual articulation which connects to an adjacent bone.

Flat and long bones both consist of compact cortical and spongy cancellous material. Cortical bone is compact and dense at about 1.3 g/cm^3 or more, while cancellous bone generally is less dense at about 1 g/cm^3. Cortical bone makes up the outer layers of the shaft of a long bone and thin outer shell at the joint. Cancellous bone exists mostly in the middle layers in bone where, by its ability to deform, it provides flexible structural support when transferring loads coming from an adjacent bone or from the external environment. At an articulating joint, the bone surface is covered by smooth, flexible cartilage, which helps to absorb transmitted shocks.

Bone is more compliant in childhood when mineralization is still relatively low: the ratio of the contents of inorganic material, mostly calcium, to organic substance is about 1:1. In contrast, the bones of the elderly are highly mineralized, with a ratio of about 7:1, and therefore are stiffer and more brittle.

Bone develops from a soft, woven-fibered material in the earliest childhood into compact, mostly lamellar material with a hard outer (cortical) shell and a spongy (cancellous) inner section. In humans, bone growth takes place until about 30 years of age, with increasing elastic modulus (indicating the stiffness of the bone) and yield strength (indicating the nominal stress at which the bone undergoes a specified

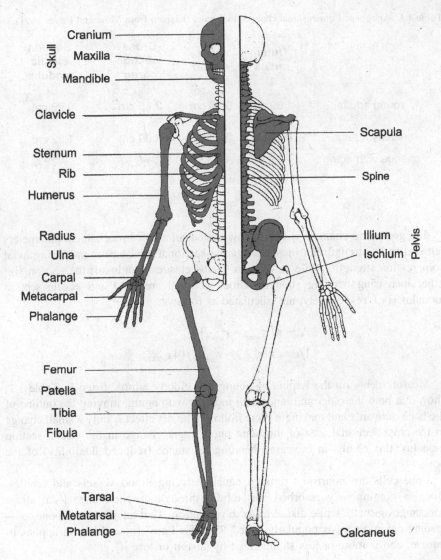

Fig. 1.1 Major skeletal bones (adapted from Langley and Cheraskin 1958; Weller and Wiley 1979; Berkow et al. 1997)

permanent deformation). Beginning after the third decade of life, bones usually become more and more osteoporotic* with larger pores (holes), which decrease the density and the structural strength of the bone. Osteoporotic long bones also tend to show increases in their outer and inner cortical diameters, which generate a more "hollowed" core; the remaining walls are more brittle than during younger age owing to an increase in mineral content in the existing bone material.

Table 1.1 Age-related dimensional changes in femurs (adapted from Mow and Hayes 1991)

	Outer radius	Inner radius	Cross-section area	Section elastic modulus
young adults	1.09 cm	0.53 cm	2.85 cm^2	0.96 cm^3
elderly adults	1.25 cm	0.76 cm	3.09 cm^2	1.32 cm^3
change with aging	+ 15 %	+ 43 %	+ 8 %	+ 38%

The geometric changes in bone allow the calculation of mechanical parameters relevant to its strength: an increase in cross-sectional area means increasing axial compression strength, while an increase of the elastic modulus of the section signifies increasing bending strength. Cross-sectional area (A) and elastic section modulus (I/c), respectively, are calculated as follows:

$$A = \pi\left(r_{outer}^2 - r_{inner}^2\right)$$
$$I/c = \pi\left(r_{outer}^4 - r_{inner}^4\right)/\left(4\, r_{outer}\right)$$

Measurements on the femurs of young and elderly adults, listed in Table 1.1, show that both the outer and inner radii increase with aging, moving the outline of the bone outwards and making it more hollow. The net effect is only a small change in the cross-sectional area of the bone but a larger change in the elastic section modulus; this results in increased bending resistance (reduced flexibility) of the bone.

Bone cells are nourished through canals carrying blood vessels and tubules. Bone is continuously resorbed and rebuilt throughout one's life; local strain encourages growth, while disuse leads to resorption. That adaptation of bone to its loading environment is popularly called "Wolff's Law" (note this also applies to muscle) and is more-or-less summarized by "use it or lose it".

1.2   Cartilage

Cartilage (Latin *cartilago*, gristle) is another major component of the human skeletal system. It consists mainly of a translucent whitish material of collagen fibers embedded in a binding substance. While structurally related to the stiffer bone material, cartilage allows considerable deformations; it is firm but elastic, flexible, and capable of rapid growth. It supplies articulating structures where required, especially at the ends of the ribs and in the joints of the limbs to absorb shocks and to facilitate smooth movements of the ends of bones against each other; and it makes up flexible structures such as the external ear and the nose.

There are three types of cartilage: hyaline cartilage, elastic cartilage, and fibrocartilage. Hyaline cartilage is smooth and glistening, and is common at the articulating surfaces of long bones such as at the knee joint; it also forms the nasal septum, the larynx (voice box), and the rib connections. Elastic cartilage is generally more flexible than hyaline, and is found at the external auditory canal and the Eustachian tube. Fibrocartilage makes up the annulus fibrosus of the intervertebral disk (discussed below in this chapter) and the meniscus in the knee.

1.3 Tendons and Ligaments

The tight tissue that wraps bundles of muscle fibers and the whole muscle condenses at the ends of the muscle to tendons (Greek *tenon*, sinew). These are mostly twisted collagen fibers that work like cables, which can bend but hardly stretch. By connecting muscle to bone, tendons serve to transfer the pull force of the muscle to the bone. To allow smooth gliding along other body structures, the tendon is usually encapsulated by a sheath (see Figs. 2.12 and 2.13 in Chap. 2): a fibrous tissue that has an inner lining, synovium, which produces a viscous fluid called synovia, that reduces friction.

Ligaments (Latin *ligare*, to bind) are like elastic straps, made of layers of parallel bundles of fiber, mostly collagen. Rings or bands of ligaments keep the tendon sheaths at the wrist and hand digits close to bones; these ligaments act as guides and pulleys during the pulling actions of the tendons and their muscles (see Fig. 2.14). Other ligaments directly connect bones across an articulation, such as in the knee joint. Ligaments also wrap the limb joints, such as the knee, or the articulating vertebrae of the spinal column.

1.4 Articulations

In adulthood, some bony joints have no mobility remaining, such as in the fibrous seams in the skull; and some have very limited mobility, such as the connections of the ribs to the sternum. However, many other articulations allow large displacements. Joints with one degree of freedom are simple hinge joints, like the elbow or the distal joints of the fingers. Some joints have two degrees of freedom, such as the wrist. Ankle, shoulder and hip joints have three degrees of freedom, as in a ball-and-socket joint. Figures 1.2 and 1.3 depict motion capabilities and the standard terms to describe them.

Simple mechanical comparisons do not describe adequately many of the more complex body articulations. For example, the wrist joint does not have a single point of rotation but, instead, consists of nearly flat surfaces that separate the forearm and wrist bones. This layout allows, mostly by sliding, flexion/extension as well as ab- and adduction displacements—see Fig. 1.2. However, the twisting supination/pronation of the hand actually takes place within the forearm—as discussed later in this chapter.

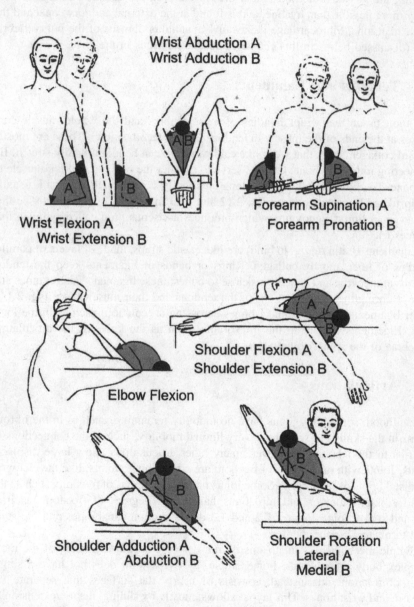

Fig. 1.2 Terms describing arm and hand motions (adapted from Van Cott and Kinkade 1972)

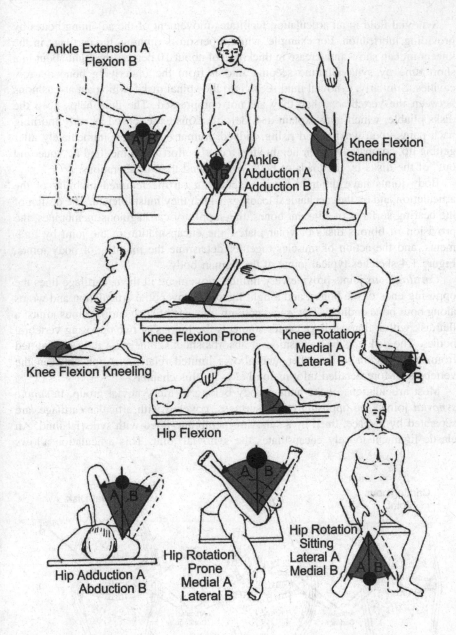

Fig. 1.3 Terms describing leg and foot motions (adapted from Van Cott and Kinkade 1972)

Synovial fluid in an articulation facilitates movement of the adjoining bones by providing lubrication. For example, while a person is running, the cartilage in the knee joint can show an increase in thickness of about 10 percent, brought about in a short time by synovial fluid seeping into it from the underlying bone marrow cavities. Similarly, synovial fluid seeps into the spinal disks, which act as cushions between the vertebrae, when they are not compressed. The fluid helps keep the disks pliable, which may explain disk deformations experienced as early morning back pains upon waking and rising. Indeed, humans are tallest immediately after getting up, compared to their height after a day's effort when the fluid is "squeezed out" of the disks by the loads of body masses and their accelerations.

Body joints have different forms, depending on the required mobility of the articulation and on the mechanical loadings which they must tolerate. The design of the bearing surfaces of adjacent bones, the supply of cartilaginous membranes, the provision of fibrous disks or volar plates, the encapsulation of the joint by ligaments, and the action of muscles together determine the mobility of body joints. Figure 1.4 sketches typical joints of the human body.

Cartilaginous joints provide very limited movement: in them, cartilage lines the opposing ends of the bones, and a tight ligament covers the articulation and wraps along both bone endings. The spinal column consists of such cartilaginous joints: a flat disk with a fibrocartilage center connects the linings of two opposing vertebral bodies. The disk acts like a cushion; it absorbs shocks, otherwise fully transmitted from one vertebra to the other, and allows limited relative motion between the vertebrae. (More detailed information below in this chapter.)

Most articulations of the human body belong to the synovial group. In *simple synovial* joints, the opposing bones surfaces, covered with articular cartilage, are separated by a space, lined by a synovial membrane filled with synovial fluid. An elastic ligament loosely encapsulates the synovial joint. This articulation allows

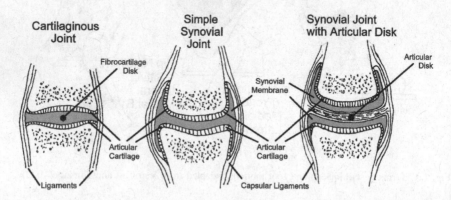

Fig. 1.4 Types of moveable body joints (adapted from Astrand and Rodahl 1977)

substantial relative angulation of the bones. In a *synovial joint with articular disk*, a fibrocartilage disk or wedge (such as the meniscus in the knee) divides the joint space, supplying additional synovial fluid and controlling its distribution and flow. This articulation provides high mobility and good shock absorption.

Articular cartilage has no blood vessels. Its nourishment is achieved through exchange of synovial fluid from the cavities in the end portion of the bone and by exchange of synovial fluid between cartilage and articular space.

1.5 Mobility

Movements in human body joints range from a simple gliding to complex displacements in several planes. For example, in the upper extremity, the rotatory pronation/supination of the hand consists of gliding-twisting between radius and ulna within the forearm (one degree of freedom); the angular swing of the forearm from the upper arm occurs in the hinge-type elbow joint (also one degree of freedom); the complex shoulder joint allows spatial circumduction (three degrees of freedom) in the shoulder. The motions themselves are limited by the shapes of the bony surfaces and by the restraining tension generated by ligaments and muscles.

Mobility, also called flexibility, indicates the range of motion that can be achieved at body articulations, as sketched in Figs. 1.2 and 1.3. Mobility is measured as the range between the smallest and largest angles enclosed by involved body segments about their common point of rotation in the articulation. However, the actual center point of rotation may move with the motion; for example, the geometric location of the axis of the knee joint displaces slightly while the thigh and the lower leg rotate about it.

Mobility varies with age, training and physical condition, and gender. It depends on whether only active contraction of the relevant muscles is used, or if gravity, an external load or perhaps even a second person contribute to achieving extreme locations. Finally, of course, the mobility is different in different planes if the articulation in question has more than one degree of freedom.

Table 1.2 provides information about voluntary (unforced) mobility in major body joints. The measurements involved 100 female US college students (ages 18–35 years), carefully controlled to resemble an earlier study on 100 male students. On each person, only one measurement was taken for each motion. The instructions were to move the limbs "as far as comfortably possible" using (if applicable) the dominant limb; 85% of the participants identified as right-dominant.

Of the 32 measurements, 24 showed statistically significant more mobility by females than by males; men had somewhat larger mobility only in wrist abduction and ankle flexion. This finding confirms earlier studies that also showed generally larger motion capability by women. Given the careful and consistent way the two studies were conducted, one may assume that the data present a realistic picture of mobility of the adult U.S. population within the working age span. Mobility commonly slightly decreases (less than about 10° in the extreme positions) in most body joints as people age into their 6th decade.

Table 1.2 Comparison of mobility measurements, in degrees, done on US college females and males (adapted from Staff 1983, Houy 1983)

Joint	Movement	5th Percentile		50th Percentile		95th Percentile		Difference*
		Female	Male	Female	Male	Female	Male	at 50th percentile
Neck	Ventral flexion	34.0	25.0	51.5	43.0	69.0	60.0	fem+8.5
	Dorsal flexion	47.5	38.0	70.5	56.5	93.5	74.0	fem+14.0
	Right rotation	67.0	56.0	81.0	74.0	95.0	85.0	fem+7.0
	Left rotation	64.0	67.5	77.0	77.0	90.0	85.0	none
Shoulder	Flexion	169.5	161.0	184.5	178.0	199.5	193.5	fem+6.5
	Extension	47.0	41.5	66.0	57.5	85.0	76.0	fem+8.5
	Adduction	37.5	36.0	52.5	50.5	67.5	63.0	NS
	Abduction	106.0	106.0	122.5	123.5	139.0	140.0	NS
	Medial rotation	94.0	68.5	110.5	95.0	127.0	114.0	fem+15.5
	Lateral rotation	19.5	16.0	37.0	31.5	54.5	46.0	fem+5.5
Elbow	Flexion	135.5	122.51	148.0	138.0	160.5	150.0	fem+10.0
Wrist	Supination	87.0	86.0	108.5	107.5	130.0	135.0	NS
	Pronation	63.0	42.5	81.0	65.0	99.0	86.5	fem+16.0
	Extension	56.5	47.0	72.0	62.0	87.5	76.0	fem+10.0
	Flexion	53.5	50.5	71.5	67.5	89.5	85.0	fem+4.0
	Adduction	16.5	14.0	26.5	22.0	36.5	30.0	fem+4.5
	Abduction	19.0	22.0	28.0	30.5	37.0	40.0	male+2.5
Hip	Flexion	103.0	95.0	125.0	109.5	147.0	130.0	fem+15.5
	Adduction	27.0	15.5	38.5	26.0	50.0	39.0	fem+12.5
	Abduction	47.0	38.0	66.0	59.0	85.0	81.0	fem+7.0
	Medial rotation (prone)	30.5	30.5	44.5	46.0	58.5	62.5	NS
	Lateral rotation (prone)	29.0	21.5	45.5	33.0	62.0	46.0	fem+12.5
	Medial rotation (sitting)	20.5	18.0	32.0	28.0	43.5	43.0	fem+4.0
	Lateral rotation (sitting)	20.5	18.0	33.0	26.5	45.5	37.0	fem+6.5
Knee	Flexion (standing)	99.5	87.0	113.5	103.5	127.5	122.0	fem+10.0
	Flexion (prone)	116.0	99.5	130.0	117.0	144.0	130.0	fem+13.0
	Medial rotation	18.5	14.5	31.5	23.0	44.5	35.0	fem+8.5
	Lateral rotation	28.5	21.0	43.5	33.5	58.5	48.0	fem+10.0
Ankle	Flexion	13.0	18.0	23.0	29.0	33.0	34.0	male+6.0
	Extension	30.5	21.0	41.0	35.5	51.5	51.5	fem+5.5
	Adduction	13.0	15.0	23.5	25.0	34.0	38.0	NS
	Abduction	11.5	11.0	24.0	19.0	36.5	30.0	fem+5.0

*Differences are listed only if significant ($\alpha < 0.5$). NS: not significant.

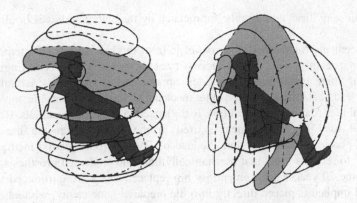

Fig. 1.5 Planes of horizontal and vertical hand reaches (adapted from Ignazi et al. 1982)

For engineering purposes, hand motion capabilities are of particular importance—see Fig. 1.5. They are the result of a combination of movements in several joints: hand, wrist, elbow, shoulder, and spine. Design of cockpits in spacecraft, airplanes and vehicles, for example, must consider the reach capabilities of the sitting operators, often under harsh working conditions such as encumbering clothing and strong accelerations.

Muscles exert primary control over joint motions by their continuous interplay between counteractive muscle groups (see Chap. 2). Moveable joints have numerous nervous connections with the muscles that act on them; this establishes local reflex circuits, which usually prevent overextensions. Control of the skeletal system is accomplished by nervous feedback and excitation signals from the central nervous system; Chap. 3 discusses these topics.

Four types of nerve endings exist in the joints. Two of them are located in the joint capsule: these are Ruffini organs that provide information about changes in joint position and speed of movement. A third kind of Ruffini organ, embedded in ligaments, signals the actual location of the joint. The fourth receptor is a free branching nerve, ending in pain-sensitive fibers. Synovial membranes and joint cartilage seem not to have nerve receptors.

1.6 Artificial Joints

Natural body joints may fail because of disease, trauma, or long-term wear and tear. For the patient, joint failure is associated with motion limitations and usually with severe pain. Damage to bone and degeneration of joint cartilage may give cause for the replacement of the articulation with an artificial, manufactured joint if conservative medical treatments are insufficient. Joint replacements in hips and knees as well as fingers have become routine: in the U.S.A. alone, current estimates are that, annually, over two million people receive joint replacements and implants, many of them elderly persons. Joint replacement usually restores function and mobility even for athletic activities* and, highly appreciated by the patient, relieves or eliminates pain.

Replacements for the ball-and-socket joint at the hip have been attempted for more than a century, with current success rates topping ninety percent. Engineering design of joint replacements includes appropriate modeling of manufactured implants* so that they can carry the mechanical loads and provide meaningful functional life. Another technical task is the selection of proper materials. Routinely successful total hip replacement started in the 1960s* with the use of the metal-on-plastic articulations and the use of PMMA (poly-methyl-methacrylate) "cement" to serve as grout that mechanically links the implanted prosthesis and the bone. Some 20 years later, cementless hip replacements were introduced wherein the metal implant is placed directly into the prepared bone cavity, without cement, either with a simple compressive press-fit and/or by coating implant surfaces with a three-dimensional surface texture. To further encourage bone in-growth or on-growth, an osteoinductive or osteoconductive chemical coating may be sprayed on the "porous" surface of the implant.

If needed, the entire articulating surfaces are replaced: in the hip, the head of the femur (thigh bone) is removed and replaced by a spherical metallic or ceramic ball on a stem, and the acetabular cup is resurfaced with a (generally) plastic or ceramic liner. In the knee, which acts as a "sloppy" hinge, the articulating surfaces on the bottom of the femur are replaced with metal, and surfaces at the top of the tibia (shin bone) and on the patella (knee cap) are resurfaced with plastic.

Both hip and knee replacements generally have the same type of design for the major load-bearing components: the metallic (or ceramic) component is convex and the plastic component is concave. The plastic generally used is an ultra-high molecular-weight polyethylene which allows for a low-friction articulation as the joint components move against each other.

Prosthetic replacements of finger joints are often a single-component molded plastic integral hinge. Use of this simple artificial joint is successful because of the low loads carried by the joints and the minimal debris generated by wear.

Other joints can also be replaced, such as the ankle and the shoulder. Although commercial implants are available, these are less common due both to the complexity of the joints (and the required surgical technique to implant an artificial replacement), and because disabilities in these joints usually have less of an impact on daily activities: one can still walk with a fused ankle, but with an immobile hip or knee, even getting in or out of bed is difficult.

The design of joint replacements is constrained by biologic and mechanical considerations. Biologically, the implanted device must be compatible with the body when it is whole and if debris particles are generated by wear and tear. The corrosive and warm environment of the body places particular requirements regarding material reactivity, mechanical strength, and wear. Of course, the device must be able to be implanted (in terms of size and complexity) and should yield near-normal range of motion and external appearance of the limb. Finally, the design of the device must consider the possibility of salvage: ideally, sufficient bone and soft tissue should remain to allow replacement of the device or fusion of the joint if needed.

As artificial replacement of hips and knees has become routine, patient expectations have continued to rise. Previously, joints were replaced primarily in older patients who placed lower physical demands (such as reduced range of motion) on the implants; now, younger and more active patients seek many decades of pain-free function. Further, patients now expect their implants to accommodate a wide range of activities. Therefore, engineers who design implants need to understand how people want to move and to work (see, for example, Fig. 11.18 in Chap. 11 on the use of anthropometric data in design). In addition, designers need to consider the ergonomics of the surgery itself (for example, surgical instruments, medical imaging, patient positioning, etc.) and the post-operative recovery of the patient (for example, rehabilitation and physical therapy).

1.7 The Hand

The hand is a complicated structure whose sections have diverse mobilities. Motion of the hand is commonly linked with arm motion. For example, supination/pronation of the total hand (see Fig. 1.2) is actually a rotation about the long axis of the forearm which takes place within the forearm: the ulna and radius twist slightly relative to each other because elastic fibers link these bones even though they do not touch.

As already mentioned, the wrist joint consists of nearly flat surfaces at the distal ends of the bones of the forearm, radius and ulna, and of similarly flat opposing surfaces on the proximal carpal bones of the hand—see Fig. 1.6. These flat surfaces allow gliding motions between arm and hand, which achieve flexion/extension as well as abduction and adduction.

The bony structure of the hand is a complex composite of 27 bones. These bones form the solid structures of the wrist (carpus), the base of the hand (metacarpus), and its five digits, the thumb and four fingers. (Actually, the tendons of the thumb contain two additional sesamoid bones of uncertain functionality.)

Adjacent to the wrist joint, the base of the hand starts with two rows of carpal bones: in the proximal row, from radial to ulnar side, scaphoid and lunate; in the distal row, trapezium, trapezoid, capitate, hamate, triquetrum, and pisiform. The carpal bones are tied together by ligaments, so only little movement can take place. The largest displacement occurs when cupping the palm as the thumb touches the little finger.

Distal to the carpals, one metacarpal bone extends toward every one of the five digits. The carpo-metacarpal (CM) joints of the four fingers are of the hinge type with limited ranges; however, the joint of the shorter and sturdier metacarpal of the thumb has a saddle design that affords two degrees of freedom. The resulting ability of the thumb to oppose all four fingers allows a remarkable variety of manipulations.

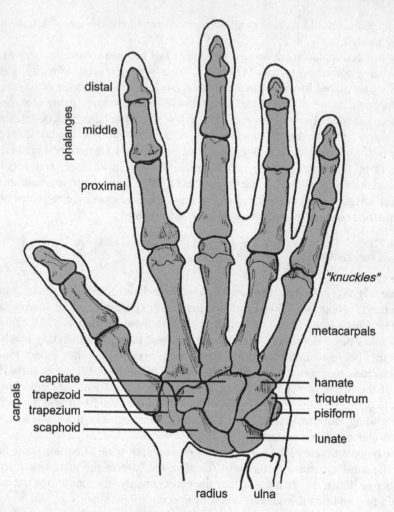

Fig. 1.6 The bones of the right hand, dorsal view

The first digit, the thumb, has distally to its metacarpus only two phalanges: the proximal and the distal phalanx. In contrast, each of the other four digits possesses three phalanges distal to the proximal ends of their metacarpals. The metacarpo-phalangeal joints (MP joints, the "knuckles" of the fingers) provide primarily flexion/extension, as in making and releasing a fist, but also some adduction/abduction, for adjoining and spreading the fingers. All the interphalangeal joints (proximal PIP, distal DIP joints) are simple hinges.

1.8 The Spinal Column

The spinal column* is a very complex, and mechanically interesting, structure. As shown in Fig. 1.7, it consists of a stack of 25 bones, the vertebrae. There are seven cervical vertebrae (C1 to C7), of which the topmost (*Atlas*) supports the skull.

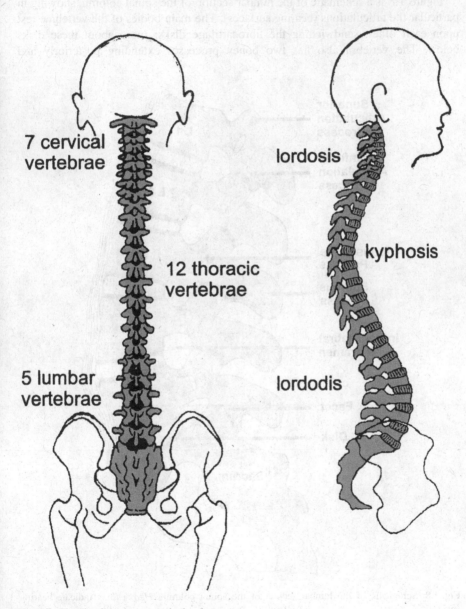

7 cervical
vertebrae

lordosis

12 thoracic
vertebrae

kyphosis

5 lumbar
vertebrae

lordodis

Fig. 1.7 Sketch of the human spinal column

The middle of the spine consists of twelve thoracic vertebrae (T1 to T12). At the bottom are five lumbar vertebrae (L1 to L5) resting on the sacrum (S1) and the "tailbone" (coccyx), which are a triangular groups of fused, rudimentary vertebrae.

In the frontal view, a healthy spine is straight; yet, seen from the side, it has a backward bend (*kyphosis*) at the thorax and two forward bends (*lordoses*) at the neck and the lower back.

Figure 1.8 is a schematic of the lumbar section of the spinal column, showing in particular the articulations (bearing surfaces). The main bodies of the vertebrae rest upon each other, sandwiching the fibrocartilage disks: more about these disks below. The vertebra also has two boney processes extending posteriorly and

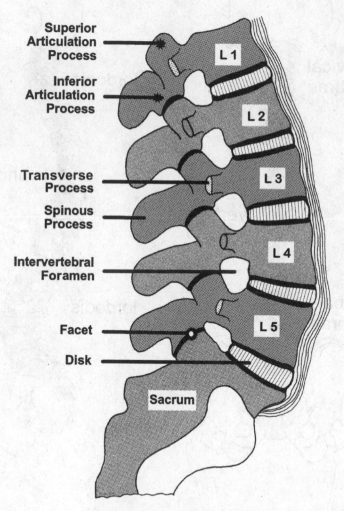

Fig. 1.8 Schematic of the lumbar section of the spinal column. *Heavy lines* indicate bearing surfaces

laterally; the top surface (*facet*) of each of these articulation processes is rounded and fits into a corresponding cavity on the underside of the articulation process of the vertebra above. The facet joints, covered with synovial tissue, influence the ability to twist the spine; this twist mostly occurs in the cervical and thoracic sections.

Figure 1.9 sketches a three-dimensional view of a typical vertebra. The front part, called the main (vertebral) body, provides two flat load-bearing surfaces, superior and inferior, for the intervertebral disks. The rear part is an arched structure: it curls around the intravertebral foramen, the opening through which the spinal cord passes. The arcs of the stacked vertebrae provide a protective tunnel for the spinal cord, which runs along the length of the spine. Each vertebral arch has five protrusions: the spinous (posterior) process, two transverse (lateral) processes, and the two articulation processes that provide four facet joints. That complex form of the vertebra provides attachment surfaces and lever arms for muscles and ligaments.

The spine is capable of withstanding considerable loads, yet flexible enough to allow a large range of postures. There is, however, a trade-off between carried load and flexibility. If there is no external load on the spine, only its anatomical structures (bone shape, joint geometry, capsules, ligaments, and muscles) restrict its

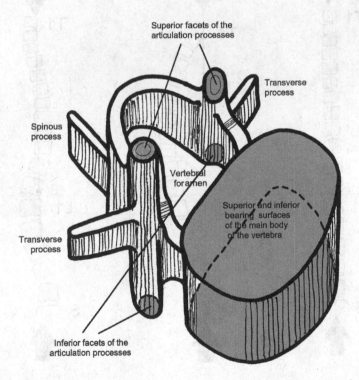

Fig. 1.9 Features of a typical vertebra: the main body, the arch with its processes and the articulation facets

mobility. Loads acting on the spinal column reduce its mobility; under heavy loading, the range of possible postures is very limited.

The spinal column is often the location of discomfort, pain, and injury because it transmits many internal and external strains. When upright, the spine transfers impacts and vibrations from the lower part of the trunk to the upper body. Conversely, forces and impacts experienced through the upper body, such as when lifting or otherwise working with the hands, are conveyed downward through the spinal column to the pelvic region and on to the seat or the floor that support the body. Thus, the spinal column must absorb and dissipate much energy, whether it is transmitted to the body from the outside or generated inside by muscles.

Modeling the spine as a straight column, as shown on the left side of Fig. 1.10, provides a simplification that allows a unique description of the relations between geometry and strain*. If the spinal column arches, shown in the center and on the right in Fig. 1.10, its load-bearing structure may fail. For a curved column to be stable, the thrust line of the load must lie within the cross sections of all

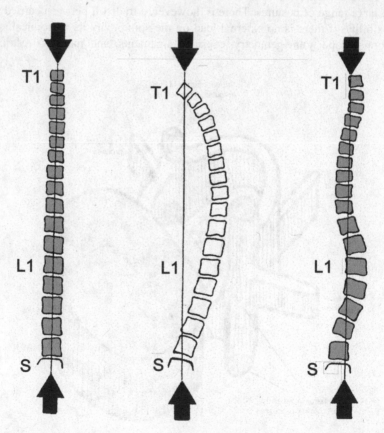

Fig. 1.10 Models of the spinal column under axial compression between T1 and sacrum (adapted from Aspden 1988)

components. If at any point the thrust line falls outside the components, either the structure buckles or a lateral force must return the spine to its stable position range. Ligaments around the spinal column and muscles attached to the transverse and spinous processes of the vertebrae can provide such lateral force; intra-abdominal pressure (IAP), which pushes in all directions within the trunk cavity, may also help in straightening the spinal curvature, so that the thrust line stays within the spinal column.

Figure 1.11 is a conceptual depiction of the major longitudinal trunk muscles that exert forces which, in concert, maintain trunk balance. Most muscles pull, more or less parallel to the spinal column, on the upper section of the body; summing their forces gives an estimate of the compression of the spinal column, which is the only "solid" load-bearing structure in the lumbar trunk section. Some of that load may also be supported by the omni-directional intra-abdominal pressure (IAP) which develops under load in the abdominal cavity.

Depending on the actual body posture, muscles (particularly the pairs of oblique and latissimus dorsi muscles) can exert significant forces in lateral and medial (anterior/posterior, also called axial) directions. These muscles generate shear,

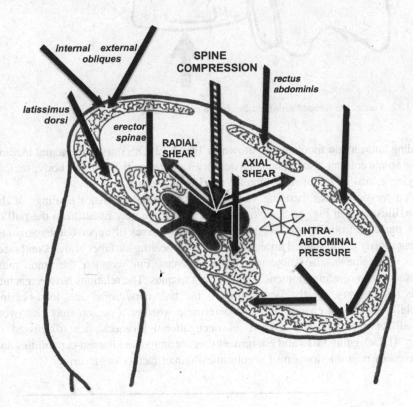

Fig. 1.11 Schematic of trunk muscle pull forces causing spine compression (adapted from Schultz et al. 1982)

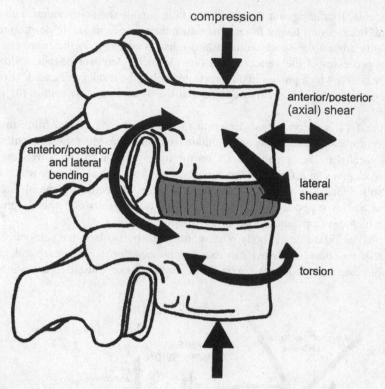

Fig. 1.12 Forces and torques acting on the spinal column (adapted from Marras 2008)

bending and torsion moments as shown in Fig. 1.12. Of course, the actual loading of the spinal column also depends on activities exerted with the upper body, such as carrying a load or performing forceful work with the hands.

Compression, shear, bending, and torsion constitute the major loadings of the spine, illustrated in Fig. 1.12. Spine compression force may result from the pull of trunk muscles, from external loads, and from the masses of upper body segments. Owing mostly to the slanted arrangement of load-bearing surfaces at disks and facet joints, the spine is also subjected to shear forces. Furthermore, the spine must withstand both bending moments and twisting torques. The relations between actual loads (which vary greatly, depending on the task conditions) and load-bearing capabilities* (which change as well) determine whether a person may be "over-loaded" or even injured—a topic of occupational biomechanics (discussed in Chap. 4). Designing tasks and equipment that accommodate human capabilities and preferences is in the domain of ergonomics/human factors engineering.

1.8.1 The Spinal Disk

The spinal intervertebral disk is able to withstand compression and deformation because of its unique structure. The core of the disk is the nucleus pulposus, which resembles a gel-filled, deformable, flattened ball: it carries most of the compressive loading transmitted along the spine and still allows the opposing surfaces of vertebral bodies to tilt with respect to each other. The annulus fibrosus wraps around the nucleus like a tire. The annulus consists essentially of collagen fibers, arranged in multiple layers that are oriented at angles to each other: this composite structure radially constrains the swelling pressure exerted by the nucleus during compression, shearing, bending, and torsion between the vertebrae below and above.

The nucleus pulposus has no blood supply of its own. It is nourished by capillaries at the junctions with the vertebral bodies as a result of osmotic pressure, gravitational force, and the pumping effects of spine movement. It also does not have sensory nerves of its own, so pain felt due to damage arises in surrounding tissue.

The disk acts as a thick-walled pressure vessel, as a multi-directional articulation, and as a shock absorber. If a disk does not function well, for example due to deterioration (such as by aging) or damage (such as by an acute injury), it can not properly transmit loads nor allow motions between adjacent vertebrae. Disk herniation occurs when part of the annulus is weakened or even cracked so that the nucleus is pushed out beyond the ring of the annulus. If displaced nucleus protrudes towards a spinal nerve, it can impinge on a nerve, affecting the transmission of nerve signals and causes pain—see Chap. 3.

Malfunction of the disk can bring about a combination of misalignment of vertebrae, displacement of vertebral or facet joints, and strain of connective tissues, muscles and ligaments. It often leads to restricted mobility and reduced ability to perform physical work, particularly the ability to lift or otherwise move loads. Low-back pain* (LBP) is common, yet often difficult to diagnose and treat. LBP results from inherent structural weakness, injury, or diseases that have been with humans since ancient times.

Aging changes the spine, just as it affects all skeletal components. Bone mass and structural strength increase with age in the young, then begin to decrease in mature adults. Children's spines are much more flexible than adult's spines, depending on the load direction. Aging and growth commonly also increase the angulation of the cervical facets and increase cervical lordosis. In the elderly, increased thoracic kyphosis and decreased lumbar lordosis are common; this is due at least in part to decreases in the height of the disks. Disk height changes stem from degeneration of the disk and changes in the curvature of the vertebral end plate, possibly from osteoporosis.

As already noted, the spinal column often is the site of overexertion and injury. This may manifest itself by deformation, displacement, strains and sprains, or damage to spinal disks, vertebrae, ligaments, cartilage, and muscle. Such injuries being both frequent and costly to businesses, much research has been directed at the causes and mechanisms involved. Since detailed discussion of problems is beyond the scope of this text, it suffices to state that uncoordinated pull of the muscles on the spinal column, particularly when associated with large external loads may (directly or indirectly, acutely or cumulatively, often in unknown and not recon-structable ways), leads to various back overexertions and injuries. Eliminating unnecessary tasks, or assigning them to machines, erases the overexertion risk. Ergonomic design of work place, equipment, and task can avoid some—but not all—risks. Even for persons with persistent back pain, work stations and procedures can be engineered to allow performance of suitable physical work*.

Notes

The text contains markers, *, to indicate specific references and comments, which follow.

Osteoporotic bones: Ostlere and Gold (1991).

Mobility is the range between the smallest and largest angles that involved body segments can enclose. Often, the actual arms of the rotation angle are not well defined: they may be the straight lines connecting the center of rotation of the enclosed articulation to the point of rotation of the next articulation (in the example of the knee joint, running distally to the ankle and proximally to the center of the hip joint); or they may be the estimated mid-axes of the adjacent body segments.

Joint replacements and implants in the USA: Total hips: >370,000 (in 2012); total knees: >680,000 (in 2012)—Steiner et al. (2012).

Joint replacement usually restores function and mobility even for athletic activities: Healy et al. (2008).

Modeling of manufactured implants: Elbert (1991).

Routinely successful total hip replacement started in the 1960s: Pioneered by John Charnley, 1911–1982.

The spinal column: Karpandji (1982), Panjabi et al. (1992) describe the geometry of vertebrae in detail. Hukins and Meakin (2000), Meakin et al. (1996), Chaffin et al. (2006), Marras (2008) and Vierra (2008) discuss biomechanics of the spine.

Spinal geometry and strain: Aspden (1988), Meakin et al. (1996), Hukins and Meakin (2000) discuss Euler buckling of the spine.

Load-bearing capabilities of the spine: Marras (2008) lists as limits: Torsion 88 Nm, Shear 1000 N (lateral and medial), Compression 3400–6400 N.

Low back pain: LBP problems have been diagnosed among Egyptians 5000 years ago (Allan and Waddell 1989) and were discussed in 1713 by Bernadino Ramazzini (translation by Wright 1993). LBP is not confined to humans; dogs and other quadrupeds often suffer from low back pain as well.

Ergonomic design for persons with back problems: For more information, see among other publications those by Violante et al. (2000), Chengular et al. (2003), Chaffin et al. (2006), Marras (2008), Vierra (2008), Oezkaya et al. (2016), Kroemer et al. (2003), Kroemer Elbert et al. (2018), Alizadeh et al. (2019).

Summary

The complex skeletal system with its bones, joints, and connective tissues allows a large range of motions, both forceful and well controlled, especially of the upper extremities. Its capabilities are highly variable; in general, they increase with use, which strengthens strained bones and lubricates loaded joints. However, damage by overloading is frequent: the spinal column, in particular, has been and continues to be the object of biomechanical and ergonomic studies. Some skeletal components can be replaced by engineered devices, including finger, hip, and knee joints.

Glossary

Acetabulum Cup-shaped cavity at the base of the pelvis (hipbone) into which the ball-shaped head of the femur fits.

Acromion Protuberance of the scapula that extends anteriorly-laterally over the shoulder joint.

Anterior In front of the body; toward the front of the body; opposed to posterior.

Articulation Joint between bones.

Atlas The top cervical vertebra, supporting the skull.

Axis Center line of an object; midline about which rotation occurs.

Bending See moment.

Biceps ("two heads") Arm muscle reducing the elbow angle.

Biceps brachii The large muscle on the anterior surface of the upper arm, connecting the scapula with the radius; flexor of the forearm.

Biceps femoris A large posterior flexor muscle of the thigh.

Bone Hard dense porous material developed from connective tissue and forming the skeleton.

Brachialis Forearm muscle connecting the mid-humerus and the ulna.

Brachioradialis Forearm muscle connecting the humerus and the radius.

Cancellous (bone) Open, latticed, porous bone.

Carpus The wrist.

Cartilage Tough, elastic, fibrous connective tissue.

Cervical Part of/pertaining to/ the cervix (neck), especially the seven vertebrae at the top of the spinal column.

Coccyx The tailbone, a triangular bone of three to five fused rudimentary vertebrae at the lower end of the spine, below the sacrum.

Collagen A protein forming the chief constituent of bone and connective tissue.

Compression The pressure (strain) generated in material caused by two opposing forces; opposite of tension.

Cortical Of/at the outside.

Degree(s) of freedom In mechanics, the number of independent linear or rotational displacements which a body can execute.

Density Mass of material per unit volume.

Diaphysis Shaft of a long bone.

Digit The thumb or any of the four fingers of the hand.

Distal Away from the center, peripheral; opposite of proximal.

Dominant The hand or foot exclusively or preferably used for certain actions.

Dorsal Toward the back or spine; also pertaining to the top of hand or foot; opposite of palmar, plantar, and ventral.

Elastic Spontaneous regaining of former shape after distortion; opposite of plastic.

Ergonomics The application of scientific principles, methods and data drawn from a variety of disciplines to the design of engineered systems in which people play significant roles.

Extend To move adjacent segments so that the angle between them is increased, as when the leg is straightened; opposite of flex.

External Away from the central long axis of the body; the outer portion of a body segment.

Facet Flat articulation surface at the upper (superior) and lower (inferior) parts of the articulation processes of a vertebra.

Fascia Layer of fibrous connective tissue enwrapping muscle, muscle bundles and bundles of fibers.

Femur The long bone of the thigh.

Flex To move a joint in such a direction as to bring together the two parts which it connects, as when the elbow is bent; opposite of extend.

Flexibility Term occasionally used instead of mobility.

Foramen An opening in a bone, such as within a vertebra (the intravertebral foramen, part of the "spinal tunnel") or between two vertebrae (the intervertebral foramen, allowing the "spinal roots" to pass).

Glycoprotein (or glucoprotein) Protein containing carbohydrate.

Ground substance In connective tissue, the amorphous gel-like material that fills the spaces between fibers (such as collagen) and cells.

Herniation Rupture; protusion of body material through an opening in its surrounding wall ("disk protrudes from its annulus").

Human (Factors) Engineering Basically the same as Ergonomics—see there.

Humerus The long bone of the upper arm.

Inferior Below, lower, in relation to another structure.

Internal Near the central long axis of the body; the inner portion of a body segment.

Knuckle The joint formed by the meeting of a finger bone (phalanx) with a palm bone (metacarpal).

Kyphosis Backward curvature of the spine; opposite of lordosis.

Lateral Lying near or toward the sides of the body; opposite of medial.

Ligament Band of firm fibrous tissue connecting bones.

Lordosis Forward curvature of the spine; opposite of kyphosis.

Lumbar Part of/pertaining to/the loin, the five vertebrae atop the sacrum.

Matrix Intercellular substance of tissue.

Medial Lying near or toward the midline of the body; opposite of lateral.

Meniscus A disk of crescent-shaped cartilage in the knee joint.

Metacarpal Pertaining to the long bones of the hand between carpals and phalanges.

Metacarpus The base of the hand (between the wrist and the digits).

Mobility The ability to move segments of the body.

Moment The product of force and its lever arm when trying to rotate or bend an object; the stress in material generated by two opposing forces that try to bend the material about an axis perpendicular to its long axis; see also Torque.

Muscle Tissue composed of bundles of fibers that can contract.

Osmosis Diffusion of a fluid through a semi-permeable membrane.

Osteoporosis Condition in which bone becomes more porous and hollow, less dense, and more brittle.

Palmar Pertaining to the palm (inside) of the hand; opposite of dorsal.

Patella The kneecap.

Pelvis The bones of the "pelvic girdle" consisting of ilium, pubic arch and ischium which compose either half of the pelvis.

Phalanges The bones of the fingers and toes (singular, phalanx).

Physiology The branch of biology that deals with the physical and chemical functions of living organisms and their parts.

Physis The body as distinguished from the mind.

Plantar Pertaining to the sole of the foot.

Plastic Able to be distorted permanently into a different shape; opposite of elastic.

Popliteal Pertaining to the ligament behind the knee or to the part of the leg behind the knee.

Posterior Pertaining to the back of the body; opposite of anterior.

Protuberance Protruding part of a bone.

Proximal The (section of a) body segment nearest the head (or the center of the body); opposite of distal.

Radius The bone of the forearm on its thumb side.

Resorb To (again) break down and assimilate something that was previously differentiated.

Sacrum A flat, triangular bone of five fused rudimentary vertebrae at the lower end of the spine, below lumbar vertebra L5.

Scapula The shoulder blade.

Spine The column of vertebrae.

Spinous (or spinal) process The posterior protuberance of a vertebra.

Sternum The breastbone.

Sub A prefix designating below or under.

Superior Above, in relation to another structure; higher.

Supra Prefix designating above or on.

Tailbone (or coccyx) Triangular bone of three to five fused rudimentary vertebrae at the lower end of the spine, below the sacrum.

Tarsus The collection of bones in the ankle joint.

Tendon Band of tough fibrous tissue connecting the end of a muscle to a bone.

Tension The strain in material generated by two opposing forces that try to stretch the material; opposite of compression.

Thoracic Part of/pertaining to the thorax (chest), especially the twelve vertebrae in the middle of the spinal column.

Tibia The medial bone of the lower leg (shin bone).

Torque A force applied at a lever arm that twists or rotates something about its central axis; see also Moment.

Torsion See Torque.

Transverse plane Horizontal plane through the body, orthogonal to the medial and frontal planes.

Triceps ("three heads") Arm muscle increasing the elbow angle.

Tuberosity A (large) rounded prominence on a bone.

Ulna The bone of the forearm on its little-finger side.

Ventral Pertaining to the anterior (abdominal) side of the trunk.

Vertebra A bone of the spine.

Volar plate Thick ligament that connects two bones.

Wolff's law "Use strengthens, disuse weakens".

References

Alizadeh M, Knapik GG, Mageswaran P, Mendel E, Bourekas EC, Marras WS (2019) Biomechanical musculoskeletal models of the cervical spine: a systematic literature review. Clin Biomech. https://doi.org/10.1016/j.clinbiomech.2019.10.027

Allan DB, Waddell G (1989) An historical perspective on low back pain and disability. Acta Orthop Scandinav 60(Suppl 234):1–23

Aspden RM (1988) A new mathematical model of the spine and its relationship to spinal loading in the workplace. Appl Ergon 19:319–323

Astrand PO, Rodahl K (1977) Textbook of work physiology, 2nd ed. Wiley, New York

Berkow R, Beers MH, Fletcher AJ (1997) The Merck manual of medical information. Merck, Whitehouse Station

Chaffin DB, Andersson GBJ, Martin BF (2006) Occupational biomechanics, 4th edn. Wiley, New York

Chengular SN, Rodgers SH, Bernard TE (2003) Kodak's ergonomic design for people at work, 2nd edn. Wiley, New York

Elbert KE (1991) Analysis of polyethylene in total joint replacement. Doctoral dissertation. Cornell University, Ithaca

Healy WL, Sharma S, Schwartz B, Iorio R (2008) Athletic activity after total joint arthroplasty. J Bone Joint Surg Am 90:2245–2252

Hukins DWL, Meakin JR (2000) Relationship between structure and mechanical function of the tissues of the intervertebral joint. Am Zool 40:42–52

Houy DA (1983) Range of joint motion in college males. In: Proceedings of the Human Factors Society 27th annual meeting. Human Factors Society, Santa Monica, pp 374–378

Ignazi G, Mollard R, Coblentz A (1982) Progress and prospects in human biometry. In: Easterby R, Kroemer KHE, Chaffin DB (eds) Anthropometry and biomechanics. Theory and applications. Plenum, New York, pp 71–98

Karpandji IA (1982) The physiology of the joints. Churchill Livingstone, Edinburgh

Kroemer Elbert KE, Kroemer HB, Kroemer Hoffman AD (2018) Ergonomics: how to design for ease and efficiency, 3rd edn. Academic Press, London

Kroemer KHE, Kroemer HJ, Kroemer-Elbert KE (2003) Ergonomics: how to design for ease and efficiency, 2nd edn. Prentice Hall, New York

Langley LL, Cheraskin E (1958) The physiology of man, 2nd ed. McGraw-Hill, New York

Marras WS (2008) The working back: a systems view. Wiley, New York

Meakin JR, Hukins DWL, Aspden RM (1996) Euler buckling as a model for the curvature and flexion of the human lumbar spine. Biol Sci 263(1375):1383–1387

Mow VC, Hayes WC (eds) (1991) Basic orthopaedic biomechanics. Raven Press, New York

Oezkaya N, Leger D, Goldsheyder D, Nordin M (2016) Fundamentals of biomechanics, 4th edn. Springer, New York

Ostlere SJ, Gold RH (1991) Osteoporosis and bone density measurement methods. Clin Orthop
    271:149–163

Panjabi MM, Goel V, Oxland T, Takata K, Duranceau J, Krag M (1992) Human lumbar vertebrae:
    quantitative three-dimensional anatomy. Spine 17:299–306

Schultz AB, Andersson GJ, Ortengren R, Haderspeck K, Nachemson A (1982) Loads on the
    lumbar spine. J Bone Jt Surg Am (64A):713–720

Staff KR (1983) A comparison of range of joint mobility in college females and males. Master's
    thesis, Industrial Engineering, A&M University, College Station

Steiner C, Andrews R, Barrett M, Weiss A (2012) HCUP projections: mobility/orthopedic
    procedures 2003 to 2012. Report # 2012-03. Agency for Healthcare Research and Quality, US
    Department of Health and Human Services, Rockville. http://www.hcup-us.ahrq.gov/reports/
    projections/2012-03.pdf. Accessed 26 Oct 2019

Van Cott HP, Kinkade RG (1972) Human engineering guide to equipment design. U.S.
    Government Printing Office, Washington

Vierra ER (2008) Low back disorders. In: Kumar S (ed) Biomechanics in ergonomics. 2nd ed.
    CRC Press, Boca Raton

Violante F, Armstrong T, Kilbom A (eds) (2000) Occupational ergonomics. Work related
    musculoskeletal disorders of the upper limb and back. Taylor & Francis, London

Weller H, Wiley RL (1979) Basic human physiology. VNR, New York

Wright WC (1993) Diseases of workers. Translation of Bernardino Ramazzinni's (1713) De
    morbis articium. OH&S Press, Thunder Bay

Chapter 2
Muscles

Overview

Skeletal muscles move body segments with respect to each other against internal and external resistances. Shortening is the only active function of the muscle. Nervous signals stimulate muscle components either to shorten dynamically, to statically retain their length, or to allow being lengthened. Various methods and techniques are available for assessing muscular control and strength. In order to use available data on body strength for design, the ergonomic engineer must first determine the critical design considerations: whether minimal or maximal exertions, static or dynamic.

The Model

In the human body, skeletal bones and connective tissues form an internal framework, moveable in its intermediate articulations. Skeletal muscles which span one or two body joints generate forces that pull on the bones, which serve as lever arms.

Skeletal muscles are "linear motors" which, when triggered by nervous signals, contract within themselves, working to shorten the distance between their opposite attachments to bones. Muscles "pull" but cannot "push" on their attachments. A muscle may be stretched by external forces, once these forces overcome the muscle's resistance.

Since muscles exist in counteracting pairs around skeletal joints, their antagonistic contractions determine and control the strength of the human body and its motions.

© Springer Nature Switzerland AG 2020
K. H. E. Kroemer et al., *Engineering Physiology*,
https://doi.org/10.1007/978-3-030-40627-1_2

Introduction

Physiology is the study of function of organisms and their parts. Within physiology, the study of muscular efforts has long been of special interest; therefore, there is a deep tradition of philosophical and experimental approaches and terminology.

Leonardo da Vinci (1452–1519) and Giovanni Alfonso Borelli (1608–1679) combined mechanical with anatomical and physiological explanations to describe the functioning of the human body. Borelli modeled the human body as consisting of long bones (links) that connect in the articulations (joints), powered by muscles that bridge the articulations. The knowledge developed by Gottfried Leibniz (1646–1716) and Isaac Newton (1642–1727) described laws governing mechanics of the human body in terms of statics and dynamics. Physiology books published until the middle of the 20th century tended to divide muscle activities into two groups: *static* and *dynamic*.

Static efforts consist of (often short) bursts of contraction that do not result in motion of the involved body segments. These "isometric" conditions have received much research attention. Consequently, there is a great deal of published information on muscle strength (and fatigue) during static efforts.

Dynamic efforts result in motion of the involved body segments. These conditions may last minutes or even hours. Work, energy, and endurance are typical research topics for dynamic activities.

2.1   Muscle Architecture

The human body has three types of muscle: *smooth, cardiac* and *skeletal (striated)*.

Smooth muscles apply pressure to vessels and organs. For example, vascular smooth muscles control the diameter (lumen) of blood vessels and hence blood flow and pressure. In both halves of the heart, cardiac muscle pumps blood through the vascular system. Electrical stimulation and contraction mechanism are similar in smooth and cardiac muscles but anatomic and physiologic characteristics of cardiac muscle resemble those of skeletal muscle. Skeletal muscles (also called voluntary muscles) control body locomotion and posture. They usually connect two body members across their common joint; some muscles cross two joints.

In ergonomics, skeletal muscles are of primary interest since they move the segments of the human body and generate energy for exertion onto outside objects. Knowledge of the characteristics of muscles is important for the design of work tasks, workplaces, and equipment.

2.1.1 Agonist-Antagonist, Co-contraction

In the human body, skeletal muscles are arranged in a "functional pair" where an opponent counteracts the contracting muscle. One muscle, or a group of synergistic muscles, flexes around an articulation while the other extends, as shown in Fig. 2.1. The active muscle is called agonist (or protagonist) and the opposing one antagonist. Co-contraction is the simultaneous contraction of two or more muscles, often of agonist and antagonist. Co-contraction serves to control the magnitude of a strength exertion or the speed of motion of limbs. Bilateral co-contraction occurs when muscle is activated that is not directly involved in a task, such as when we tighten muscles in the left arm when those in the right arm execute a strong effort.

2.1.2 Components of Muscle

There are several hundred skeletal muscles in the human body, known by their Latin names. (The often-used Greek prefix *myo* means muscle). Many muscles are spindle-shaped (fusiform) with a wide "belly" and narrow ends where the wrapping tissue combines to form cable-like tendons, which attach to bones: *origin* is the proximal attachment, *insertion* the distal attachment. By weight, muscle consists of 75% water and 20% protein; the remaining 5% of weight include fats, glucose and glycogen, pigments, enzymes, and salts.

Fibrous connective tissue, *epimysium*, wraps each muscle; it is called *fascia* when it is smooth and allows the muscle to slide against its surrounds. As sketched in Fig. 2.2, every muscle actually consists of bundles of muscles, fascicles (*fasciculi*), wrapped in a sheath of specialized connective tissue (*perimysium*). In turn, every fascicle is made up of separate bundles of long muscle cells, called muscle fibers; again, connective tissue (*endomysium*) wraps each bundle of cells. A large number of fibers lie side-by-side. Some muscle cells are as long as the fascicle; shorter fibers often appear in series, head-to-head. The muscle fiber is made up of fibrils (also called *myofibrils*), discussed below.

A complex membrane, the *sarcolemma* (also called plasmalemma) envelopes each muscle fiber. Inside that envelope are *myofibrils* and a gelatin-like substance, *sarcoplasm*, which fills the spaces within and between the fibrils. The sarcoplasm contains dissolved proteins, minerals, glycogen, fats, and myoglobin.

Figure 2.3 shows the intricate network of small tubes (*tubules*) which run parallel to the myofibrils and also loop around them: this is the sarcoplasmic reticulum. Another network has interconnected transverse tubes, T-tubules. They start as indentations (*invaginations*) of the sarcolemma and then penetrate the muscle fiber sideways. Thus, they are extensions of the sarcolemma and provide pathways for extracellular fluid.

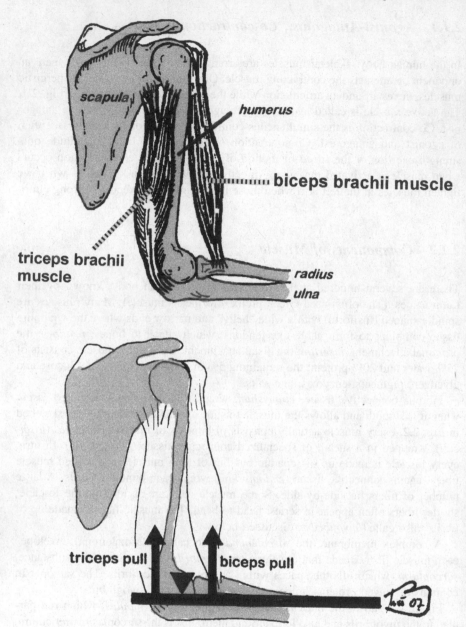

Fig. 2.1 Biceps and triceps muscles as antagonistic pair control elbow flexion and extension. Not shown are the brachialis muscle (attaching to humerus and ulna) and the brachioradialis muscle (connecting humerus with radius), which act together with the biceps as a synergistic flexor group

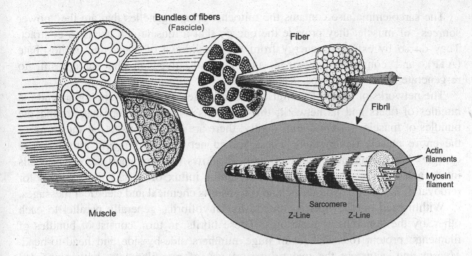

Fig. 2.2 Major components of muscle (adapted from Astrand et al. 2003; Wilmore et al. 2008)

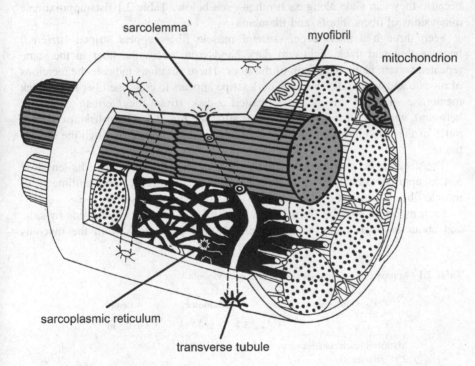

Fig. 2.3 Schematic of major components of muscle fiber (adapted from Astrand et al. 2003; Wilmore et al. 2008)

The sarcolemma also contains the mitochondria, organelles that are the "power sources" of muscle: they provide the energy that a muscle cell needs to contract. They do so by extracting energy from the breakdown of adenosine triphosphate (ATP), and converting glycogen and fats, dissolved in the sarcoplasm, to re-generate ATP—see Chaps. 7 and 8.

The networks of tubules of a muscle fiber connect with the wrapping around the bundles of fibers and its network; this connects with the membranes around the bundles of muscles (*perimysium*); finally, there are connections to the sheathing of the entire muscle (*epimysium*). These joined networks within the muscle are a complex "plumbing and control" system. It allows fluid transport among the cells inside and outside the muscle; it provides for influx of energy carriers and for removal of metabolic byproducts; and it transmits chemical and electrical messages.

Within each muscle fiber are thread-like myofibrils, generally parallel to each other, by the hundreds or thousands. These fibrils, in turn, consist of bundles of filaments: protein rods that lie in huge numbers side-by-side and head-to-head. *Myosin* and *actin* are the two primary types of myofilaments, both elongated polymerized protein molecules. These are the "movers" of muscular contraction because they can slide along each other—see below. Table 2.1 lists approximate dimensions of fibers, fibrils, and filaments.

Seen through a microscope, skeletal muscle fibers appear striped (*striated*) because thin and thick, light and dark bands run across the fiber in the same repeated pattern along the length of the fiber. These striations indicate the locations of myosin and actin rods. One such dark stripe appears to cross the fiber like a thick membrane or disk: this is the so-called z-disk (from the German *zwischen*, between), where actins originate on either side; see Fig. 2.4. The z-disk also carries much of the "plumbing and control" networks, just mentioned, through the muscle tissue.

The distance between two adjacent z-disks defines the sarcomere. Its length at rest is approximately 250 Å (1 Å = 10^{-10} m); accordingly, one millimeter of muscle fiber length can contain about 40,000 sarcomeres in series.

Each myofibril contains from 100 to 2500 myosin filaments, lying side by side, and about twice as many actin filaments. Small projections from the myosins

Table 2.1 Approximate dimensions of muscle components

Component	Diameter	Length
Fiber	5×10^5 to 10^6 Å	Up to 0.5 m
Myofibril (each contains up to 2500 filaments)	10^4 to 5×10^4 Å	
Actin myofilament Myosin myofilament	50 to 70 Å 100 to 150 Å	10^4 Å 2×10^4 Å

1 Å (Ångstrom) = 10^{-10} m

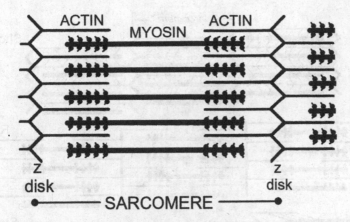

Fig. 2.4 Schematic of a sarcomere with its myofilaments myosin and actin

(resembling miniature hooked golf clubs), called cross-bridges, protrude towards neighboring actins. The actin filaments (twisted double-stranded protein molecules) wrap in a double helix around the myosin molecules. In cross-section, six actin rods surround each myosin rod in a regular hexagonal array. That combination of actin and myosin is the contracting microstructure (also called the *elastic element*) of the muscle.

2.1.3 Muscle Contraction

Shortening is the only active action a muscle can do; passively, it can be stretched by external forces.

By convention, one distinguishes between three lengths of muscle: the relaxed (resting) length, and its contracted (shortened) or its elongated (stretched) lengths. Figure 2.5 sketches these three conditions. The sarcomere can shorten to about 60% in contraction or it may be stretched to about 160% of its resting length without damage. At around 200% stretch, muscle fibers, tendons, or tendon attachments are likely to break.

In a relaxed muscle, attractive forces between the actin and myosin filaments are chemically neutralized, but an incoming nervous signal (see Chap. 3) initiates filament activity. Cross-bridges are established and released as the actin "ratchets" along the myosin, sliding the opposing heads of actin rods towards each other*. This pulls the z-disks towards the tails of the myosin filaments: the sarcomere (and with it the muscle) shortens. After the contraction, the muscle returns to its resting length, primarily through a recoiling-type action of its shortened filaments, fibrils, fibers and of other connective tissues.

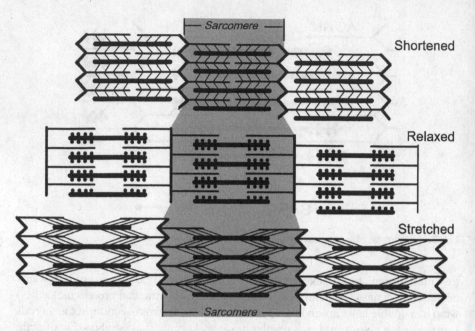

Fig. 2.5 Schematic of sarcomeres in contracted, relaxed and extended muscle

Careless use of the term "contraction" can lead to confusion:

- "Contraction" literally means "pulling together", as in the sliding of actin on myosin filaments, which pulls the z-disks together and hence shortens the sarcomere. This action is called "concentric".
- During muscular shortening, tension develops in the muscle only to the extent that resistance against the shortening exists. Therefore, the event of a "contraction" does not necessarily imply a forceful effort by the muscle.
- During a "static contraction", the sarcomere length appears unchanged (isometric).
- In an "eccentric contraction", the sarcomere is actually lengthened.

To avoid misleading implications and contradictions in terms, it is often better to use the terms activation, effort, or exertion, instead of contraction*.

2.1.4   Relations Between Muscle Length and Tension

Muscle can contract to about 60% of its resting length. In this condition, the actins proteins curl completely around the myosin rods and the z-disks are as close as attainable. This is the shortest possible length of the sarcomere, at which the muscle can just barely develop any active contraction force.

External force can stretch the muscle beyond its resting length. This slides actin and myosin fibrils along each other. When the muscle is extended to 120–130% of resting length, the cross-bridges between actin and myosin attain their best positions to generate contact for contractile resistance. If the elongation continues even farther, the cross-bridge overlap diminishes; at about 160% of resting length, so little overlap remains that the muscle cannot develop any active resistance anymore. With further elongation, passive tissue resistance to stretch grows, like in a rubber band, until tissue snaps.

Below resting length, tension in the muscle results from the active contraction effort of the muscle. Above resting length, the total tension in the muscle is the summation of active and passive strains. Accordingly, the curve of *active* contractile tension developed within a muscle is zero at approximately 60% resting length, about 0.9 at resting length, at unit value at about 120–130% of resting length, and then falls back to zero at about 160% resting length. (These values apply to an isometric twitch contraction, discussed below.) Figure 2.6 shows, schematically, the combination of active and passive tensions in a muscle. Because of that combination, the muscle becomes more forceful as its length extends.

The summation effect explains why we instinctively pre-stretch (that is, preload) muscles for a strong exertion, as in bringing the arm behind the shoulder before throwing a ball forward.

In engineering terms, muscles exhibit viscoelastic qualities*: "viscous" because their behavior depends both on the amount by which they deform and on the rate of deformation; and "elastic" because, after deformation, they return to the original length and shape. These behaviors, however, are not purely in the muscle, because it is non-homogeneous, anisotropic, and discontinuous in its mass. Nevertheless, nonlinear elastic theory and viscoelastic rules can describe major features of muscular performance.

Viscoelasticity also helps to explain why the highest possible muscle tension is developed isometrically ("statically", while resisting eccentric stretch), whereas

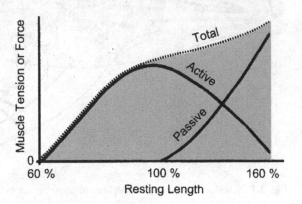

Fig. 2.6 Active, passive and total tension within a muscle at different lengths

muscle tension is decidedly lower in active shortening ("dynamically", during concentric movement). The higher the velocity of muscle contraction, the faster actin and myosin filaments slide by each other and the less time is available for cross-bridges to develop and hold. This principle of speed-related reduction in tension capability of the muscle holds true for both concentric and eccentric activities. In eccentric activities, however, the total tension resisting the stretch increases with larger length owing to the summing of active and passive tensions within the muscle.

2.1.5 The "Motor Unit"

The axon endings of nerve fibers permeate muscles; they are the final branches of the efferent (motor) part of the peripheral nervous system (discussed in Chap. 3). The area where axon and sarcolemma of the muscle contact is called the motor endplate, and is sketched in Fig. 2.7. Each nerve fiber innervates several, usually hundreds or even thousands of muscle fibers through as many motor endplates. These fibers under common control are a motor unit: the same signal stimulates them all. However, the muscle fibers of one motor unit lie usually not close to each other but are spread throughout the muscle. Thus, "firing" one motor unit does not cause a strong contraction at one specific location in a muscle but rather a weak contraction throughout the muscle.

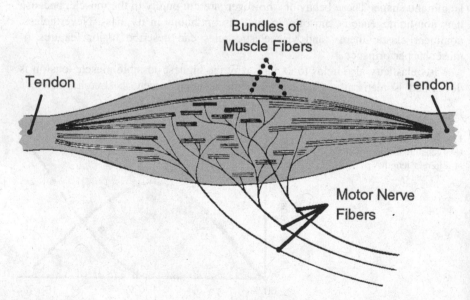

Fig. 2.7 Schematic of the motor endplates of three nerves (adapted from Guyton 1979)

The anecdotal "all-or-none principle" suggests that all muscle fibers of one motor unit are either fully relaxed or fully contracted. This principle does not describe the condition of a muscle in the initial phase of twitch buildup or during the return to the resting state. However, once the stimulus exceeds a threshold potential, the muscle as a whole will generally provide a complete response (contraction).

2.1.5.1 Muscle Twitch

Twitch is a single contraction of a motor unit; it starts upon receipt of a nervous stimulus and ends with complete relaxation. A twitch lasts up to 220 ms in skeletal muscle, yet only about 10 ms in ocular muscle. As Table 2.2 shows, twitches begin with a latent period; followed by the phase of shortening, then a period of relaxation, and finally there is time for recovery—together about 130 ms for a typical fast twitch muscle fiber (see below for more information on fast and slow fibers). When a fiber contraction is not yet completely released by the time the next stimulation signal arrives, summation (superposition) of twitches occurs. The new contraction builds on a level higher than if the fibers had been completely relaxed; consequently, the contraction achieves higher contractile tension in the muscle. Such "staircase" effect takes place when excitation impulses arrive at frequencies of ten or more per second. When muscle stimulation is above a critical frequency, 30–40 stimuli per second, successive contractions fuse together to a maintained contraction, called tetanus. In such superposition of twitches, the tension generated may be double or triple as large as in a single twitch; a full tetanus may have up five times the single-twitch tension.

2.2 Muscle Fatigue

Muscular fatigue may be defined operationally and simply as "a state of reduced physical ability, which can be restored by rest". The feeling of fatigue indicates that the body is becoming unable to continue or repeat an effort. A benefit of the signals of fatigue is prevention of exhaustion and perhaps serious injury to muscles. An everyday example is the muscle fatigue that one painfully experiences when working with raised arms, such as adjusting a fixture overhead, depicted in Fig. 2.8. Increasing discomfort is followed by pain in the shoulder muscles (which keep the arms up) or in the neck muscles (which keep the head bent backwards); this makes it impossible to go on with the work, even though nerve impulses still arrive at the neuromuscular junctions and the resulting action potentials continue to spread over the muscle fibers.

Table 2.2 Single twitch of a fast-twitch muscle fiber

Period, duration	Muscle action	Energy metabolism (see Chapter 7)
Latent, about 10 ms	No reaction yet of the muscle fiber to the motor neuron stimulus	none
Shortening, about 40 ms	Cross-bridges between actin and myosin vibrate and "ratchet" the heads of the actin rods towards each other along the myosin	Energy for this process is freed from the ATP complex, mostly anaerobically
Relaxation, about 40 ms	Cross-bridges stop oscillating, bonds between myosin and actin break, the muscle returns to its original length	ATP is re-synthesized by ADP
Recovery, about 40 ms	none	Aerobic metabolism oxidizes glucose and glycogen, final regeneration of ATP and phosphocreatine

Fig. 2.8 Fatiguing work overhead (adapted from Nordin et al. 1997)

The reasons for fatigue are complex. They may relate to energy delivery in the muscle, to accumulation of metabolic byproducts such as lactate, to overexertion of muscular contraction mechanisms, even to events in the nervous system. Occurrence of fatigue depends on the type and intensity of the effort, on the fiber type of the involved muscles, on the person's fitness and training, and on the individual's motivation. Often, several of these factors combine to cause fatigue. So-called central fatigue, associated with one's sense of effort and motivation, may occur also in the nervous control system; a strong emotional drive to perform, as in competition, can temporarily overcome muscular fatigue.

Sufficient arterial blood supply to the muscle and unimpeded blood flow through the intricate pathways that penetrate muscle are crucial for muscle activities. Blood flow determines the ability of contractile and metabolic processes to continue, since blood supplies the needed oxygen and energy carriers to rebuild ATP. Equally important, blood removes metabolic byproducts, particularly lactic acid and potassium as well as the heat, carbon dioxide and water liberated during metabolism—see Chaps. 6 and 7.

An artery enters the muscle usually at about its mid-length and branches profusely from there. Some of the small arteries and their arterioles transverse muscle fibers while other blood vessels parallel the fibers; many crosswise linkages exist among them. This forms a complex network of blood vessels (the "capillary bed", see Chap. 6) permeating the muscle tissue. On the exit side, venules gather blood laden with metabolic wastes and channel it into veins. This capillary network is particularly well developed at the motor endplates.

The abundance of capillaries in the muscle provides good facilities for the supply of oxygen and nutrition to the muscle cells and for the removal of metabolic byproducts (see Chap. 7). However, a strongly contracting muscle generates strong pressure inside itself, as can be felt by tapping a tightened biceps or calf muscle. By this pressure, the muscle compresses its own fine blood vessels. This impedes, and may even shut off, the muscle's circulation in spite of a reflexive increase in systolic blood pressure. Complete interruption of blood flow through a muscle leads to disabling muscle fatigue within a few seconds, forcing relaxation of the muscle's contraction and allowing resumption of blood flow.

If repeated and/or strong efforts of the muscle severely diminish its supply of blood, lactic acid can remain as a byproduct of anaerobic glycolysis. If so much lactate builds up that it blocks the breakdown of ATP, the muscle quickly loses its ability to function. These biochemical events perturb the coupling between nervous excitation and muscle contraction—the muscle experiences "fatigue" and must rest. Fortunately, simply taking a break, thus avoiding the cause of the problem, leads to complete recovery.

The stronger the exertion of a muscle, the shorter the period during which this strength can be maintained. Figure 2.9 shows this relation between isometric strength exertion and endurance schematically: a maximal exertion can be maintained for just a few seconds; 50% of strength is available for about one minute; but less than 20% can be applied for long periods.

Fig. 2.9 Endurance times of isometric muscle efforts

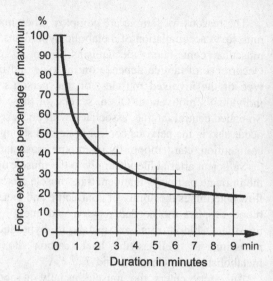

Fatigue depends on the magnitude and duration of effort compared to the capability of the involved muscle; hence, physical training and skill development can be effective to some extent to counteract fatigue. However, the proper ergonomic approach is minimize the risk of fatigue to design out any work requirements that generate fatigue: the engineering solution is "fitting the task to the human".

2.3   Activities of Entire Muscles

2.3.1   Control of Muscle

Generally, one cannot voluntarily contract more than two-thirds of all fibers of a muscle at once; this seems to be a safeguard against overstraining the muscle-tendon structure. However, contraction of all fibers at the same time can occur as a result of a reflex; this may strain the muscle (or associated tendon) to its total structural tensile capacity, resulting in a tear.

The activities of a whole muscle, which comprises many motor units (discussed above), are controlled by *recruitment coding*: this determines how many and which motor units turn on at any given instant. The cooperative effort of all participating motor units determines the tension in the whole muscle and hence the force that it transmits via tendons to the skeletal bones.

Exact control of the tension in a muscle depends on the number of muscle fibers innervated by one nerve axon (see Chap. 3): the smaller the number of fibers per axon, the finer the muscle control. For example, in eye musculature, one nerve controls only seven muscle fibers, for an innervation ratio of 1:7, whereas the quadriceps femoris extending the knee has a ratio of approximately 1:1000.

The nervous action potential spreads along the muscle at speeds of 1–5 m/s to initiate either recruitment or rate control: *rate control* determines the number twitches per unit time.

2.3.2 Muscle Fiber Types

The proportion of fiber types in an individual seems to be genetically determined, but the total number of fibers, their size, contractile properties, and metabolic and fatigue characteristics can change with training or lack of use.

Muscle fibers may be of the slow-twitch type (also called *Type I* or red fiber). Relatively low-rate signals trigger them; therefore they are often called low-threshold fibers. Slow fibers have small motor neurons and contract relatively leisurely, taking 80–100 ms until peak is reached. They produce comparatively low forces and are well suited for activities of light to moderate intensity even over extended time because they resist fatigue.

Other muscle fibers consist of fast-twitch fibers (also called *Type II* or white or high-threshold fibers), which have relatively short contraction times of about 40 ms. Fast fibers have large motor neurons and quick contraction–relaxation speed. They exist in two main kinds: fast fatigue-resistant (FR) fibers are capable of high force production, whereas fast fatiguable (FF) fibers can produce the largest force and the greatest contraction–relaxation speed but (as the name suggests) fatigue easily.

Slow fibers are mostly recruited for finely controlled actions, while strong efforts generally involve fast fibers. For activities that require rapid, high force and/or high power, FR units are recruited first, followed by FF units*.

2.4 Strength of Muscles and Body Segments

The term *strength* is often used in confusing ways: it may refer to

- the tension (or force) *within* a given muscle; or to
- the internal transferral of muscle effort across links and joints (*internal transmission*); or to
- a body segment's exertion of force or moment to an outside object (*external exertion*).

Employing the following definition and proper use of the related terms should provide clarity:

"**Muscle strength**" is the maximal force, or tension, that muscle can develop voluntarily between its origin and insertion—the vectors "M" in Fig. 2.10. Use of the proper terms "force" (in N) or "tension" (in N/mm^2 or N/cm^2) helps to avoid uncertainty.

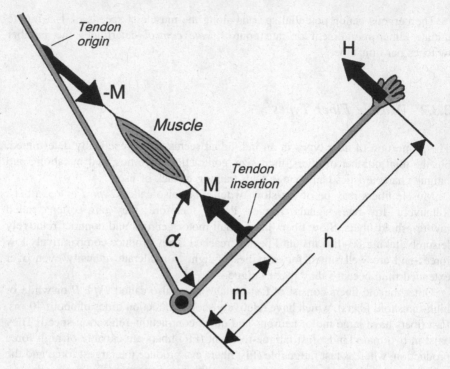

Fig. 2.10 The muscle-tendon unit exerting pull force M to bone links at origin and insertion

"**Internal transmission**" is the manner in which muscle effort (usually in terms of moments, in Nm) transfers inside the body along links and across joint (s). If the internal transmission path of the moment (in Nm) traverses several link-joint structures in series, each transfers the arriving moment (by the existent ratio of lever arms, such as m and h in Fig. 2.10) to the next link-joint until resistance is met, usually at the point where the body interfaces with an outside object.

"**Body strength**" (often **Segment Strength**) is the force, moment or torque that a body segment can apply to an object external of the body. The segment is usually hand, elbow, shoulder, back, or foot—for example, "Hand Strength H" in Fig. 2.10.

2.4.1 Muscle Strength

The "strength" of a muscle is properly described as the maximal force, or tension, that the muscle can develop voluntarily between its origin and insertion. Structurally, it depends primarily on the number of muscle fibers in use, and on their thickness. Strength training increases the thickness of fibers, but probably not their number. Endurance training also increases capillary density and mitochondrial volume. Either kind of training also improves the overall coordination of motor unit activation in the central nervous system (CNS, more on this in Chap. 3) because "muscles are the slaves of their motoneurons"*.

Within the muscle, filament contraction in the longitudinal direction of the muscle fiber generates tension, as discussed above. The tensions in each filament combine to a resultant tension of the muscle. Its magnitude is proportional to the cross-sectional thickness of the muscle. Maximal tensions measured in human skeletal muscle are within the range of 16–61 N/cm^2: 30 N/cm^2 is a typical value, occasionally called "specific (human muscle) tension"*. So, if the muscle cross-section area is known (such as from cadaver measurements or from magnetic resonance imaging (MRI) scans), one can estimate a resultant muscle force; however, this calculation relies on assumptions about cross-section and specific tension values. A technique suitable to measure the muscle force directly in the living human is still lacking.

2.4.2 Internal Transmission of Muscle Strength

The tension developed by a muscular contraction (whether concentric, isometric, or eccentric) tries to pull together two points: the *origin* of the muscle (that attachment closest to the center of the body) and *insertion* at the opposite end of the muscle-tendon unit (away from the center of the body)—see Fig. 2.10.

The origin (origins if the muscle has two "heads" such as the biceps, or three such as the triceps) is commonly at the surface of a bone. From there, usually a short tendon runs to the muscle "belly" with its contractile elements. On the other end of the muscle, a tendon reaches outward toward the insertion. The tendon may be quite long; for example, the tendons that connect the muscles in the forearm with the digits of the hand are around 20 cm in length. The tendon insertion usually attaches to bone, but may also end in strong connective tissues, such as in some of the fingers. (More about the hand below).

In engineering terms, muscle acts like a "linear motor" that generates pull force along its long axis, attempting to shorten its length. A muscle cannot push. The muscle pulls on long bones (biomechanically, the structural "links" between articulations) which serve as solid lever arms: that muscle pull generates a moment about a body joint. The moment developed depends, hence, not only on the strength of the muscle but also on its lever arm, and on the pull angle as well—compare Figs. 2.1 and 2.10.

The forearm link articulates in the elbow joint about the distal end of the upper arm link. The flexor muscle connects both links. It has its origin near the proximate end of the upper arm link* and it inserts on the forearm link at the relatively short distance m from the elbow. The flexor generates a muscle force M, which pulls on the forearm at the angle α (which depends on the existing elbow angle). Hence, the product of M and sin α is the component of M that is perpendicular to the forearm; it generates a moment T about the elbow. The magnitude of T depends on force M, its lever arm m and its pull angle α:

$$T = m\,M \sin \alpha$$

Moment T transforms into hand force H, perpendicular to the forearm link at its lever arm h:

$$T = Hh$$

Solving for hand force H yields

$$H = T/h = (m\,M \sin \alpha)/h$$

If H, h, m, and α are measured, one can compute the muscle force M:

$$M = (H\,h)/(m \sin \alpha)$$

(See also Chap. 4 for further use of biomechanical techniques).

This simple model shows that the amount of force (here, H) available at the interface of the body with an external object depends on the

- internal muscle force (here, M);
- internal lever arms (here, m and h);
- internal pull angle (here, α) which, in turn, depends on the angle between the two links.

By experience, we acquire the skill to position our body segments so as to achieve those lever arms and pull angles that permit the best use of our muscles to generate body strength needed to lift a heavy load, to push a big object, or to squeeze the handles of a hand tool.

The calculation of internally transferred moments is fairly simple for static efforts but can become quite complicated in dynamic exertions, for three reasons:

1. angles of force vectors change,
2. muscle functions change with motion, and
3. accelerations and decelerations of body masses require force expenditures.

2.4.3   Body (Segment) Strength

As a muscle contracts, its pulling force is internally transmitted to the approximate body member, being transformed in magnitude and direction during that transmission. This results in a force or torque that the body can apply to an external object, such as

- by hand to a tool, or to a handle on a box as in load lifting;
- by shoulder or back in pushing or carrying; or
- by the feet in operating pedals, or in walking or running.

The quality and quantity of the force, moment, or torque that the body can apply to an outside object depends on mechanical and physical conditions, especially on

- the body segment employed, for example hand or foot.
- the type of body object attachment, such as a simple touch or a surrounding grasp.
- the coupling type, by friction or with interlocking.
- the direction of the force/moment/torque vector.
- the needs for caution and control in task execution.
- static or dynamic exertion.

Consideration and proper selection of these conditions are critical tasks for the designer and ergonomist.

2.4.4   Exerting Strength with the Hand

The hands are our most-used work implements in daily life, able to both exert large forces and to perform delicate manipulations. Anatomically, they are complex structures based on 27 bones, as shown in Chap. 1, Fig. 1.6. The set-up of bones and the arrangement of muscles and tendons is a complicated example of internal transmission, just discussed.

Within the hand are the so-called intrinsic muscles. They mostly control the thumb (thenar muscles) and the little finger (hypopthenar muscles), and they execute adjustments of the metacarpals (interosseus muscles). However, major motions of the digits and the execution of large forces are primarily done by the extrinsic muscles, located in the forearm.

Since the extrinsic hand muscles are so far away from the digits, they possess long tendons, all of which must cross the narrow wrist. Each digit has two flexor tendons on the palmar side, one attaching to the distal phalanx and the other to a closer phalanx. The flexor muscles of each digit are on the underside of the forearm. At the dorsal side of the forearm are the extensor muscles: these straighten the digits.

Figure 2.11 provides a simplified view of a cut across the proximal part of the hand: the carpal bones form a shallow channel, which the transverse carpal ligament covers, thus making it the "carpal tunnel". This tight compartment must accommodate all the flexor tendons coming from the forearm as well as the median nerve and blood vessels.

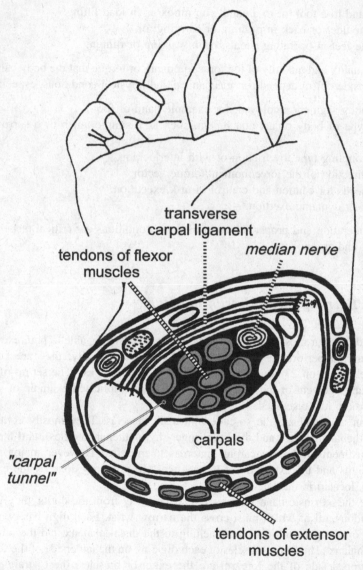

Fig. 2.11 Schematic view of the cross-section of the right hand near the wrist. The flexor tendons pass through the carpal tunnel near the palm. The extensor tendons lie side-by-side on the dorsal side of the carpal bones (adapted from Kroemer and Kroemer 2001)

On the other (dorsal) side of the carpal bones, the extensor tendons of the forearm muscles lie side by side, usually visible just under the skin. The thumb, the index and the little finger each have two extensor tendons; the other fingers each have one extensor. After passing the wrist, several of the extensor tendons interconnect at the back of the hand which can cause fingers to move simultaneously, especially the ring and little fingers.

Tendon sheaths envelope both flexor and extensor tendons. The sheaths are tough and flexible tubes that guide and keep moving tendons in place—see Figs. 2.12 and 2.13. That guidance is essential when tendons must change direction, such as when following a curved wrist or bent digit. Sheath structures contain fibrous ligament tissue, often in ring-like (*annular*) or crossed (*cruciform*) arrangements, which provide tubular loops functioning like pulleys so that tendons can bend digits instead of bow-stringing across joints—see Figs. 2.12 and 2.14.

The synovial layer of a sheath normally provides enough lubrication to ease the movements of the tendons within their enclosures; at the wrist, the displacement can amount to several centimeters. However, overuse can overload the lubricating capabilities of synovial fluid; examples are forceful repeated hammering or rapid

Fig. 2.12 Flexor tendons and their sheaths in the palm side of the hand (adapted from Kroemer and Kroemer 2001)

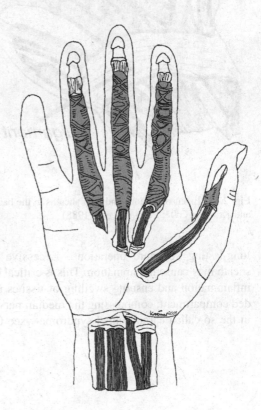

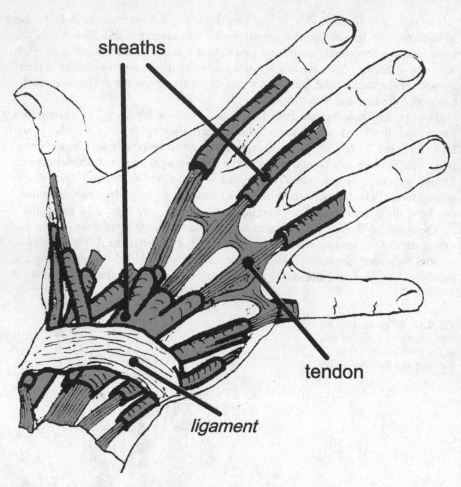

Fig. 2.13 Extensor tendons and their sheaths in the back side of the hand (adapted from Kroemer and Kroemer 2001; Putz-Anderson 1988)

long-lasting keyboard operation*. Excessive friction between a tendon and its sheath may cause inflammation. This is critical especially in the carpal tunnel where inflammation and ensuing swelling of tissues increases pressure within that crowded compartment; compressing the median nerve can make it malfunction, resulting in the so-called carpal tunnel syndrome—see Chap. 3.

Fig. 2.14 Schematic of the guidance provided by a tendon sheath (adapted from Kroemer 2017)

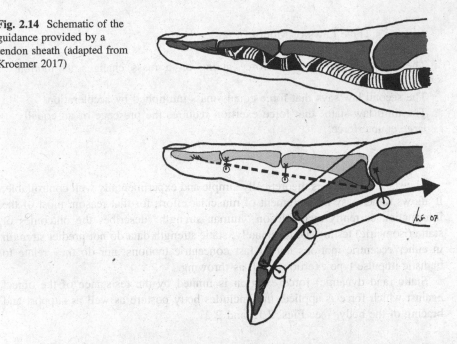

2.5 Static and Dynamic Exertions

In some muscle efforts, no segment movement and no visible change in muscle length occur; in traditional physiologic terminology, this is called an *isometric* exertion: from the Greek *iso*, meaning the same; and *metrein*, referring to the length of the muscle. (Yet, muscles and their tendons have elasticity: thus, they do stretch somewhat with the pull of a muscle. Accordingly, even in a so-called isometric effort, muscle length may change by a small amount).

Most efforts, however, are in motion (*variometric*): concentric when the muscle shortens, eccentric when the muscle becomes lengthened by an external force.

2.5.1 Static Strength

By definition, during an isometric effort, there is no change in muscle length and therefore involved body segments do not move. All forces acting within the system are in equilibrium, as the first of Newton's laws requires. Therefore, the physiological "isometric" case is about equivalent to the "static" condition in physics.

Newton's Three Laws of Motion:

The first law explains that force acting on a mass changes its motion condition.
The second law says that force equals mass multiplied by acceleration.
The third law states that force exertion requires the presence of an equally large counterforce.

The static condition is theoretically simple and experimentally well controllable. It allows rather easy measurement of muscular effort; for that reason, most of the information currently available on "human strength" describes the outcomes of static (isometric) testing. Unfortunately, static strength data do not predict strength in either eccentric motions nor in fast concentric motions, nor do they relate to ballistic-impulse type exertions such as throwing.

Static (and dynamic) force exertion is limited by the resistance of the object against which force is applied; this includes body posture as well as support and bracing of the body—see Figs. 2.15 and 2.21.

2.5.2 Dynamic Strength

Changing muscle length produces segment motion, a *dynamic* condition in mechanics terms. (Within the physics field of dynamics, *kinematics* describe the features and properties of motion; *kinetics* concern the forces that cause motion). Body movement incorporates changes in muscle configuration, muscle length, muscle force vector direction and leverage.

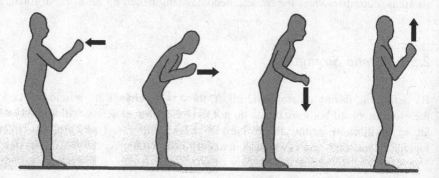

Fig. 2.15 Force exerted with the hand depends on body posture alone if bracing against solid support is not available

Given the arrangement of human skeletal muscles, especially their pull angles and lever arms (mechanical advantages), the displacement generally is small at the muscle but amplified at the point of exertion to the outside; Figs. 2.1 and 2.10 illustrate this for arm-hand travel.

The time derivatives of displacement (velocity, acceleration, and jerk) are important both for the muscular effort and the external effect: for example, acceleration (the change in velocity) determines force and impact, as per Newton's second law.

Definition and experimental control of dynamic muscle exertions are much more complex tasks than in static testing. Many schemes for experimentation using various independent and dependent experimental variables can be developed*; most attempts to measure dynamic efforts use equipment that makes the subject's hand or foot move on a certain path, generally linear or circular:

2.5.2.1 Control of Velocity

Control of velocity as the independent test variable: If velocity is set to a constant value, one speaks of an *isokinematic* (*isovelocity*) measurement. (Note that occasionally this condition is mislabeled "isokinetic"). Force and/or repetition are generally the chosen dependent test variables while mass is controlled. The limitation of this approach is that the constant velocity is demanded at the equipment interface (usually a handle or pedal), not at the muscle of interest. In this case, the actual rate of muscle length change is not isokinematic.

2.5.2.2 Control of the Amount of Muscle Tension

Control of the amount of muscle tension (force) as the independent test variable: force (moment, torque) is set to a constant value. In this *isotonic** test, mass properties and displacement (and its time derivatives) are likely to be controlled independent variables, and repetition is a dependent variable. Again, the limitation of this approach is that the constant effort is demanded at the equipment interface (such as handle or pedal), not at the muscle of interest. In this case, the actual muscle effort usually is not isotonic.

2.5.2.3 Concurrent Control of the Amount of Effort and Motion

Concurrent control of the amount of effort and motion means that the *isoforce* condition is often combined with an *isometric* condition, such as in holding a load motionless. This is obviously not a dynamic but a static exercise.

2.5.2.4   Controlling the Mass upon Which Muscle Acts

Controlling the mass upon which muscle acts by keeping it constant is the mark of an *isoinertial* test. Repetition of moving such constant mass (as in lifting or lowering, throwing, jumping, running) may either be a controlled independent or a dependent variable. Further, displacement and its derivatives may become dependent outputs. Force (or moment) applied is liable to be a dependent value, according to Newton's second law. The obvious limitation of this approach is that controlling the amount of external mass to be moved does not control muscle tension.

This complexity explains why, in the past, dynamic approaches have found few applications outside training and rehabilitation. Dynamic tests of muscle capabilities have not yet yielded many data applicable in ergonomics—the one exception is isoinertial testing which has found wide use in setting guidelines for "safe" lifting and lowering*. If one is allowed to perform as one pleases, such as in the "free dynamic tests" common in sports, there is no appreciable experimental control over the actual use of muscles; however, the experimenter can measure outcomes such as distance thrown, height jumped, weight lifted, running times achieved.

Mechanically, muscles and tendons show elasticity under tension: they are not rigid in length. Interestingly, this elastic property is generally acknowledged as allowing eccentric muscle stretch, but the existence of elasticity is commonly ignored in concentric muscle effort and usually disregarded in tendons. Neglecting to consider elastic properties can lead to misunderstandings regarding muscle strength measurements. Table 2.3 lists often used terminology in strength testing. All "iso-terms" can be grossly misleading because they generally describe equipment characteristics, not muscle events*.

Table 2.3 Terms describing muscular efforts

Term	Meaning	Same as	Opposite of
concentric	muscle shortens against resistance		eccentric
dynamic	muscle changes length and force		static
eccentric	lengthening of a resisting muscle by external force		concentric
isoinertial	effort moves a constant mass		
isokinetic	force remains constant	isotonic	
isokinematic	velocity remains constant		velocity changes
isometric	position remains constant	static	dynamic
isotonic	force remains constant	isokinetic	force (tension) varies
static	muscle length (and force) remain constant	isometric	dynamic

2.5.3 Relationships Between Measurements of Static and Dynamic Strength

Tension inside a muscle changes with the velocity of its shortening or lengthening. When length is not changing (the isometric condition), nervous activation signals can generate very strong muscle tension, as discussed earlier. When shortening, with increasing concentric velocity there is less and less time to establish forceful cross-bridges between actin and myosin filaments until, at the highest possible concentric speed, the force approaches zero. A similar effect occurs with increasing eccentric velocity: establishing cross-bridges becomes more difficult with increasingly fast lengthening, but the passive resistance of muscle-tendon tissues to stretch elevates the overall level of internal muscle tension.

Very slow motions may be interpreted as a sequence of static exertions, done one-by-one at every point along the expected path of motion*. Accordingly, static measurements of muscle strength also reflect its force capabilities in very slow motion. However, as the speed of motion increases, static data become increasingly less indicative of dynamic capabilities. Eccentric muscle efforts are usually rather slow; in this case, static measurements provide approximate information about eccentric muscle strength. In contrast, many concentric motions are quite fast, such as when running, throwing, shoveling or hammering. In such rapid muscle contractions, the dynamic muscle output is likely to be rather different from statically generated muscle strength.

2.6 Regulation of Strength Exertion

The conceptual model in Fig. 2.16 helps to understand the factors involved in the exertion of voluntary strength. It shows a series of feedforward commands, arising in the central nervous system (CNS, see Chap. 3), to generate muscular activities aimed to execute a given task. Throughout the steps of ensuing activities in the body, several feedback loops report on the status of activities, so that new commands can adjust the actions in order to achieve the task efficiently and safely.

2.6.1 Feedforward

The control initiatives of the central nervous system start by calling up an "executive program" which, innate or learned, exists for all common muscular activities, such as walking, hammering or lifting objects. "Subroutines" modify that general program to make it appropriate for the specific case, such as walking downstairs, hammering carefully, or lifting with caution. Another modifier is motivation, which determines how much of the structurally possible strength a person will exert under the given conditions. Table 2.4 contains a listing of circumstances that may increase or decrease one's willingness to exert.

Fig. 2.16 Schematic of
generation and control of
muscle effort

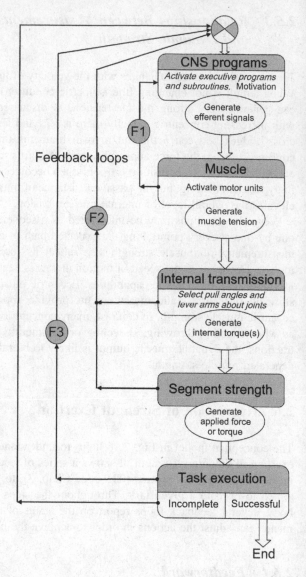

The results of these complex interactions are the excitation commands which the
efferent nervous pathways transmit to the motor units involved; there, the signals
trigger muscle contractions. The actual tension developed in the muscle depends on
the rate and frequency of the signals received, on the size of the muscle, and the
number of motor units involved.

The existing biomechanical conditions, especially lever arms and pull angles,
modify the output of the muscle effort. These conditions are constant in a static
effort but they change in the course of dynamic activities.

The internal transmission of the torque (s) generated around body joints (for
example from shoulder over elbow and wrist to the hand) alters the effects of

Table 2.4 Factors likely to increase (+) or decrease (–) muscular performance

Circumstances	Likely Effect
Feedback of results to subject	+
Instructions on how to exert	+
Arousal of ego involvement/aspiration	+
Drugs	+
Subject's outcry, startling noise	+
Hypnotic commands to perform strongly	+
Setting of goals, incentives	+
Competition, contest	?
Spectators	?
Fear of injury	–

muscle output until they arrive at the body segment that applies force/moment/torque to a tool or other object or to a measuring instrument.

2.6.2 Feedback

Figure 2.16 shows several feedback loops: they serve to monitor, control, and modify muscular exertion and body segment positions and motions. The first feedback loop, F1, is primarily a reflex-like arc that originates at receptors (proprioceptors, see Chap. 3) in the joints signaling location, in the tendons indicating changes in muscle tension, and in the muscles indicating their lengths. These interoceptors send their signals to the signal generator in the spinal cord very quickly—but how they do it is not well understood. Feedback loops F2 and F3 originate at exteroceptors and pass their signals through a "comparator", which compares the actual conditions reported by the body sensors to those conditions expected according to the feedforward signals. If actual and expected conditions differ, new commands are generated to eliminate the discrepancy.

Feedback loop F2 starts at receptors sending kinesthetic signals, reporting on events related to touch, pressure, and body position. When lifting an object, for example, body position and motion are monitored continuously along with the sensations of pressure in the hand and of the object rubbing against one's body.

Feedback loop F3 begins at other exteroceptors and provides information on sounds and vision related to the effort. For example, this may be the sounds or movements generated in equipment by the exertion of strength; it may be the pointer of an instrument that indicates the force applied; or it may be the experimenter or coach giving feedback and exhortation to the subject, depending on the status of the effort—see Table 2.4.

2.7   Measuring Muscle Strength

2.7.1   The "Maximal Voluntary Effort"

Ethically, people may not be subjected to tests that can damage them. Thus, tests of human body strength do not probe structural integrity of muscles and tendons; instead, the subject him/herself assesses what magnitude of effort is tolerable or suitable under the experimental conditions. That personal evaluation determines the magnitude of effort that the subject decides to execute. This fact is expressed in the term "maximal voluntary effort" (MVE), often called "maximum voluntary contraction" (MVC).

Accordingly, all human strength tests are self-controlled by the test subject. An individual's maximal voluntary effort is not only dependent on status, fitness and skill of the musculoskeletal system: rather, the MVE is also the result of instructions, feedforward and feedback signals, and experimental conditions. MVE depends decidedly on situational motivation, as listed in Table 2.4.

Instructions given to the subject, and comments made during the experiment, can have major impact on the magnitude of the MVE. An appropriate approach is for the experimenter to instruct the subject to exert maximal voluntary effort according the subject's self-assessment of suitability, safety, or avoidance of fatigue: typical for this approach are "psychophysical tests" regarding lifting or lowering loads where the subject receives the instruction to imagine that the task would have to be done continuously for hours*.

2.7.2   Measurement Opportunities

For a discussion of measurement techniques, it is advantageous to separate the model shown in Fig. 2.16 into three sections. First, the *feedforward* part contains the excitation signals, the contracting muscle and the internal transmission of muscle strength. The second part contains sensory *feedback*. The third part concerns the *output*, the force/torque/power that the hand (or other body segment) exerts to an external object, usually a work tool or measuring instrument. Each section provides opportunities for measurement.

2.7.2.1   Measuring via Feedforward

When considering the feedforward section of the model, it becomes apparent that there are still no suitable means to measure the executive programs, the subroutines, or the effects of will or motivation on the excitation signals, which the central nervous system generates. In spite of decades of experience in recording electroencephalograms (EEGs), we still can glean only rather general and quantitative information from them.

However, techniques are available which allow the observation and recording of efferent excitation impulses that travel along the motor nerves to the motor endplates of the involved muscle. Implanted (intrusive) or surface (non-invasive) electrodes can pick up these electric events, which are then recorded in electromyograms (EMGs). Their evaluation requires complex computer programs and skilled expertise* to conclude useful results. Experimental techniques to measure the tensions within filaments, fibers, or muscles need more development to become practical*.

It is relatively easy to observe and record the positions and relative angles of body segments. This information allows some general estimation of mechanical advantages (lever arms) during the internal transmission of muscle exertion. However, the internal mechanical advantages are complex even in a static posture and change during motions: consider the varying lever arms of tendon attachments and the pull angles as sketched in Fig. 2.10. Further development of biomechanical models with fine details of the human body—see Chap. 4—should lead to useful anthromechanical assessments.

2.7.2.2 Measuring via Feedback

In theory, the feedback loops in the nervous system offer some interesting possibilities for measurements. Yet, the afferent pathways from interoceptors are anatomically and functionally associated with the feedforward paths for the efferent impulses. Current technology makes it nearly impossible to distinguish the electric events associated with feedback signals from those associated with feedforward impulses. Until advanced measurement and interpretation techniques are available, the first two feedback loops (F1 and F2) remain not useful for strength measurements.

Yet, it is common to use the third feedback loop (F3), which starts at the subject's eyes and ears as exteroceptors, to control feedback by providing (or withholding) information about strength exertions*. Trainers and coaches routinely use such feedback manipulation to exhort enhanced performance.

2.7.2.3 Measuring Output

The strength exerted by a body segment is the final result of a complex system of feedforward and feedback signals, controlled elements, and modifying conditions. In an anthromechanical context, Figs. 2.15 and 2.21 illustrate that during efforts (static and dynamic) that involve several body segments, the magnitude of exertion depends on temporary body stances and on bracing against solid body supports—and, of course, on muscular capabilities.

During dynamic exertions, the arrangement of the chain of force vectors (examples in Figs. 2.20 and 2.24) changes with time and motion. Identification of the crucial link in the chain can be difficult even in static efforts. It is much more laborious in dynamic exertions where changing motion changes the biomechanical strain in the body links, and can add links to the chain or remove them.

Identification of the muscles involved is challenging: currently, EMG technology provides limited but incomplete help. Identification of those muscles which are critical to the effort is an arduous enterprise, particularly so during motion. Innovative procedures and instrumentation are still needed, especially for measuring individual muscle (tendon) force in vivo.

Figures 2.19 and 2.21 provide examples of "practical shortcuts" to measure body strength such as by hand-arm, shoulder, foot-leg, or back. Like in sports (referred to earlier), they do not explicitly identify the involved body links, do not describe their specific arrangements, and do not verify the critical muscles employed:

An instrument placed at the interface between the body and a resisting structure records unambiguously the combined output of all components in the loop. This leads to an operational definition: *"Maximal voluntary body (segment) strength is what is measured externally".*

Using that simple definition allows employing commonly available measuring devices to record the amount and direction of the force vector (or moment, torque, impulse, power) exerted over time. All external conditions are easy to describe: location of the interface between body and measuring device, position and support of the body, and so forth.

Simply measuring the resulting output does not satisfy the desire to understand and control the complicated system that generates the result. It does not provide information on specific muscles. However, there is one good reason for proceeding: the measured strength output is exactly the practical information that trainer, coach, physician, engineer and subject want.

2.8   The Strength Test Protocol

A carefully devised experimental protocol is essential for proper measurements. Such a protocol specifies the selection of subjects, their protection and their information; the control of the experimental conditions; the use, calibration, and maintenance of the measurement devices; and other important facts, such as (the avoidance of) training and fatigue effects. The protocol must be so exact and comprehensive that, by following it, another experimenter can replicate the measurements.

Regarding the selection of subjects, care must be taken to ensure that the subjects participating in the tests are in fact a representative sample of the population about which data are to be gathered. Regarding the management of the experimental conditions, the control over motivational aspects is particularly difficult. Outside of sports and medical testing, it is widely accepted that the experimenter should not give exhortations and encouragements to the subject.

Table 2.5 is a listing of items in a test protocol derived from the so-called Caldwell Regimen*, which was originally meant for isometric tests but minor amendments have adapted it for dynamic testing.

Table 2.5 Basics of strength test protocols

The following items are those of primary importance. Others should be added as appropriate. Items marked with ^ apply to both static and dynamic testing.

1^. Description of the subjects:
  (a) Population and sample selection.
  (b) Current health and status: medical examination and questionnaire are recommended.
  (c) Gender.
  (d) Age.
  (e) Anthropometry (at least height and weight).
  (f) Training and experience related to the strength testing.

2^. Information to the subjects about the test purpose and procedures.
  (a) Subjects shall avoid overexertion but give the best effort.
  (b) Instructions to the subject should be kept factual and not include emotional appeals.
  (c) Inform the subject during the test session about his/her general performance in qualitative, non-comparative, positive terms. Do not give instantaneous feedback during the exertion.
  (d) Rewards, goal setting, competition, spectators, fear, noise and the like can affect the subject's motivation and performance; therefore, should be avoided.

3. Static strength is measured according to the following conditions:
  (a) Static strength is assessed during a steady exertion sustained for four seconds.
  (b) The subject should be instructed to "increase to maximal exertion (without jerk) in about one second and maintain this effort during a four second count."
  (c) The transient periods of about one second each, before and after the steady exertion, are disregarded.
  (d) The strength datum is the mean score recorded during the first three seconds of the steady exertion. (The "peak" strength observed during the effort is up to one third larger than the average measure over an exertion period of three seconds.)

(continued)

Table 2.5 (continued)

4^. The minimal rest period between related efforts should be two minutes; more if symptoms of fatigue are apparent.

5^. Description of the conditions existing during testing:
(a) Body parts and muscles chiefly used.
(b) Body position (body motion in dynamic testing) related to the effort.
(c) Body support/reaction force available.
(d) Coupling of the subject to the measuring device (to describe location of the strength vector).
(e) Strength measuring and recording devices used.

6^. Data reporting:
(a) Mean (median, mode).
(b) Standard deviation.
(c) Skewness.
(d) Minimum and maximum values.
(e) Sample size.

2.9   Designing for Body Strength

The engineer or designer wanting to consider human strength faces a number of decisions. These include:

1. Is the exertion static or dynamic?

Most available body segment strength data concern static (isometric) exertions. These data (examples follow below) provide reasonable guidance also for slow motions, although they are probably too high for concentric motions and perhaps too low for eccentric motions. Of the little information available for dynamic strength exertions, most is limited to isokinematic cases. In dynamic exertions, other considerations often are important, such as physical endurance (circulatory, respiratory, metabolic) capabilities of the operator, and prevailing environmental conditions*.

2. Is the exertion by hand, by foot, or with other body segments?

If a choice of exertion modes is possible, the selection should rely on physiologic and ergonomic considerations to achieve the safest, least strenuous and most efficient performance. Usually, foot motions consume more energy, are less accurate and slower but stronger than hand movements over the same distance. Specific design information is available for hand and foot exertions and for use of other body segments—see the examples below.

3. **What are body posture and support during the strength exertion?**

Exertion of strength does not depend only on the size of the muscle mass involved, but also on the reaction force stabilizing the body—as per Newton's third law: force exertion requires the presence of an equal counterforce. The counterforce comes from support surfaces: the floor to stand upon, the seat to sit on, the backrest or wall to brace against. Strength transmission between the points of exertion and support depends on the positions and masses of involved body parts, which is especially important in dynamic tests. Figure 2.21 illustrates how body bracing and posture affect hand force exertion.

4. **Is a maximal or a minimal strength exertion the critical design factor?**

Maximal user output usually determines the structural strength of the object—even the strongest operator should not be able to break a handle or a pedal, for example. Therefore the design value is, with a safety margin, above the highest perceivable strength application.

Minimal user output is that strength expected from the weakest operator which still yields the desired result. Hence, it sets the operational force or torque limits at which a device can be successfully operated or a heavy object moved.

The minimal and maximal expected strength exertions establish the range of foreseen strength applications. Average (or median) user strength has usually no design value because half of the values (or users) are weaker and all others are stronger.

2.9.1 *Proper Statistical Use of Strength Data*

Measured strength data are commonly reported as averages (means) and standard deviations. This allows the use of common computational techniques to determine data points of special interest to the designer. In reality, however, strength data generally occur in a skewed dispersion rather than in a bell-shaped "normal" cluster. Treating them as if they stemmed from a normal distribution is a statistical mishap (more on this in Chap. 11) which is generally not of great concern to the designer because, usually, the data points of design interest are at the extremes of the distribution: either brute maximal forces or moments that the equipment must be able to bear without breaking, or minimal exertions that even "weak" persons are able to generate.

In the interest of safety, it is prudent to expect the strongest exertions to lie well above the strongest measured exertion because "panic" exertions can be much larger those measured in an experiment. Even in well-controlled strength tests on the same subject, measurements may vary considerably: standard deviations may easily vary by 10 or more percent*. Minimal strengths are selected at the low end of the distribution, often as 5th percentile values, as in Fig. 2.19.

2.9.2   Designing for Hand Strength

The human hand is able to perform a wide variety of activities, ranging from fine control to large forces:

- Fine manipulation of objects, requiring little displacement and force. Examples are writing by hand, assembly of small parts, adjustment of controls.
- Fast movements to an object, with moderate accuracy to reach the target but fairly small force exertion there. An example is the movement to a switch and its operation.
- Frequent movements between targets, usually with some accuracy but little force; such as in an assembly task, with parts taken from bins and assembled.
- Forceful activities with little or moderate displacement, such as in many assembly or repair activities, for example when turning a hand tool against resistance.
- Forceful activities with large displacements, for instance yielding an axe or sledge hammer.

Of the digits of the hand, the thumb is the strongest and the little finger the weakest. The whole hand, all digits in action combined with the palm, can exert large gripping and grasping strengths, but their execution depends on the coupling between the hand and the handle—see Fig. 2.17.

Muscles in the arm and shoulder can develop fairly large moments and torques. Flexion strength about the elbow joint follows in principle the distribution depicted in Fig. 2.18. Note that the data show the results of isolated static measurements, each done one after the other, not in one continuous sweep as the continuous curves connecting the static results may misleadingly suggest—see also Fig. 2.27 through Fig. 2.29.

Examples of static hand forces at different arm positions are shown in Fig. 2.19. Whereas these forces are applied with the hand, they are generated by arm and shoulder muscles and therefore depend on arm posture and general body support.

The flow of strength vectors between the points of application and support of the body depends on the positions of involved body segments, especially on their joint angles—see Fig. 2.19. Accordingly, the strength of exertions with the extremities, shoulder, backside or other, is determined both by body posture and by body support via friction or bracing. An example: a person standing on slippery ground may find it impossible to push a heavy object horizontally with hand or shoulder.

The amount of strength available for exertion onto an object outside the body depends on the weakest part in the chain of strength-transmitting body parts. Hand force, for instance, may be limited by finger strength, or shoulder strength, or low back strength; and in every case, by the reaction force available to the body. Figures 2.20 helps in determining where the weak link is in the chain of strength transmission.

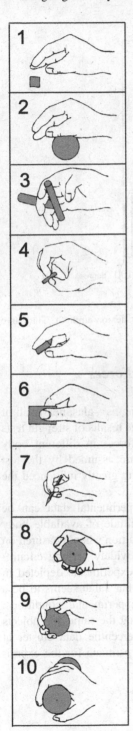

1. Digit Touch. One digit touches the object without holding it.

2. Palm Touch. Some part of the inner surface of the hand touches the object without holding it.

3. Finger Palmar Grip (Hook Grip). One finger or several fingers hook(s) onto a ridge or handle; thumb counteraction is not needed.

4. Thumb-Fingertip Grip (Tip Grip). The thumb opposes one fingertip.

5. Thumb-Finger Palmar Grip (Pinch or Plier Grip). Thumb pad opposes the palmar pad(s) of one finger or of several fingers near the tips. This grip evolves easily from coupling # 4.

6. Thumb-Forefinger Side Grip (Lateral Grip or Side Pinch).Thumb opposes the radial side of the forefinger.

7. Thumb-Two-Finger Grip (Writing Grip). Thumb and two fingers (often forefinger and index finger) oppose each other at or near the tips.

8. Thumb-Fingertips Enclosure (Disk Grip). Thumb pad and the pads of three or all four fingers oppose each other near the tips. The grasped object does not touch the palm.

9. Finger-Palm Enclosure (Collet Enclosure). Most or all of the inner surface of the hand is in contact with the object, enclosing it. This enclosure evolves easily from enclosure # 7.

10. Power Grasp. All of the inner surface of the hand grasps the object, which is often a nearly cylindrical handle arranged parallel to the knuckles and protruding on the side(s).

Fig. 2.17 Couplings between hand and handle (adapted from Kroemer 1986)

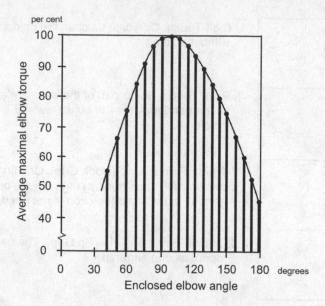

Fig. 2.18 Scheme of the relation between elbow angle and elbow flexion strength. The curve connects the results of single static tests

2.9.3   Using Tables of Exerted Moments and Forces

The literature contains many sources for data on body strengths—alas, nearly all of them for static efforts. Figure 2.21, for example, displays the results of strength tests on 45 young men who exerted static horizontal push forces in different body postures with various kinds of body bracing. The posture assumed by the test subject and the availability of reaction forces due to bracing greatly influenced the outcomes of the strength tests.

Deriving design recommendation from published experimental data can be challenging task: while the data indicate orders of magnitude of available body strength, the exact numbers should be viewed with great caution when measured on various subject groups, on rather few people, and under widely varying circumstances. Figure 2.22 graphs the scatter of all push force experiments depicted in Fig. 2.21. All 5th percentile data are close to 500 N; so that value seems to be a reasonable estimate of minimal forces exerted in all experimental conditions. Figure 2.23 similarly shows the scatter of all push data that the same 45 subjects exerted with only their preferred hand. Here, the 5th percentile data cluster at 250 N, which one may accept as a reasonable minimum estimate for the experimental conditions.

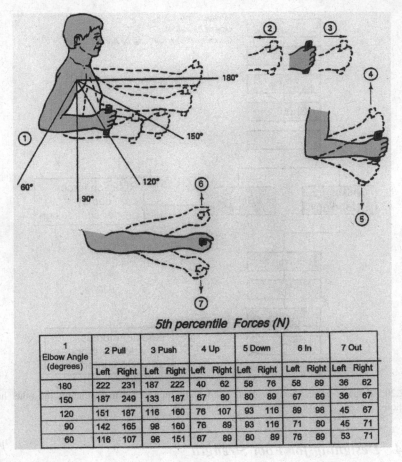

5th percentile Forces (N)

1 Elbow Angle (degrees)	2 Pull		3 Push		4 Up		5 Down		6 In		7 Out	
	Left	Right	Left	Right	Left	Right	Left	Right	Left	Right	Left	Right
180	222	231	187	222	40	62	58	76	58	89	36	62
150	187	249	133	187	67	80	80	89	67	89	36	67
120	151	187	116	160	76	107	93	116	89	98	45	67
90	142	165	98	160	76	89	93	116	71	80	45	71
60	116	107	96	151	67	89	80	89	76	89	53	71

Fig. 2.19 Hand forces exerted by sitting men: Fifth-percentile values, in Newton(adapted from MIL HDBK 759 1981)

The information in Table 2.6 derives from the same set of empirical data. They were extrapolated to show the effects of (a) location of the point of force exertion, (b) body posture, (c) body support and (d) friction at the feet on the horizontal push (and pull) forces, which male soldiers could exert. If the conditions or subjects are different, different strength values should be expected as well. To verify that a new design is operable, it is advisable to take body strength measurements on a sample of the intended user population, under conditions similar to actual use.

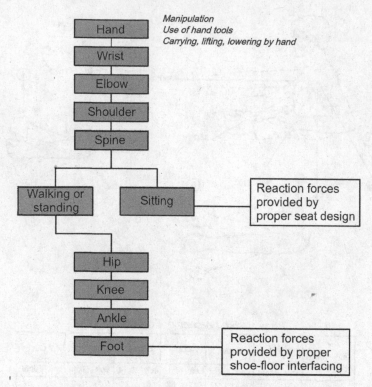

Fig. 2.20 Chain of critical body segments and body support for manipulating and other hand activities

2.9.4   Designing for Foot Strength

If a person stands at work, operation of foot controls should not be required because the operator would have to balance standing on only the other leg. For a seated operator, such as in a vehicle, use of foot controls is much easier because the seat supports the body. This allows the feet to move more freely. Given suitable conditions, the seated operator's feet can exert large forces and energies, for which the seat provides the needed reactions, as Figs. 2.24 and 2.25 indicate.

When riding a bicycle, the legs transmit energy through the feet to the pedals. Normally, the pedals are directly underneath the body, so that the body weight above them provides the reactive forces to the forces transmitted to the pedals. Placing the pedals more forward makes body weight less effective for generation of reaction force to the pedal effort; so, a recumbent bicycle needs a backrest against which the buttocks and low back press while the feet push forward on the pedals.

Height (1) of force plate	Distance (2)	Force, N	
		Mean	SD
 Percent of shoulder height*		With both hands	
50	80	664	177
50	100	772	216
50	120	780	165
70	80	716	162
70	100	731	233
70	120	820	138
90	80	625	147
90	100	678	195
90	120	863	141
 Percent of shoulder height*			
60	70	761	172
60	80	854	177
60	90	792	141
70	60	580	110
70	70	698	124
70	80	729	140
80	60	521	130
80	70	620	129
80	80	636	133
 Percent of shoulder height*		With both hands	
70	70	623	147
70	80	688	154
70	90	586	132
80	70	545	127
80	80	543	123
80	90	533	81
90	70	433	95
90	80	448	93
90	90	485	80
 100 percent of shoulder height*	Percent of thumb-tip reach*	With both hands	
	50	581	143
	60	667	160
	70	981	271
	80	1285	398
	90	980	302
	100	646	254
		With preferred hand	
	50	262	67
	60	298	71
	70	360	98
	80	520	142
	90	494	169
	100	427	173
 100 percent of shoulder height*	Percent of span*		
	50	367	136
	60	346	125
	70	519	164
	80	707	190
	90	325	132

Fig. 2.21 Horizontal push forces (means and standard deviations, in N) exerted by 45 young males with their hands, the shoulder and the back. Legend: (1) Height of the center of the force plate, 20 cm high and 25 cm wide; (2) Horizontal distance between the surfaces of the force plate and the opposing bracing structure; (*) Anthropometric definitions in Chap. 11 (adapted from Kroemer and Robinson 1971)

Fig. 2.22 Forces exerted with the hands, the preferred shoulder, or the back while braced against a footrest or wall—see Fig. 2.21 (adapted from Kroemer and Robinson 1971)

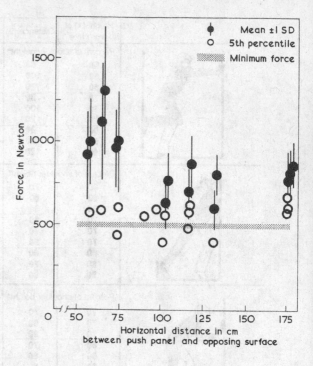

Fig. 2.23 Forces exerted with the preferred hand while braced against a footrest or wall—see Fig. 2.21 (adapted from Kroemer and Robinson 1971)

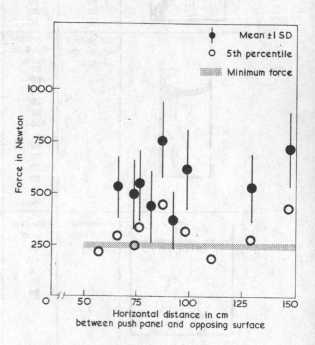

Table 2.6 Horizontal push and pull forces (in N), which male soldiers can exert intermittently or for short periods of time (adapted from Kroemer and Robinson 1971)

Horizontal Force* at least	Applied with (see Fig. 2.21)	Condition**
100 N, push or pull	Both hands, or one shoulder, or the back	With low traction, 0.2 $< \mu >$ 0.3
200 N, push or pull		With medium traction, μ about 0.6
250 N push	One hand	Braced against a vertical wall located 0.5 to 1.5m from and parallel to the push plate; OR Anchoring the feet on perfectly non-slip ground
300 N, push or pull	Both hands, or one shoulder, or the back	With high traction, μ about 0.9
500 N, push or pull		Braced against a vertical wall located 0.5 to 1.5m from and parallel to the push plate; OR Anchoring the feet on perfectly non-slip ground
750 N push	The back	Braced against a vertical wall located 0.5 to 1.5m from and parallel to the push plate; OR anchoring the feet on perfectly non-slip ground

* Force may be nearly doubled for two operators, nearly tripled for three operators
** μ is the coefficient of friction at the ground

Fig. 2.24 Chain of critical body segments and body support for foot actions

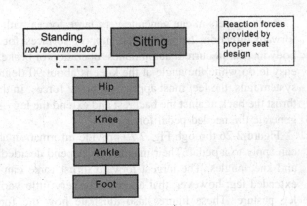

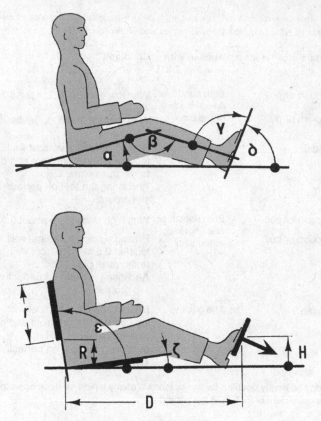

Fig. 2.25 Conditions affecting pedal force: body angles (upper illustration) and work space dimensions (adapted from Kroemer 1971)

A sitting person can generate very large forces with nearly extended legs in the forward direction, helped by the support surfaces for buttocks and back and by body inertia. In current automobiles, operation of brake or clutch pedal is normally easy to do while the angle at the knee is about 90 degrees. But if the power-assist system fails, the feet must apply high brake forces: in this case, the operator should thrust the back against the backrest and extend the leg—as in Fig. 2.25—in order to generate the needed pedal force.

Figure 2.26 through Fig. 2.29 provide information about the forces that the foot can apply to a pedal. Their magnitudes depend decidedly on body support and hip and knee angles. The largest forward thrust force can be exerted with the nearly extended leg: however, that allows only very little variation in pedal position and leg posture. These figures also illustrate how the force that the foot can exert depends on the chain of strength transmission between seat and pedal. Bad seat design and adjustment; frail muscles; deficient joints at hip, knee or ankle; and/or low friction between foot (shoe) and pedal may all make for a "weak kick".

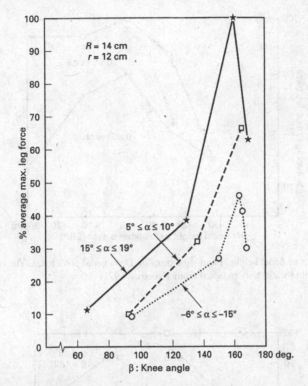

Fig. 2.26 Effects of thigh angle α and knee angle β on pedal push force. The curves connect the results of single static tests (adapted from Kroemer 1971)

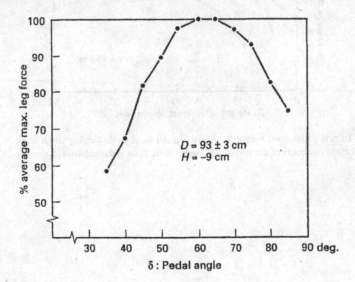

Fig. 2.27 Effects of ankle (pedal) angle δ on foot force generated by ankle rotation with fixed leg extension D and pedal height H. The curve connects the results of single static tests (adapted from Kroemer 1971)

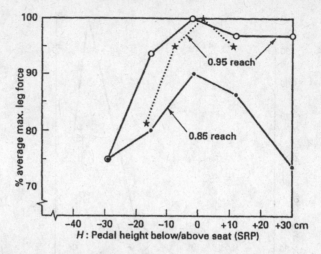

Fig. 2.28 Effects of pedal height H and leg extension D on pedal push force. The curves connect the results of single static tests (adapted from Kroemer 1971)

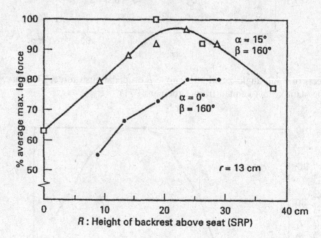

Fig. 2.29 Effects of backrest height R on pedal push force with varying thigh angle α and knee angle β. The curves connect the results of single static tests (adapted from Kroemer 1971)

Notes

The text contains markers, *, to indicate comments and references, which follow:

The actin "ratchets" along the myosin: Huxleys' "Sliding Filament" theory: Herzog (2008).

The term contraction: Cavanagh (1988).

Viscoelastic qualities: in engineering terms, a viscous material resists shear flow and deforms over time when load is applied. An elastic material immediately deforms when load is applied, and returns to its original shape once the load (or stress) is removed.

FR units are recruited first, followed by FF units: Kumar (2004, p 18).

"Muscles are the slaves of their motoneurons": Basmajian and DeLuca (1985, p. 431).

"Specific human muscle tension": Enoka (1988).

The flexor muscle has its origin near the proximate end of the upper arm link: This is a simplification: in fact, the biceps has to two origins ("heads", as the name indicates) one of which crosses the shoulder joint.

Long-lasting keyboard operation: Kroemer (1989, 2001, 2017), Kroemer (2017).

Strength tests using various independent and dependent experimental variables: Kroemer et al. (1989), Kumar (2004, 2008).

Isotonic: The term *isotonic* often has been applied wrongly. Some older textbooks described lifting or lowering of a constant mass (weight) as typical for isotonics. This is physically false for two reasons: first, according to Newton's laws of motion, changes in acceleration and deceleration of a mass require application of changing (not constant) forces. The second fault lies in overlooking the changes that occur in the mechanical conditions (pull angles and lever arms) under which the muscle functions during the activity (see Chap. 4 for more detail). Hence, even if there were a constant force to be applied to a moving external object, the changes in mechanical advantages still would cause changes in muscle tension.

Guidelines for "safe" lifting and lowering: Snook (2005), Waters (2008).

Terms using "iso-" generally describe equipment characteristics, not muscle events: Significant gaps in understanding muscle functions exist; terminology describing testing procedures is inexact and uses misnomers: Kumar (2004).

A sequence of static exertions, done one-by-one at every point along the expected path of motion: The common graphic style of connecting the data points with an uninterrupted curve misleadingly implies a continuous motion, whereas in reality the strength measurements were done statically, point by point.

Psychophysical tests: Dempsey (2004), Snook (2005), Waters (2008).

Evaluations of electromyograms (EMGs): Basmajan and de Luca (1985), Biasiucci et al. (2019), Merletti et al. (2004), Sommerich and Marras (2004).

Experimental techniques to measure the tensions within filaments, fibers, or muscles: Bell and Syrotuik (2008), Fukashiro et al. (1995), Komi et al. (1996), Schuind et al. (1992).

Control feedback by providing (or withholding) information about strength exertions: Kroemer and Marras (1980).

Caldwell Regimen: Caldwell et al. (1974), Dempsey (2004), Gallagher et al. (2004).

Strength assessments: Astrand et al. (2003), Wilmore et al. (2008).

Dynamic exertions and environmental conditions: Physiologic and ergonomic texts provide such information; for example by Astrand et al. (2003), Chaffin et al. (2006), Kroemer Elbert et al. (2018), Chengular et al. (2003), Gallagher et al. (2004), Winter (2004), Cavagna (2017).

Standard deviations of static strength may easily vary by 10 or more percent: Astrand et al. (2003).

Literature on body strength: Compiled, for example in standards, by ISO, NASA and the Military; by Chengular et al. (2003), Kroemer et al. (2003), Kroemer Elbert et al. (2018), Kumar (2004, 2008), Marras and Karwowski (2006).

Summary

Muscle contraction is brought about by active shortening of muscle sub-structures. Elongation of the muscle is done by external forces.

Excitation signals from the central nervous system control muscle contraction. Each specific signal affects those fibers that form a motor unit, of which a muscle contains many.

An efferent stimulus causes a single twitch contraction of the motor unit. A rapid sequence of stimuli can lead to a superposition of muscle twitches, which may fuse together into a sustained contraction.

Prolonged strong contraction leads to muscular fatigue, which hinders and cuts short the muscle effort. Hence, a maximal voluntary contraction can last only a few seconds.

During an isometric contraction, (in theory) muscle length remains constant; this establishes a static condition for the affected body segments. In an isotonic effort, the muscle tension remains constant, which usually coincides with a static (isometric) effort.

Dynamic activities result from changes in muscle length, which bring about motion of body segments. In an isokinematic effort, body segment speed remains unchanged. In an isoinertial test, external mass properties remain constant. Note that terms using the prefix "iso" are often misnomers.

Maximal muscle tension depends on the individual's muscle size, muscle length, and motivation.

Individual body (segment) strength depends on the person's muscular properties, on the person's motivation, and on situational factors, especially posture and body support.

Individual body (segment) strength is measured routinely as force (or moment or torque) applied to an instrument external to the body.

Measurement of strength necessitates carefully controlled experimental conditions.

Design of equipment and work tasks for human body segment strength capabilities requires

- *determining whether the exertion is static or dynamic*
- *establishing with what body part force or moment or torque is exerted*
- *arranging the best suitable body posture*
- *selecting the strength percentile (minimum and/or maximum) that is critical for the operation.*

Glossary

Acceleration Second time-derivative of displacement.

Actin Muscle filament capable of sliding along myosin (see there).

Activation of muscle See contraction.

Agonist The muscle performing an intended action—same as protagonist. See also antagonist.

Antagonist The muscle opposing the action of an agonist.

Anthromechanics Biomechanics applied to the human body.

Body (segment) strength The ability to exert force (see there) or moment (see there) or torque (see there) to an external object.

Co-contraction Simultaneous contraction of two or more muscles.

Concentric (muscle effort) Shortening of a muscle against a resistance.

Contraction (of muscle) Literally, "pulling together" the z-disks delineating the length of a sarcomere, caused by the sliding action of actin and myosin filaments. Contraction develops muscle tension only if the shortening is resisted. During an isometric "contraction" no change in sarcomere length appears; in an eccentric "contraction" the sarcomere is actually lengthened. To avoid such contradiction in terms, it is often better to use the terms activation, effort, or exertion.

Dependent variable In experiments and tests, the variable whose value shows the effects, if any, of the controlled (independent) variable(s).

Displacement Distance moved (in a given time).

Dynamics A subdivision of mechanics that deals with forces and bodies in motion.

Eccentric (muscle effort) Lengthening of a resisting muscle by external force.

Effort (of muscle) See contraction.

Elastic Spontaneous regaining of former shape after distortion; opposite of plastic.

Electromyogram EMG, graphic record of the electric activity of a muscle.

EMG See Electromyogram,.

Energy (The capacity to do) work. Proper units are the joule (J) and calorie (cal).

Endomysium Connective tissue enwrapping a muscle fiber.

Epimysium Connective tissue enwrapping muscle.

Ergonomics The application of scientific principles, methods and data drawn from a variety of disciplines to the design of engineered systems in which people play significant roles.

Exertion (of muscle) See contraction

Fascia Smooth band or sheet of connective tissue separating muscles and internal organs, for example as epimysium.

Fascicle Bundle of muscles, fascilulus.

Fast-twitch fiber Muscle fiber with relatively short contraction time (about 40 ms) and a large motoneuron. Also called type II or white or high-threshold fiber.

Fiber See muscle.

Fibril See muscle fibers.

Filament See muscle fibers.

Force In physics, a vector that can accelerate a mass. As per Newton's second law, the product of mass and acceleration; the proper unit is the Newton, with $1 \text{ N} = 1 \text{ kg m s}^{-2}$. On Earth, 1 kg applies a (weight) force of 9.81 N (1 lb exerts 4.44 N) to its support due to gravity. Muscular force often is described as tension (stress) multiplied with the transmitting cross-sectional area.

Free dynamic In this context, an experimental condition in which neither displacement and its time derivatives, nor force are manipulated as independent variables.

Independent variable In experiments and tests, the variable whose value is intentionally manipulated to show effects, if any, on the observed (dependent) variable(s).

In situ In the original position/place.

In vivo Within the living organism/body.

Iso (Greek) Prefix meaning equal; unchanged, constant.

Jerk Third time-derivative of displacement.

Kinematics A subdivision of dynamics that deals with the motions of masses, but not with the causing forces. Compare with kinetics.

Kinetics A subdivision of dynamics that deals with forces applied to masses. Compare with kinematics.

Lever arm One component of the formula for moment (or torque): "force *times* lever arm". Also called mechanical advantage, it represents the distance from a fulcrum at which the force vector acts.

Mechanical advantage In this context, the lever arm (see there) of a force vector.

Mechanics The branch of physics that deals with forces applied to bodies and their ensuing motions.

Mitochondrium Organelle in a (muscle) cell, able to produce ATP—see Chap. 7.

Moment The product of force and its lever arm when trying to rotate or bend an object; the stress in material generated by two opposing forces that try to bend the material about an axis perpendicular to its long axis; see also Torque.

Motor endplate Contact area of axon and sarcolemma of the muscle.

Motor unit All muscle filaments under the control of one efferent nerve axon.

Muscle A bundle of fibers, able to contract or be lengthened. In this context, striated (skeletal) muscle that moves body segments about each other under voluntary control.

Muscle contraction The result of contractions of motor units distributed through a muscle so that the muscle shortens its length.

Muscle fiber Element of muscle, containing fibrils which consist of filaments.

Muscle fibril Element of muscle fiber, containing filaments.

Muscle filaments Muscle fibril elements (actin and myosin, both polymerized protein molecules) capable of sliding along each other, thus shortening the muscle and, if doing so against resistance, generating tension.

Muscle force The product of tension within a muscle multiplied with the transmitting muscle cross-section.

Muscle strength The ability of a muscle to generate and transmit tension in the direction of its fibers. See also body strength.

Muscle tension The pull within a muscle expressed as force divided by transmitting cross-section

Myo (Greek) Prefix referring to muscle.

Myosin Muscle filament (see there) along which actin (see there) can slide.

Mys (Greek) Prefix referring to muscle.

Perimysium Connective tissue enwrapping bundles of muscle fibers.

Physiology The branch of biology that deals with the physical and chemical functions of living organisms and their parts.

Plasmalemma Membrane of a muscle cell, also called sarcolemma.

Power Work (done) per unit time.

Protagonist The muscle performing an intended action – same as agonist. See also antagonist.

Rate coding The time sequence in which nerve signals arrive at a specific motor unit and cause it to contract.

Recruitment coding The time sequence in which nerve signals arrive at different motor units and cause them to contract.

Repetition Performing the same activity more than once. (One repetition indicates two exertions).

Sarcolemma Membrane of a muscle cell, also called plasmalemma.

Sarcoplasmic reticulum The "plumbing and control" system of the muscle.

Slow-twitch fiber Muscle fiber triggered by relatively low-rate signals; therefore often called a low-threshold fiber. It has a leisurely contracting time (80–100 ms) and small motoneurons. Also called type I or red fiber.

Statics A subdivision of mechanics that deals with bodies at rest.

Strength See body strength and muscle strength.

Tension Force divided by the cross-sectional area through which it is transmitted.

Torque A force applied at a lever arm that twists or rotates something about its central axis; see also Moment

Type I fiber Slow-twitch muscle fiber, see there.

Type II fiber Fast-twitch muscle fiber, see there.

Variometric Of differing length.

Velocity First time-derivative of displacement.

Viscoelastic Exhibiting both viscous behavior (deformation response to load is time-dependent) and elastic behavior (deformation response to load is instantaneous and recoverable)

Work The product (integral) of force and distance moved; a measure of the energy expended to apply a force to move an object.

References

Astrand PO, Rodahl K, Dahl HA, Stromme SB (2003) Textbook of work physiology. Physiological bases of exercise, 4th edn. Human Kinetics, Champaign

Basmajian JV, DeLuca CJ (1985) Muscles alive, 5th edn. Human Kinetics, Baltimore

Bell GJ, Syrotuik DG (2008) Physiology and biochemistry of strength generation and factors limiting strength development in skeletal muscle (Chapter 2). In: Kumar S (ed) Biomechanics in ergonomics, 2nd edn. CRC Press, Boca Raton

Biasiucci A, Franceschiello B, Murray, MM (2019) Electroencephalography. Curr Biol 29(3): R80–R85 (2019). https://doi.org/10.1016/j.cub.2018.11.052

Caldwell LS, Chaffin DB, Dukes-Dobos FN, Kroemer KHE, Laubach LL, Snook SH, Wasserman DE (1974) A proposed standard procedure for static muscle strength testing. Am Ind Hyg Assoc J 35(4):201–206

Cavagna G (2017) Physiological aspects of legged terrestrial locomotion. Springer, Heidelberg

Cavanagh PR (1988) On "muscle action" vs "muscle contraction". Biomechanics 21:69

Chaffin DB, Andersson GBJ, Martin BJ (2006) Occupational biomechanics, 4th edn. Wiley, New York

Chengular SN, Rodgers SH, Bernard TE (2003) Kodak's ergonomic design for people at work, 2nd edn. Wiley, New York

Dempsey PG (2004) Psychophysical aspects of muscle strength (Chapter 7). In: Kumar S (ed) Biomechanics in ergonomics, 2nd edn. CRC Press, Boca Raton

Enoka RM (1988) Neuromechanical basis of kinesiology. Human Kinetics, Champaign

Fukashiro S, Komi PV, Jaervinen M, Miyashita M (1995) In vivo achilles tendon loading during jumping. Eur J App Physiol 71:453–458

Gallagher S, Moore JS, Stobbe TJ (2004) Isometric, isoinertial, and psychophysical strength testing devices and protocols (Chapter 8). In: Kumar S (ed) Biomechanics in ergonomics, 2nd edn. CRC Press, Boca Raton

Guyton AC (1979) Physiology of the human body, 5th ed. Saunders, Philadelphia

Herzog W (2008) Determinants of muscle strength (Chapter 4). In: Kumar S (ed) Biomechanics in ergonomics, 2nd edn. CRC Press, Boca Raton

Komi PV, Belli A, Huttunen V, Bonnefoy R, Geyssant A, Lacour JR (1996) Optic fibre as a transducer of tendomuscular forces. Eur J Appl Physiol 72:278–280

Kroemer AD, Kroemer KHE (2017) Office ergonomics, 2nd edn. CRC Press, CRC Boca Raton

Kroemer KHE (1971) Foot operation of controls. Ergonomics 14(3), 333–361

Kroemer KHE (1986) Coupling the hand with the handle: an improved notation of touch, grip, and grasp. Hum Factors 28:337–339

Kroemer KHE (1989) Cumulative trauma disorders: their recognition and ergonomic measures to avoid them. Appl Ergon 20(4):274–280

Kroemer KHE (2001) Keyboards and keying. An annotated bibliography of the literature from 1878 to 1999. Int J Univ Access Info Soc UAIS 1(2):99–160

Kroemer KHE (2017) Fitting the human. Introduction to ergonomics, 7th edn. CRC Press, Boca Raton

Kroemer KHE, Kroemer AD (2001) Office ergonomics. Taylor & Francis, London

Kroemer KHE, Kroemer HJ, Kroemer-Elbert (2003) Ergonomics: how to design for ease and efficiency, 2nd edn. Prentice Hall, New York

Kroemer Elbert KE, Kroemer HB, Kroemer Hoffman AD (2018) Ergonomics: how to design for ease and efficiency, 3rd edn. Academic Press, London

Kroemer KHE, Marras WS (1980) Toward an objective assessment of the 'maximal voluntary contraction' component in routine muscle strength measurements. Europ J Appl Physiol 45:1–9

Kroemer KHE, Robinson DE (1971) Horizontal static forces exerted by men standing in common working positions on surfaces of various tractions. AMRLTR-70-114. Wright-Patterson AFB, Aerospace Medical Research Laboratory

Kumar S (ed) (2004) Muscle strength. CRC Press, Boca Raton

Kumar S (ed) (2008) Biomechanics in ergonomics, 2nd edn. CRC Press, Boca Raton

Marras W, Karwowski W (eds) (2006) The occupational ergonomics handbook, 2nd edn. CRC Press, Boca Raton

Merletti R, Farina D, Rainoldi A (2004) Myoelectric manifestations of muscle fatigue (Chapter 18). In: Kumar S (ed) Biomechanics in ergonomics, 2nd edn. CRC Press, Boca Raton

MIL HDBK 759 (1981) Human factors engineering design for army material. U.S. Army Missile Command. Naval Publications and Forms Center, Philadelphia

NASA (1989) Man-systems integration standards, Revision A and following, NASA-STD 3000. LBJ Space Center, Houston

Nordin M, Andersson GBJ, Pope MH (1997) Musculoskeletal disorders in the workplace: principles and practices. Mosby, St. Louis

Putz-Anderson V (1988) Cumulative trauma disorders. Taylor & Francis, London

Schuind F, Garcia-Elias M, Cooney WP, An KN (1992) Flexor tendon forces: in vivo measurements. J Hand Surg 17:291–298

Snook SH (2005) Psychophysical tables: lifting, lowering, pushing, pulling, and carrying (Chapter 13). In: Stanton N, Hedge A, Brookhuis K, Salas E, Hendrick H (eds) Handbook of human factors and ergonomics methods. CRC Press, Boca Raton

Sommerich CM, Marras WS (2004) Electromyography and muscle force (Chapter 17). In: Kumar S (ed) Biomechanics in ergonomics, 2nd edn. CRC Press, Boca Raton

Waters TR (2008) Revised NIOSH lifting equation (Chapter 20). In: Kumar S (ed) Biomechanics in ergonomics, 2nd edn. CRC Press, Boca Raton

Wilmore JH, Costill D, Kenney WL (2008) Physiology of sport and exercise, 4th edn. Human Kinetics, Champaign

Winter DA (2004) Biomechanics and motor control of human movement, 3rd edn. Wiley, New York

Chapter 3
Neuromuscular Control

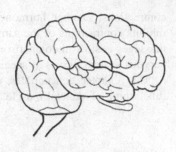

Overview

The central nervous system is one of the control and regulation networks of the body. It collects inputs from various sensors that respond to internal and external stimuli. The cerebrum, the cerebellum, and the spinal cord contain the main integration and regulation functions concerning motor activities. The pathways for incoming and outgoing signals are the neurons, which possess the ability to facilitate or inhibit the transmission of impulses.

The Model

The nervous system transmits information about events inside and outside the body from various sensors along its afferent pathways to the brain. Here, decisions about appropriate actions and reactions are made, and feedforward signals are generated and sent along the efferent pathways to muscles.

Introduction

The overall purpose of the human regulatory and control systems is to maintain equilibrium (*homeostasis*) on the cellular level and throughout the body despite changing demands on the body generated by varying external environments and task requirements. Temporary but often instantaneous action signals must be generated in response to acute demands on the body.

The human body is under the control of two corresponding organizations: the endocrine system and the nervous system. Endocrine glands produce hormones, which circulate with the blood stream to carry chemical messages to those cells programmed to receive them. For example: one set of hormones (*norepinephrine*) stimulates smooth muscle in some organs, but in others inhibits the muscular contractions; another hormone (*acetylcholine*) has just the opposite effect and inhibits contractions of the same smooth muscles. The response time to hormonal messages is slow in comparison with the reaction to electrical impulses, which the nervous system directs at specific receptors, for example in finger flexor muscles.

This chapter concentrates on the nervous system, particularly as it affects functioning of skeletal muscle.

3.1   Organization of the Nervous System

The human nervous system and its functions can be described by function or by anatomical location.

3.1.1   Organization by Function

Functionally, two major divisions of the human nervous system are the autonomic (visceral) system and the somatic system. The *autonomic* nervous system, ANS, has two functional branches: the sympathetic division stimulates the "fright, flight or fight" response whereas the parasympathetic division controls the unconscious relaxation into "rest and digest" or "feed and breed". The *somatic* (or *somesthetic*: Greek *soma*, body) nervous system, SNS, provides voluntary control of mental activities, skeletal muscle, and conscious actions and body movements.

3.1.2   Organization by Location

Anatomically, the nervous system has three major subdivisions. The first subdivision is the *central* nervous system, CNS, which includes brain and spinal cord; it has primarily control functions. The second subdivision is the *autonomic* nervous system, described above; it generates instinctive attitudes and regulates involuntary functions such as cardiac and smooth muscle, blood flow, digestion, or glucose release in the liver. The third subdivision is the *peripheral* nervous system, PNS, which includes the cranial and spinal nerves; it has no control functions but transmits signals to and from the brain.

Figure 3.1 depicts this organization of the human nervous system. CNS and PNS, the central and peripheral nervous systems, are of particular importance to the ergonomist.

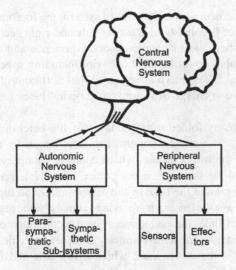

Fig. 3.1 Organization of the human nervous system

3.2 The Central Nervous System: Brain and Spinal Chord

The brain of an adult human weighs between 1200 and 1400 g and has approximately 100 billion nerve cells. Figure 3.2 shows the customary division of the brain into several sections. Its left and right halves, called hemispheres, are grossly anatomically similar mirror-images of each other.

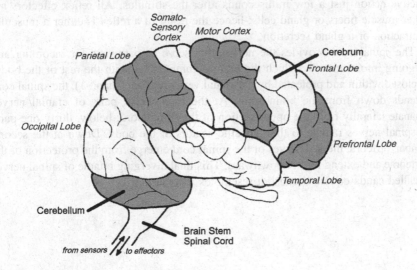

Fig. 3.2 The human brain seen from the right side

With respect to the neuromuscular control system, the forebrain's *cerebrum* is of particular importance: it consists of the two (left and right) cerebral hemispheres, each divided into lobes, and manages sensory experience and consciousness. The cerebrum's frontal lobes control skilled behavior including speech, mood, thought, planning; the parietal lobes interpret sensory input and control body movements; the occipital lobes are in charge of vision; the temporal lobes generate memory and emotions.

The *cortex*, the many-folded outmost layer of the cerebrum, is important for attention, perception, awareness, thought, memory, language, and consciousness. Specific regions are the motor cortex, which controls voluntary movements of the skeletal muscles, and the somato-sensory cortex, which interprets sensory inputs.

Deep within the brain, at the base of the cerebrum and on top of the brain stem, are the thalamus (associated with relaying signals and regulating alertness), hypothalamus (which links the endocrine system to the nervous system), and the basal ganglia (associated with semi-voluntary complex activities such as walking).

In the hindbrain is the *cerebellum*, which distributes and integrates impulses from the cerebral association centers to the motor neurons in the spinal cord. In humans, the cerebellum helps regulate voluntary movement by contributing to coordination, balance, precision, and accurate timing.

The *brain stem* connects the main brain with the spinal cord, and controls basic vital body functions such as breathing, heart rate, and consciousness. The human brain stem includes the midbrain, pons, and medulla oblongata.

The *spinal cord* is an extension of the brain: it coordinates fast actions, particularly reflex limb movements such as the "knee jerk". A reflex begins with a stimulation of a sensory receptor, which triggers a signal to the spinal cord. There, it evokes an immediate response and impulse to the appropriate muscles. Since no time-consuming higher brain functions are involved, an effector can execute a reactive action just a few milliseconds after the stimulus. All reflex effectors are either muscle fibers or gland cells; hence the result of a reflex is either a muscular contraction or a gland secretion.

The spinal cord provides the nerves that serve as pathways for incoming and outgoing information by which the brain communicates with the rest of the body. Enclosed within and protected by the spinal vertebrae (see Chap. 1), the spinal cord extends down from the brain stem. At the top, twelve pairs of cranial nerves emanate laterally from the upper section of the spinal cord; below, thirty-one pairs of spinal nerves radiate to their specific sectors of the body. Down at the second lumbar vertebra, the last nerves of the spinal cord emerge from the protection of the vertebrae and extend farther downward. This final diverging bundle of spinal nerves is called cauda equina because it resembles a horse tail.

3.3   Sensors and Effectors of the Peripheral Nervous System

The peripheral nervous system, PNS, has an *afferent* (sensory) division and an *efferent* (motor) division. The sensory division transmits information from body sensors to the CNS, and therefore is sometimes called the feedback section of the PNS. It carries signals concerning the outside from external receptors *(exterocep- tors)*, and from internal receptors *(interoceptors)* reporting on changes within the body. Since all of these sensations come from various parts of the body, external and internal receptors together are also called somatic sensors.

Internal receptors include the *proprioceptors* (Latin *proprius*, "one's own"). Among these are the muscle spindles, nerve filaments wrapped around muscle fibers; they detect the amount of stretch of the muscle. Golgi organs are associated with muscle tendons and detect their tension, hence report to the central nervous system information about the strength of contraction of the muscle. Ruffini organs are kinesthetic receptors located in the capsules of articulations. They respond to the degree of angulation of the joints (joint position), to slow changes in join position, and to sustained strain.

Another set of interoceptors, called *visceroceptors*, reports on events within the visceral (internal) structures of the body, such as organs of the abdomen and chest, as well as on events in the head. The usual modalities of visceral sensations are pain, burning sensations, and pressure. Similar sensations may also come from external receptors; since the pathways of visceral and external receptors are closely related, information about the body is often integrated with information about the outside.

External receptors provide information about the interaction between the body and the outside: sight (vision), sound (audition), taste (gustation), smell (olfaction), temperature, chemical agents, and touch (taction). Several of these are of particular importance for the control of muscular activities: in particular, the sensations of touch, pressure, and pain can be used as feedback to the body regarding the direction and intensity of muscular activities transmitted to an outside object.

The sensors in the *vestibulum* (see Fig. 3.3) of the inner ear are also proprio- ceptors: they detect and report the position of the head in space and respond to sudden changes in its attitude. This is done by sensors in the three semi-circular canals, each located in another orthogonal plane. Signals from proprioceptors in the neck, triggered by displacements between trunk and head, allow to relate the position of the trunk to that of the head.

Figure 3.4 sketches the location of common receptors at the body surface: Krause's end bulbs react to cold; Ruffini, Meissner's, and Pacinian corpuscles respond to warmth, touch, and pressure, respectively. Merkel's endings are

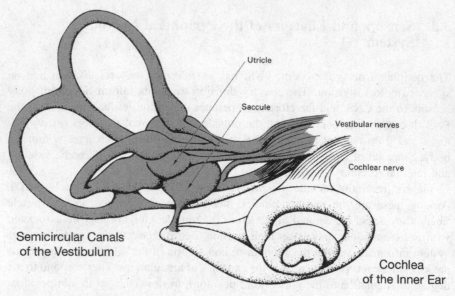

Utricle

Saccule

Vestibular nerves

Cochlear nerve

Semicircular Canals
of the Vestibulum

Cochlea
of the Inner Ear

Fig. 3.3 The vestibulum

mechanoreceptors, responding to touch. Furthermore, free nerve endings abound throughout the skin of the body; they respond to light and heavy touch, to heat and cold, and probably are chiefly responsible for reporting pain sensations.

The types of receptors, and their densities, vary throughout the body: tongue, lips, fingers and feet are the most sensitive regions of the body. Since nerve pathways interconnect extensively, the sensations reported are not always specific to a modality; for example, very hot or cold sensations can be associated with pain, which may also be caused by hard pressure on the skin.

Almost all sensors respond vigorously to a change in the stimulus but, if the load remains, report less and less during the next seconds or minutes. This adaptation makes it possible to live with continued pressure, such as from clothing. The speed of adaptation is specific to sensors. Furthermore, the speeds with which the sensations are transmitted to the central nervous system are quite different for different sensors: light and sound cause the fastest responses whereas pain transmission is usually the slowest.

The efferent (motor) division of the PNS transmits information from CNS through the nerves and ganglia. Many decisions made in the CNS that are based on the evaluation of incoming signals in turn generate action signals which the PNS's motor division transmits to "effector" muscles. Hence, the motor division is also often called the feedforward section of the PNS. The main effects (as pertain to this book) of the action signals are to make muscle motor units contract (for more on muscles, see Chap. 2).

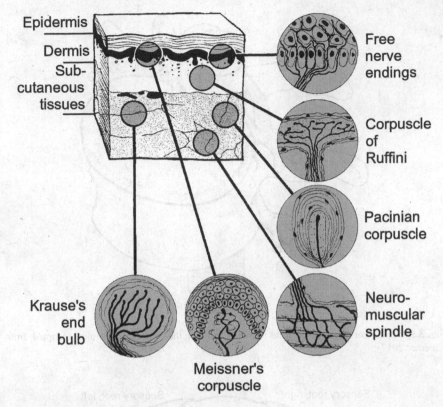

Epidermis

Dermis

Sub-
cutaneous
tissues

Free
nerve
endings

Corpuscle
of
Ruffini

Pacinian
corpuscle

Krause's
end
bulb

Neuro-
muscular
spindle

Meissner's
corpuscle

Fig. 3.4 Location of common receptors near the body surface (adapted from Langley and Cheraskin 1958)

3.3.1 Nervous Pathways of the Peripheral Nervous System

The spinal cord runs from the brain stem down to the second lumbar vertebra. The stacks of bone arches of the vertebrae (*intra*-vertebral foramcna, see Figs. 1.1–1.9 in Chap. 1) form a protective tunnel (occasionally called a channel) for the soft and fragile cord. Between adjacent vertebrae, two nerve bundles emerge from the cord to the left and the right, as depicted in Figs. 3.5 and 3.6; each passes through a lateral opening between adjacent vertebrae (*inter*-vertebral foramen). These lateral nerve bundles are often called nerve roots (which is what they look like as they emerge from the stem). The nerves contain the fibers of both the sensory and motor tracts; they are the pathways by which the brain and spinal cord communicate with the rest of the body.

The uppermost section of the spinal cord contains twelve pairs of *cranial* nerves, which serve structures in the head and neck as well as the lungs, heart, pharynx and larynx, and many abdominal organs. They control eye, tongue, and facial movements, and the secretion of tears and saliva. The main inputs are from the eyes, the

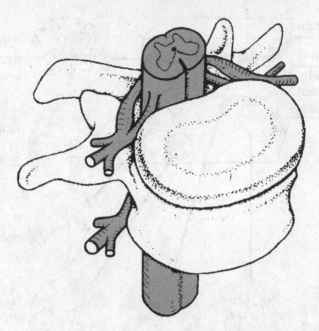

Fig. 3.5 Spinal cord and nerve roots passing through the vertebral foramina (adapted from Kroemer 2017)

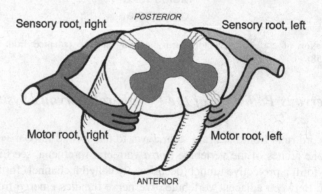

Fig. 3.6 Sensory and motor roots (adapted from Kroemer 2017)

taste buds in the mouth, the nasal olfactory receptors, and touch, pain, heat, and cold receptors of the head.

Below the neck, thirty-one pairs of *spinal* nerves emanate between the thoracic and lumbar vertebrae and serve defined sectors of the rest of the body. Figure 3.7 shows specific areas of the skin (dermatomes) innervated by distinct nerves and lists where their roots start in the spinal column.

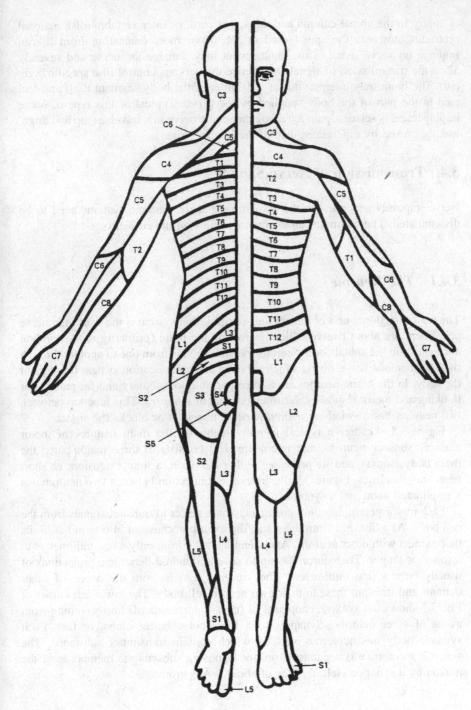

Fig. 3.7 Sensory dermatomes with the spinal nerve roots: C means cervical, T thoracic, L lumbar, S sacral

Injury to the spinal column and the spinal cord, or inter-vertebral disk material protruding towards the spinal cord or the nerve roots emanating from it, can impinge on nerve tissue. This impingement may damage the nerve and severely affect the transmission of signals and hence the nervous control of a specific body part. The brain may interpret this as a disorder of the body segment itself and feel pain in the part of the body supplied by the nerve. Typical of this type of nerve impingement is sciatica, pain felt along the sciatic nerve that traverses hip and thigh, usually caused by a herniated disk of the lumbar spine.

3.4   Transmission of Nerve Signals

Nerve impulses generated in the autonomic and the somatic systems need to be disseminated. The communication structure relies on linked neurons.

3.4.1   The Neuron

The basic functional unit of the nervous transmission system is the *neuron*, a nerve cell. There are about twelve billion neurons in the brain (primarily in the cerebral cortex) and in the spinal cord. Neurons transmit signals from one to another through their filamentous nerve fibers, which serve as a communication system throughout the body. In the brain, neurons are also responsible for memory and for patterns of thinking and motor responses. Neurons connect in *synapses*. This junction between two neurons has a switching ability: it either transmits or blocks the signal.

Figure 3.8 sketches a typical neuron combining the main features of motor neurons, sensory neurons, and interneurons. It consists of three major parts: the main body (soma), and its processes: either the axon, a long extension, or short branching dendrites. Figure 3.9 illustrates the connection between two neurons via a myelinated axon and a synapse.

Each motor neuron has only one axon, which serves to transmit signals from the cell body. At a distance from the soma, the axon branches out into terminal fibrils that connect with other neurons. Axon lengths range from only a few millimeters to a meter or longer. The neuron has up to several hundred dendrites, projections of usually only a few millimeters. They receive signals from the axons of other neurons and transmit these to their own neuron cell body. The lower left corner of Fig. 3.7 shows the synapse endpoints of (usually hundreds of) fibrils coming from axons of other neurons. Synapses look like knobs, bulbs, clubs, or feet. Each synaptic body has numerous vesicles which contain transmitter substance. The synaptic membrane is separated from the opposing subsynaptic membrane of the neuron by a synaptic cleft, a space of about 2 Ångstrom.

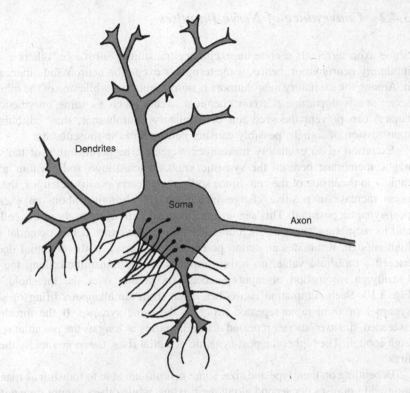

Fig. 3.8 Typical neuron components: soma, dendrites, axon. Synapses are sketched only in the lower left part of the figure

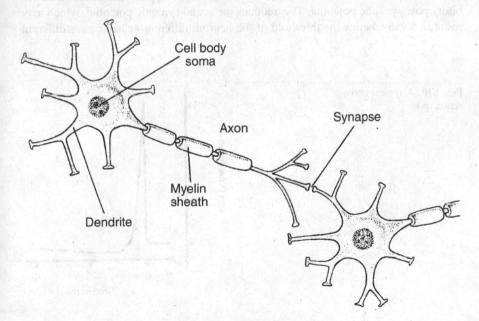

Fig. 3.9 Two motor neurons connected by axon and synapse

3.4.2   Conveyance of Nerve Impulses

Some axon terminals secrete an excitatory transmitter substance, others carry an inhibitory neurohumor; hence, some terminals excite the neuron and others inhibit it. Among the excitatory neurohumors is norepinephrine, while one of the inhibitory secretions is dopamine. Certain chemical agents (such as some anesthetics and curare) can prevent the secretion of excitatory neurohumor, thus inhibiting the transmission of signals, possibly causing paralysis of the musculature.

Excretion of an excitatory transmitter increases the permeability of the subsynaptic membrane beneath the synaptic knob, which allows sodium ions to flow rapidly to the inside of the cell. Since sodium ions carry positive charges, the result is an increase in positive charge inside the cell, bringing about an excitatory post-synaptic potential. This sets up an electrical current throughout the cell body and its membrane surfaces, including the base of the axon. If this potential spikes high enough, it initiates an action potential in the axon. If the potential does not exceed a threshold value, no action impulse will be transmitted along the axon. Usually, a summation of inputs succeeds in getting over the threshold, as in Fig. 3.10. Such summation can either result from simultaneous firing of several synapses or from rapid repeated firing of the same synapse. If the threshold is exceeded, the axon carries repeated action impulses as long as the potential remains high enough. The higher the post-synaptic potential rises, the more rapidly the axon fires.

Depending on their type and size, some neurons are able to transmit as many as a thousand impulses per second along their axons, while others cannot transmit more than about twenty-five signals per second.

Inhibitory transmitters in a synapse create a negative potential called the inhibitory post-synaptic potential. This reduces the actual synaptic potential, which may result in a value below the threshold of the neuron: different neurons have different

Fig. 3.10   A typical nerve action spike

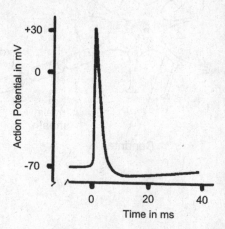

thresholds. Thus, the inhibitory synaptic knobs may prevent neuron discharge and stop transmission.

Most nervous impulses are not carried by a single neuron from receptor to destination but follow a chain of linked neurons. Once developed, the action potential propagates itself point-by-point along a nerve fiber, not needing further stimulation. At each successive point along the fiber, the action potential rises to a maximum and then rapidly declines. Because of this shape, the action potential is called a spike potential. The height of the spike, its strength, does not vary with the strength of the stimulus, nor does it weaken as it progresses along the axon. The stimulus is either strong enough to elicit a response, or there is no potential; this is an all-or-none phenomenon similar to the one discussed in the contraction of skeletal muscle.

While nerve fibers hardly fatigue, synapses do; some very rapidly, others rather slowly.

The velocity of the nerve impulse is a constant for each nerve fiber, ranging from 0.5 m/s to about 150 m/s. The particular speed does not change along a specific fiber; it correlates with the diameter of the fiber, being faster in a thick fiber than in a thin strand. In the peripheral parts of the nervous system, axons and dendrites are sheathed along their length. The envelope consists of myelin, a white material composed of protein and phospholipid. The presence of a myelin sheath allows a greater speed of signal conduction. Skeletal muscles are served by thick myelinated axons terminating at the motor endplates, while pain fibers are the thinnest and not myelinated.

3.5 Control of Muscle

Reflex actions can take place without involvement of higher brain centers, but most complex voluntary muscular activities need fine regulation through efferent feed-forward and afferent feedback control, sketched in Fig. 3.11. Muscle actions differ, so they require variable involvement of the higher brain centers, such as the cerebral cortex, the basal ganglia and the cerebellum. Apparently, the motor cortex controls

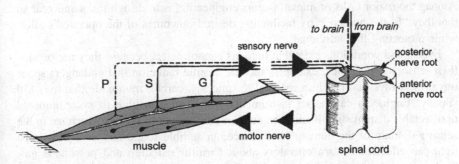

Fig. 3.11 Motor and sensory nerves controlling muscle action: pain signal P; signal from muscle spindle S; signal from Golgi organ G (adapted from Kroemer 2017)

mostly very fine, discrete muscle movements, whereas everyday gross movements, such as walking, running or posture control, are tasks done in the basal ganglia with their large pools of neurons. This may be the locus for "executive programs" mentioned in Chap. 2.

Activation of a muscle follows this sequence:

Step 1: An excitation signal travels along the efferent nervous pathways to the axon of the signal-carrying motor nerve, which terminates at a motor endplate located in a z-disk: this is the myoneural junction for the motor unit of the muscle (discussed in Chap. 2).

Step 2: The excitation signal stimulates a de-polarizing action of the muscle cell membrane. This allows spread of the action potential along the sarcoplasmic reticulum.

Step 3: The potential triggers the release of calcium ions into the sarcoplasmic matrix surrounding the filaments of the motor unit.

Step 4: This removes tropomyosin, the hindrance for interactions between actin and myosin filaments through chemical, mechanical, and electrostatic actions.

Step 5: Opposing heads of actin rods move towards each other, sliding along the myosin filaments: their heads may meet, bunch up, even overlap. This is a contraction of the sarcomere. Shortening of many fibrils at the same time shortens the whole muscle. (Note that a muscle can attempt to contract but, instead, may become lengthened by an overwhelming stretching force—see Chap. 2.)

Step 6: Rebounding of calcium ions in the sarcoplasmic reticulum switches the contraction activity off, which allows the filaments to relax.

A common way to record the occurrence of efferent signals is to implant a wire or needle electrode, or to attach a surface electrode to the skin, near the motor endplates of the innervated motor unit of the muscle of interest (see Chap. 2) and to record the electrical activities in an electromyogram, EMG*.

3.6   Ergonomic Engineering to Facilitate Control Actions

Among the major tasks of human factors engineering is to design tasks and gear so that they "fit the human"* by facilitating desired outcomes of the operator's effort while protecting the individual.

Events and conditions exist that humans cannot detect because they are outside their sensory capabilities; examples include cosmic radiation that endangers space travelers, x-rays that can hurt medical technicians, carbon monoxide that may kill silently. Engineers can select instruments which are sensitive to such humanly undetectable distant stimuli and then convert their output to signals which are in the sensory domain of the human: for instance, an audible sound and a visible strobe light can effectively warn operators about harmful radiation and poisonous gas. Such technical conversion is useful if distal stimuli are fully incompatible with human sensors or if stimuli are not conspicuous enough to be perceived securely.

Fig. 3.12 Converting distal signals into proximal stimuli that can serve as inputs to the human processor (adapted from Kroemer 2006)

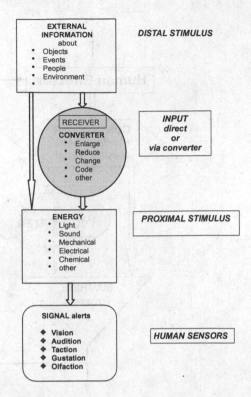

Figure 3.12 illustrates the principle of transforming imperceptible distal signals into proximal stimuli which humans can process.

Converting human physical output as needed to control complex machinery, such as automobiles, ships, airplanes and spacecraft, is a similar ergonomic task—see Fig. 3.13. One of the technical issues concerns insufficient human strength: various ways to augment muscular capabilities are currently used, such as "power(ed) steering" and "power(ed) brakes" in many vehicles. One of the allied tasks is to provide appropriate sensory feedback so that, in spite of the converting device, the operator retains "feel for the behavior" of the machinery.

A related task is the prediction of the future response of a compound system following current control inputs; examples are the course that a large ship will travel depending on a current rudder setting, or the future trajectory of a plane or spacecraft in response to present control actions. Generating immediate feedback regarding future system performance requires not only knowledge of the converters used to transmit human inputs to the system, but also of the dynamic properties and future behavior of the system: often, appropriately programmed computers provide that information.

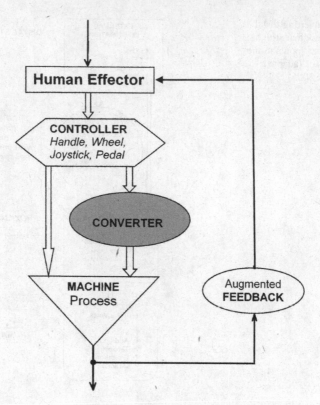

Fig. 3.13 Transforming human effector output (adapted from Kroemer 2006)

Injury to an efferent nerve reduces the ability to transmit signals to innervated muscle motor units and hence impedes the control of activities of muscles, for instance to organize and regulate specific force or torque for application to tools, equipment, and work objects. Likewise, impairment of afferent nerves reduces the information that sensors can relay to the central nervous system. Sensory feedback is very important for many activities because it contains information about force and pressure applied, position assumed, and motion experienced.

Sensory nerve impairment usually brings about sensations of numbness, tingling, or pain in the associated body part. Impairment of an autonomic nerve reduces the ability to control such functions as sweat production in the skin. Common signs of reduced autonomic function are dryness and shininess of skin areas controlled by that nerve.

Spinal cord injuries often cause severe impairment of nervous pathways. Falls from a height and automobile accidents, for example, may cause sudden displacement of the head with respect to the trunk. Too often, this results in dislocation or breakage of vertebrae which, in turn, damages the spinal cord passing through the vertebral foramen. Damage to the cord may render it incapable of transmitting feedforward signals to body areas innervated by nerves below the injury point, and also make it unable to conduct feedback signals from these body parts to the brain. As a result, the victim has no control over or sensation in the innervated body segments.

To the engineer, the task is preventing such accidents, or at least to design protective devices to mitigate the consequences: so far, little success has been achieved for climbers of ladders, whereas persons in automobiles have benefitted from improved protective shells (automobile body designs) and from shock-protecting devices inside the cabin, such as air bags, head rests, and body restraints, which can minimize whiplash and body impact on car structures.

Acute trauma can affect nerve functioning, but so can the cumulative effects of often repeated light trauma, even if each event by itself is not harmful. A typical case is the "carpal tunnel syndrome" (CTS) in the hand: it is the final episode in a chain of events. The common causes are overly repetitive finger motions, as in long-lasting in piano playing, keyboarding, knitting, gardening and many other hand activities at work or leisure.

Carpal tunnel syndrome (CTS) is among the best known cumulative trauma disorders. It has been reported in the medical literature starting in the middle and late 19th century, initially called palsy, spasm, or cramp, and later called myalgia and tenosynovitis. In 1959, Tanzer described several CTS cases resulting from cumulative trauma: two of his patients had recently started to milk cows on a dairy farm, three worked in a shop in which objects were handled on a conveyor belt, two had done gardening with considerable hand weeding, and one had been using a spray gun with a finger trigger. Two other patients had been working in a large kitchen where they stirred and ladled soup twice daily for about 600 students.

In 1966 and 1972, Phalen described the typical gradual onset of numbness in the thumb and in the three adjacent fingers of the hand. In 1975, Birkbeck and Beer described the results of a survey they made of the work and hobby activities of 658 patients who suffered from CTS: about 80% of them did work that required light yet highly repetitive movements of the wrists and fingers. In 1976, Posch and Marcotte analyzed 1,201 cases of CTS*. Yet, even today CTS remains a common problem, affecting millions of adults (and more women than men) each year.

Excessive repetition of finger flexor tendons motions in their sheaths causes increasing friction and, often, inflammation of these tissues. The associated swelling leads to pressure inside the narrow space between the carpal bones of the hand and the transverse carpal ligament, the "carpal tunnel" shown in Fig. 2.11 of Chap. 2. Continuing the repetitive actions worsens inflammation, swelling and pressure. That strain compresses the tendons in their sheaths, blood vessels and the median nerve, which all pass through the carpal tunnel. The median nerve inner-vates thumb, index and middle fingers and the near side of the ring finger. (The little finger and the outside of the ring finger are under control of the ulnar nerve; it passes through Guyon's canal and may suffer from ulnar tunnel syndrome.) Excessive pressure on the nerve impairs all its functions, autonomic, sensory and motor. Usually, the victim first has the feelings of tingling and numbness in the innervated digits, then pain and reduced control of finger motions. The medical diagnosis of CTS often relies on the measurement of reduced conduction velocity in the afflicted median nerve. The solution is straight-forward: a surgeon (partly) cuts the restricting transverse carpal ligament to relieve the pressure and give the tissues space to heal.

Of course, it is best to avoid the problem by not performing excessive repetitions of finger flexions. Accordingly, the fundamental solutions are in the expertise of ergonomic engineers who can design equipment and work tasks to suit human capabilities. Since the 19th century, overuse of the Morse telegraph key, of the keyboard of the piano, and of Sholes' 1878 QWERTY typewriter keyboard all have been known to lead to cumulative repetition trauma*, yet only the Morse key has fallen by the wayside.

Technology that today seems at the brink of practical use, especially nan-otechnology, may provide means to pick up nervous signals and transmit these to effectors, and to provide feedback to the CNS. Such bioengineering achievement could provide immeasurable help to persons whose natural body control system has been damaged.

Notes

The text contains markers, *, to indicate specific references and comments, which follow.

Electromyogram: Sommerich, Marras (2004, 2008), Marras and Karwowski (2006), Biasiucci et al. (2019).

Design tasks and gear so that they "fit the human": for example, Kroemer (2006, 2017), Kroemer and Kroemer (2017), Kroemer Elbert et al. (2018).

Carpal tunnel syndrome, CTS: Tanzer (1959); Phalen (1966, 1972), Birkbeck, Beer (1975), Posch and Marcotte (1976), Pfeffer et al. (1988), Arndt and Putz-Anderson (2001).

Keyboard and key use leading to cumulative repetition trauma: Kroemer (2001, 2017), National Research Council (1999).

Summary

The body must continuously control muscle functions according to information about conditions and events reported from its many and varied sensors. The information feedback to the central nervous system (where decisions are made and actions initiated) and the feedforward signals to the muscles flow along the peripheral nervous system.

Afferent and efferent signals are transmitted along neurons, consisting of soma, dendrites, and axons. At the neuron, nerve fibrils (from other neurons) end in synapses which serve as selective switches. Depending on the strength of the incoming signal, it is either transmitted or not transmitted across the synaptic membrane.

With many sensors reacting to the same stimulus, and many afferent pathways transmitting the signals at different intensities and speeds, the peripheral nervous system serves as a filter or selector for the central nervous system.

Damage to efferent or afferent nerves usually results in loss of motor control and of sensory feedback.

Glossary

Adaptation Responsive adjustment, here diminished reaction.

Afferent Carrying inward, toward the CNS, see there.

Autonomic (visceral) nervous system Part of the nervous system that regulates involuntary actions; has sympathetic and parasympathetic subsystems.

Axon Long process of a nerve fiber, conducting impulses away from the nerve soma.

Carpal tunnel syndrome (CTS) An impairment of the median nerve.

Central nervous system (CNS) Brain and spinal cord together.

Cerebellum (Latin, small brain) Part of the brain that regulates complex voluntary muscle actions, maintains body posture and balance.

Cerebrum Largest portion of the brain, controls and integrates motor and sensory functions as well as higher functions such as thought, reason, memory and emotions.

CNS Central nervous system, see there.

Cortex The outer layer of grey matter of the brain covering the cerebral hemisphere.

Cranial nerve Nerve that serves structures in the head, neck, lungs, heart, pharynx, larynx, and many abdominal organs.

CTS Carpal tunnel syndrome, see there.

Cumulative trauma Damage that increases by successive addition of small injuries.

Dendrite Short process of a nerve fiber that conducts impulses inward to the nerve soma.

Dermatome Skin area innervated by the sensory fibers of a single spinal nervenerve.

Distal stimulus Stimulus (see there) outside the body.

Efferent Carrying outward from the CNS to an effector, usually a muscle, gland or organ.

Electromyogram (EMG) Graphic record of the electric activity of a muscle.

EMG Electromyogram, see there.

Ergonomics The application of scientific principles, methods and data drawn from a variety of disciplines to the design of engineered systems in which people play significant roles.

External receptors Receptors that respond to signals outside the body; same as exteroceptors.

Exteroceptors See external receptors.

Feedback Information carried inward toward the CNS from body sensors.

Feedforward Information carried outward from the CNS to an effector, often a muscle.

Foramen Opening or passage in a bone.

Innervate To supply a body part with nerves.

Internal receptors Receptors that respond to signals inside the body; same as interoceptors.

Interoceptors See internal receptors.

Kinesthesia Sense that detects body position, and movements of joints, muscle, tendons.

Motor endplate Contact area of axon and sarcolemma of the muscle; see Chap. 2.

Motor unit All muscle filaments under the control of one efferent nerve axon; see Chap. 2.

Muscle A bundle of fibers, able to contract or be lengthened. In this context, striated (skeletal) muscle that moves body segments about each other under voluntary control; see Chap. 2.

Muscle contraction The result of contractions of motor units distributed through a muscle so that the muscle shortens its length; see Chap. 2.

Myelin An envelope around an axon that consists of protein and phospholipid.

Nerve root Nerve bundle that emanates laterally from the spinal cord.

Neuron A nerve cell.

Parasympathetic system The part of the autonomic nervous system (see there) that opposes the physiological effect of the sympathetic system.

Peripheral nervous system (PNS) The cranial and spinal nerves nerves outside the CNS, see there.

Plasmalemma Membrane of a muscle cell, also called sarcolemma.

PNS Peripheral nervous system, see there.

Proprioceptors Sensors that report on the status of the body (Latin *proprius*, "one's own").

Proximal stimulus Stimulus (see there) that arrives at a body sensor.

Reflex Involuntary action produced automatically in response to a stimulus.

Root See nerve root.

Sciatica Pain felt along the sciatic nerve that traverses hip and thigh.

Soma Body, such as of a cell (Greek *soma*, body).

Somatic Of, relating to, or affecting the body.

Spinal cord Strand of nerve tissues extending from the medulla down through the foramina of the spinal column.

Spinal nerves Bundles of nerve that emanate between the thoracic and lumbar vertebrae and serve defined sectors of the rest of the body, see nerve roots.

Stimulus Agent, action or condition that elicits (or should elicit) a physiological or psychological response.

Sympathetic system The part of the autonomic nervous system (see there) that opposes the physiological effect of the parasympathetic system (see there).

Synapse Junction between neurons, see there.

References

Arndt B, Putz-Anderson V (2001) Cumulative trauma disorders. 2nd ed. Taylor & Francis, London

Biasiucci A, Franceschiello B, Murray MM (2019) Electroencephalography. Curr Biol 29(3):R80–R85. https://doi.org/10.1016/j.cub.2018.11.052

Birkbeck MQ, Beer TC (1975) Occupation in relation to the carpal tunnel syndrome. Rheum Rehabil 14(4):218–221

Kroemer KHE (2001) Keyboards and keying. An annotated bibliography of the literature from 1878 to 1999. Int J Univ Access Info Soc UAIS 1(2):99–160

Kroemer KHE (2006) "Extra-ordinary" ergonomics: how to accommodate small and big persons, the disabled and elderly, expectant mothers and children. CRC Press, Boca Raton

Kroemer KHE (2017) Fitting the human. Introduction to ergonomics, 7th edn. CRC Press, Boca Raton

Kroemer AD, Kroemer KHE (2017) Office ergonomics. Taylor & Francis, London

Kroemer Elbert KE, Kroemer HB, Kroemer Hoffman AD (2018) Ergonomics: How to design for ease and efficiency, 3rd edn. Academic, London

Langley LL, Cheraskin E (1958) The physiology of man. 2nd ed. McGraw-Hill, New York

Marras WS (2008) The working back: a systems view. Wiley, New York

Marras WS, Karwowski K (eds) (2006) The occupational ergonomics handbook, vol 1 & 2, 2nd edn. CRC Press, Boca Raton

National Research Council (1999) Work-related musculoskeletal disorders: Report, workshop summary, and workshop papers. National Academy Press, Washington

Pfeffer GB, Gelberman RH, Boyes JH, Rydevik B (1988) The history of carpal tunnel syndrome. J Hand Surg 13B(1):28–34

Phalen GS (1966) The carpal-tunnel syndrome—seventeen years' experience in diagnosis and treatment of six-hundred fifty-four hands. J Bone Jt Surg 48-A(2):211–228

Phalen GS (1972) The carpal-tunnel syndrome. Clinical evaluation of 598 hands. Clin Orthop Relat Res 83:29–40

Posch JL, Marcotte DR (1976) Carpal tunnel syndrome. An analysis of 1201 cases. Orthop Rev 5:25–35

Sommerich CM, Marras WS (2004) Electromyography and muscle force (Chap. 7). In: Kumar S (ed) Biomechanics in ergonomics, 2nd edn. CRC Press, Boca Raton

Tanzer RC (1959) The Carpal-tunnel syndrome—a clinical and anatomical study. Am J Ind Med 11:343–358. Also in J. Bone Jt Surg 41A:4, 626–634

Chapter 4
Anthromechanics

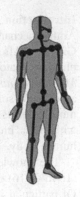

Overview

Anthromechanics is the discipline of biomechanics applied to the human: biomechanics explain characteristics, actions and responses of a biological system in mechanical terms. The discipline of anthromechanics incorporates methods and techniques from physics, especially mechanics; mathematics, including computer sciences; anatomy, particularly anthropometry; and, of course, biology and physiology. In this chapter, anthromechanical considerations explore and explain the action of muscles bridging body joints, the assessment of mass properties of the human body, and the transmission of forces and moments along a chain of body segments.

The Model

The body, consisting of solid, flexible and fluid components, possesses mass and other material properties that obey physical laws. The body is built on and around a solid skeleton of links (the long bones) with sections that articulate in joints. Muscles are the "engines" that power the body. Using this approach, mechanical terms and computational techniques can describe many functions of the human body.

Introduction

Anthromechanics* is not a new science. Leonardo da Vinci (1452–1519) and Giovanni Alfonso Borelli (1608–1679) combined the then-existing knowledge of physics, anatomy, and physiology. In his book *De Motu Animalium (About the Motion of Animals)*, Borelli developed a model of the human skeleton consisting of a series of solid links (long bones) joined at their articulations and powered by muscles bridging the articulations. This concept of a "stick person" embellished with mass properties still underlies many current biomechanical models of the human body.

The knowledge developed by Isaac Newton (1642–1727) and Gottfried Leibniz (1646–1716) explained the physical relationships between force, mass, and motion. Of particular importance are Newton's laws of motion:

– the first explains that force acting on a mass changes its motion condition;
– the second states that force equals mass multiplied by acceleration;
– the third makes it clear that force exertion requires the presence of an equally large counter force.

4.1   Treating the Body as a Mechanical System

Beginning in the late 1800s, researchers determined the masses of body segments, investigated the interactions between mass distribution and external impulses applied to the human body, and discussed statics and mechanics of the human body. Since then, anthromechanic research has addressed, among other topics, the response of the body to vibrations and impacts; human strength and motion of the whole body and its components; functions of the spinal column; hemodynamics and the cardiovascular system; and prosthetic devices.

Treating the human body solely as a mechanical system means to make many and gross simplifications, for example, disregarding mental functions. Furthermore, mechanical analogies also simplify reality, such as:

- Anthropometric Data: dimensions obtained from statistical manipulation of data measured on groups of people.
- Articulations: joints and bearing surfaces usually with simplified geometries, with lubricants and often considered frictionless.
- Bones: lever arms, central axes, structural members.
- Contours: surfaces of geometric bodies.
- Flesh: volumes, masses.
- Muscles: motors, dampers or locks.
- Nerves: control and feedback circuits.

- Organs: generators or consumers of energy.
- Tendons: cables transmitting muscle forces.
- Tendon Sheaths: pulleys and sliding surfaces, often considered frictionless.
- Tissue: elastic load-bearing surfaces, springs, contours.

4.1.1 Stress and Strain

In engineering terms, *stress causes strain*. For example: the weight of the upper body stresses the spinal column, generating strain in its structures. Engineering stress is a measure of the force applied over a unit area, and has units of pressure (N/mm^2). Engineering strain can be ascertained as deformation (often as a change in length relative to the original length, and thus expressed in unitless percent) corresponding to the applied stress. (Detailed discussions of stress analysis are beyond the scope of this text. However, there are many basic mechanics books available for further information.)

In the 1930s, Hans Selye introduced another concept: *stressor causing stress* (and distress). He was borrowing an engineering term to describe a psychologic/ physiologic condition. Unfortunately, the use of the term "stress" as either being cause (in engineering) or result (in psychology) can create much confusion: What is, for example, "job stress"? To avoid ambiguity, the engineering terminology will be used in this text: **stress produces strain**.

4.2 Mechanical Bases

Mechanics is the study of forces and of their effects on masses. *Statics* considers masses that are at rest or in equilibrium because balanced forces act on them. *Dynamics* concerns the motions of masses, where the (change in) motion results from the action of external forces on the mass.

Dynamics, also called *kinesiology* when applied to the human body, subdivides into two fields. *Kinematics* considers the motions (displacements and their time derivatives velocity, acceleration, and jerk) but not the forces that bring these about; in contrast, *kinetics* is concerned with just these forces.

Newton's three laws, cited earlier in the introduction to this chapter, are fundamental to biomechanic analyses. The second law sheds light on one important factor in biomechanics: force. Force is defined by its ability to change the velocity of a mass by generating acceleration. A force can cause a mass to speed up or slow down, and to change its direction of motion. Mathematically, force is a vector since it has both magnitude and direction.

Force is not a basic unit of the international metric measuring system but a derived one, calculated from the linear acceleration of a mass. (Acceleration is the first time-derivative of velocity.) Relatedly, no device exists that measures force directly. All devices for measuring force rely on other physical phenomena, usually either measurement of displacement (such as bending of a metal beam) or acceleration experienced by a mass.

The correct unit for force measurement is the Newton (N)*. One Newton of force accelerates one kilogram of mass by one meter per squared second: $1 \text{ N} = 1 \text{ kg} \times 1 \text{ m s}^{-2}$. On Earth, one kg has a weight (note that weight is a force!) of 9.81 N.

Moment is a vector that can change the angular velocity of a mass by generating an angular acceleration. Moment is calculated as the product of force and its lever arm (distance) to the articulation about which it acts. By definition, the lever arm is at a right angle to the direction of the force; in kinesiology, the lever arm often is called the mechanical advantage. Torque is also a moment but usually tries to twist ("torque") an object about its long axis.

4.2.1   Free-Body Diagram

To calculate the forces or moments that act on an object of interest, we establish a space around it; within its fictional boundaries, we enforce Newton's laws. If there is no motion, or no change in motion, then the object is in static equilibrium: *static* since there is no new motion and *equilibrium* because the forces and moments are balanced. When balanced, the sum of the forces, and the sum of the moments, must both be zero:

$$\sum \text{forces} = 0$$
$$\sum \text{moments} = 0$$

Example: On Earth, establish a space around a person standing on a weight scale. Within that space, all forces (and moments) sum up to zero. Accordingly, the force with which the scale is pushing up on the person's feet must balance the downward force, the weight, due to the mass of the person. The weight (mass × acceleration, Newton's second law) equals the reaction force (Newton's third law).

Considering the person on a scale is an example of establishing fictional boundaries which seem to cut a person or object from its environs. This is called a "free body": depending on where boundaries are placed, that may be the entire human pushing on an object, or is can be just arm or hand of a person manipulating a tool. Such free-body diagrams are useful tools for engineering or physics: free-body diagrams depict the object under investigation isolated but with vector forces and moments applied to it that represent the effect of the environs on that object in a given situation.

4.2.2 Static Equilibrium

Figure 4.1 illustrates how forces and moments produced by the arm muscle can change with arm position. Biceps brachii, the primary flexing muscle, pulls with force M on the forearm. That biceps force M has a lever m around the elbow joint. For calculating the moment T about the elbow joint, the lever must be perpendicular to the direction of vector M:

$$T = m\,M$$

As Fig. 4.1 shows, the length of the lever changes with the elbow angle α: the lever m is longest at a nearly right elbow angle; the leverage gets smaller the more the forearm is flexed or extended with respect to the upper arm.

Figure 4.2 sketches the same condition in more detail and more realistically. The actual direction of the muscle force vector M usually is not at a right angle to the lever arm m, as it was in Fig. 4.1. The vector P (which is perpendicular) depends on the angle ß between P and M, the relationship between the two vectors is

$$P = M\cos\!\text{ß}$$

Vector P generates a moment T around the elbow joint:

$$T = m\,P$$

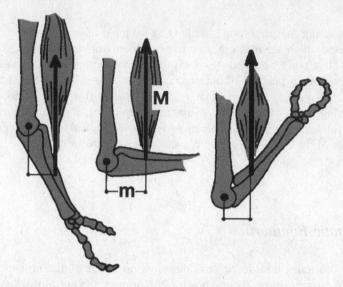

Fig. 4.1 Increasing or decreasing elbow angle α changes the lever arm m of force vector M.

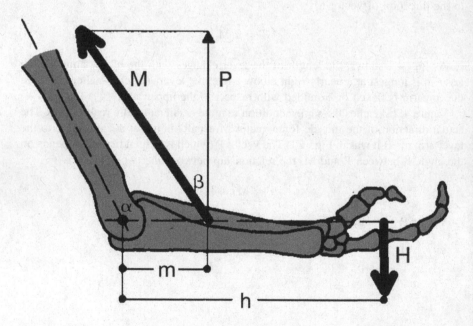

Fig. 4.2 Interaction between hand force H and muscle force M

If all forces and moments are in balance, in every direction, the sum of the forces must be zero, as must be the sum of the moments. This requires counteraction to the moment T by an equally large but opposing moment which may be generated by a force vector H (parallel to P) pulling down on the hand. With its perpendicular lever arm h, H establishes equilibrium if

$$h\,H = m\,P = m\,M\cos\beta$$

This yields an equation for calculating the actual muscle force vector M:

$$M = (h/m)(H/\cos\beta)$$

This equation confirms that, to generate a hand force vector H, muscle force M varies with the ratio between the lever arms (h/m) and with the pull angle ß. Measuring the external force H, the angle ß, and the lever arms h and m, provides all the information needed to determine the magnitude of the muscle force vector M.

Note that in this example the arm segments were considered massless and static. However, real-life conditions are a bit more complex. It can be rather difficult to determine the insertion point of the biceps muscle force vector M on the radius bone because encapsulating ligaments restrain the muscle in a groove on the humerus bone to near the elbow bend. This makes the angle ß steeper than assumed in Fig. 4.2. Furthermore, the biceps muscle has two heads, which split along the upper part of the humerus and attach in different locations to the shoulder blade. Moreover, the brachialis muscle helps in flexing the elbow; it originates at about half the length of the humerus on its anterior side and inserts on the ulna. In some arm positions, the brachioradialis muscle, connecting radius and ulna, can also contribute to elbow flexion.

Figure 4.3 illustrates the elbow flexor conditions more realistically (but still only in a 2-dimensional plane): M is again the contractile force vector of the biceps, (90° − ß) is its pull angle with respect to the long axis of the forearm, and P = M cosß its moment-generating force about the elbow joint at the lever arm m. A similar analysis for the brachialis force vector N shows its component force Q perpendicular to the lever arm n. Angle γ (which changes with the elbow angle α) is the difference between the directions of N and Q, hence Q = N cosγ. The concurrent action of M and N is opposed by moment h H (assuming a right angle between hand force H and its lever arm h: if this is not the case, a vector analysis of H determines its angular components).

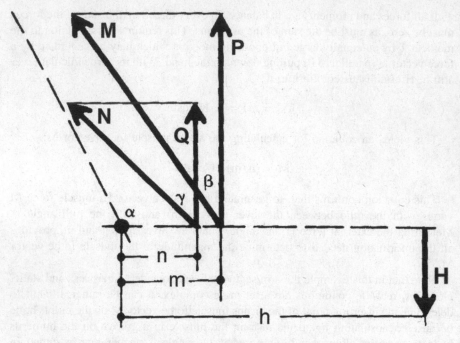

Fig. 4.3 Interaction between hand force H and two muscle forces, M and N

This is now a (two-dimensional) free-body diagram*: fictional boundaries seem to cut the forearm from the upper arm. Establishing such fictional boundaries around forearm-hand, or likewise around the upper arm or other segments, allows convenient mechanical modeling and computational procedures.

In order to keep the forearm-hand segment linked to the upper arm, horizontal and vertical reaction forces and a restraining moment must exist at the elbow boundary. A free-body analysis of the upper arm proximally traverses the free-body boundary at the elbow. In this cross-over, the joint reaction forces and moment in the forearm analysis transfer in their opposite directions (Newton's third law) to the distal end of the upper arm free-body diagram. Then the analysis can proceed in similar fashion as done for the forearm-hand section.

Static equilibrium in a "free body" system requires (in two dimensions):

- all horizontal forces sum to zero; and
- all vertical forces sum to zero; and
- all moments sum to zero.

Equilibrium requires the presence of joint reaction forces*: here, a horizontal force J and a vertical reaction force V, which counteract the horizontal components of M and N (namely M sinß and N sinγ) and their vertical components; without them, the forearm would move sideways and up or down. Likewise, a joint-reaction moment T assures that the elbow angle α does not change. (Forces and moment may be generated by shoulder muscles acting on the upper arm—this is not sketched in Fig. 4.3.)

With J, all forces in the horizontal direction sum to zero:

$$M \sin \beta + N \sin \gamma + J = 0$$

With V, all forces in the vertical direction sum to zero:

$$P + Q + H + V = 0$$

With T, all moments about the elbow joint sum to zero:

$$m P + n Q + h H + T = 0$$

When these three equations are satisfied, the forearm is at rest with respect to the upper arm. (Note that the actual magnitudes of J, V, T and their actual signs (+ or −) become evident when the equilibrium equations are solved. Mathematically, this requires establishing a coordinate system.)

This discussion shows that it is fairly easy to measure the resultant output of all concurring muscular forces (such as the hand force H in Figs. 4.2 and 4.3) but that overall output does not indicate which muscles contribute how much. Furthermore, muscles may be active which were not considered: for example, the triceps muscle may also be active in order to control and regulate the effort; this would add to the moment generated by H while counteracting the biceps and brachialis muscles, making them increase their efforts. Thus, the anthromechanical approach, as used here, calculates only "minimal net results": the smallest total of muscle forces and moments needed to achieve the desired effect, such as H. The actual efforts of individual muscles* may be much higher than the equations indicate, especially if antagonistic muscles counteract each other. With current technology, electromyograms can help to assess the activities of specific muscles—see Chaps. 2 and 3.

4.2.3 Dynamic Analyses

Mechanical analyses are useful for many ergonomic and engineering assessments. Anthromechanics can identify physical efforts of everyday activities such as when lifting loads. They are useful for the evaluation of assist devices for walking (canes, crutches or walkers). They can determine the forces applied at body joints (data needed for the design of artificial joints), and the efficacy of external protective equipment such as back belts for use during lifting or of knee braces for support and protection during strenuous activities.

Traditional dynamic analyses follow the same approach as used in the static example above:

- Chose the analysis space;
- Construct the appropriate free-body diagram;
- Identify forces and moments crossing the free-body boundaries; and
- Solve the equations to determine the unknown values.

However, dynamic equilibrium must incorporate the ˋeffects of changes in velocities (that is, accelerations) as per Newton's second law. Accordingly, dynamic forces derive from "mass × linear acceleration" and dynamic moments from "mass-moment-of-inertia × angular acceleration". With these modifications, the equations for equilibrium (above) remain valid. Yet, determining the mass properties of all components and their linear or rotational accelerations (such as centrifugal, tangential and Coriolis—all of which may be changing over time) can become daunting tasks. New computerized analyses can assist in the use of efficient paradigms and procedures in anthromechanics.

4.3   Anthropometric Inputs

Anthromechanics rely much on anthropometric data, often adapted and simplified to fit the mechanical approach. Figure 11.3 in Chap. 11 identifies the reference planes used originally in anthropometry and subsequently in anthromechanics: the medial (mid-sagittal) plane divides the body into right and left halves; the frontal (coronal) plane establishes anterior and posterior sections of the body; the transverse (horizontal) plane cuts the body in superior and inferior parts. Their common intersection establishes the origin of a Cartesian xyz coordinate system. However, anatomy defines only the location of the medial plane while the frontal and transverse planes need to be fixed by consensus. When the human stands upright (in the so-called anatomical position), convention is to assume that the three planes meet in the center of mass (CM) of the body, which is in the pelvic region. If the body is not in this anatomical position, one may decide to establish a new origin depending on the given conditions.

4.3.1   Links and Joints

Simplifying the human skeletal system into a relatively small number of straight-line links (representing long bones) and joints (representing major articulations) can make treatment and computation simple. Figure 4.4 shows a typical link-joint system. In this example, hands and feet are not subdivided into their components, and the spinal column consists of only three links. Clearly, such simplification does not represent the true design of the human body—yet, depending on the desired application, it may suffice to stand in for certain mechanical properties.

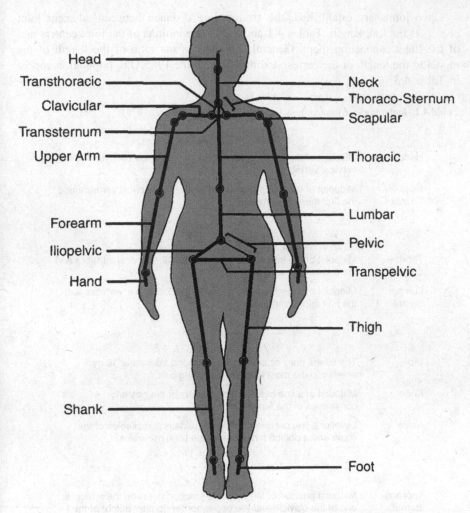

Fig. 4.4 Typical link-joint system (adapted from NASA/Webb 1978)

Determination of the location of joint centers of rotation is relatively easy for simple articulations, such as the hinge-type joints in fingers, elbows, and knees. However, this is much more difficult for complex joints with several degrees of freedom, such as in the hip and, even more difficult, in the shoulder. Approximations may suffice: in Fig. 4.4, the shoulder joint consists of two artic-ulations with an intermediate scapular link. In many cases, carefully simplifying location and properties of a joint can be adequate to meet modeling requirements. However, it is important to understand the limitations imposed by the simplifying assumptions made in such models: model joint characteristics may reflect the true body articulation only partially or merely within a limited range of motion.

Once joints are established, the straight-line distance between adjacent joint centers is the link length. Tables 4.1 and 4.2 list definitions of the joint centers and of the links connecting them. General estimates of the ratio of the length of the model to the length of the corresponding bone for the 1985 U.S. population appear in Table 4.3.

Table 4.1 Definitions of joint centers (adapted from NASA/Webb 1978)

HEAD and NECK	
Head-Neck	Midpoint of the interspace between the occipital condyle and first cervical vertebra.
Neck-Thorax	Midpoint of the interspace between the 7th cervical vertebra and the first thoracic vertebra.

TRUNK	
Thorax-Lumbar	Midpoint of the interspace between the 12th cervical vertebra and the first lumbar vertebra.
Lumbar-Sacral	Midpoint of the interspace between the 5th lumbar vertebra and the first sacral vertebra.

LEG	
Hip	The lateral point at the tip of the femoral trochanter 10 mm anterior to the most laterally projecting part.
Knee	Midpoint of a line between the centers of the posterior convexities of the femoral condyles.
Ankle	Level of a line between the tip of the ;lateral malleolus of the fibula and a point 5 mm distal to the tibial malleolus.

ARM	
Thoraco-sternal	Midpoint position of the palpable junction between the proximal end of the clavicle and the upper border (jugular notch) of the sternum.
Clavi-Scapular	Midpoint of a line between the coracoid tuberosity of the clavicle (at the posterior border of the bone) and the acromial-clavicular articulation (or the tubercle at the lateral end of the clavicle): the point should be visualized as on the underside of the clavicle.
Gleno-Humeral (Shoulders)	Midregion of the palpable bony mass of the head and tuberosities of the humerus; with the arm abducted about 45° relative to the vertebral margin of the scapula, a line approximately bisects the joint when dropped perpendicularly to the to the long axis of the arm from the outermost margin of the acromion.
Elbow	Midpoint on a line between the lowest palpable point, the medial epicondyle, of the humerus, and a point 8 mm above the radiale (radio-humeral junction).

Table 4.2 Definition of body links (adapted from NASA/Webb 1978)

HEAD and NECK	
HEAD	This straight line between the occipital condyle/C1 interspace center and the center of mass of the head.
NECK	The straight line between the occipital condyle/C1 and C7/T1 vertebral interspace joint centers.

TRUNK	
TORSO (total)	The straight-line distance from the occipital condyle/C1 interspace joint center to the midpoint of a line passing through the right and left hip joint centers.
THORAX sublinks	• *Thoraco-sternum*: a closed linkage system composed of three links. The right and left transthoraxes are straight-line distances from the C7/T1 interpace to the right and left sterno-clavicular joint centers. The transsternum link is a straight-nine distance between the right and left sterno-clavicular joint centers. • *Clavicular*: straight-nine distance between the right and left sterno-clavicular and the claviscapular joint centers. • *Scapular*: straight line between the claviscapular and the gleno-humeral joint centers. • *Thoracic:* The straight line between the C7/T1 and the T12/L1 vertebral interspace joint centers.
LUMBAR	The straight line between the T12/L1 and L5/S1 vertebral interspace joint centers.
PELVIS	A linkage system composed of three links. The right and left ileo-pelvic links are straight lines between the L5/S1 interspace joint center and a hip joint center. The transpelvic link is a straight line between the right and left hip joint centers.

LEG	
THIGH	The straight line between the hip and knee joint centers of rotation.
SHANK	The straight line between the knee and ankle joint centers of rotation.
FOOT	The straight line between the ankle joint center and the center of mass of the foot.

ARM	
UPPER ARM	The straight line between the gleno-humeral and elbow joint centers of rotation.
FOREARM	The straight line between the elbow and wrist joint centers of rotation.
HAND	The straight line between the wrist joint center of rotation and the center of mass of the hand.

Table 4.3 Ratios (in %) of link length to bone length (adapted from NASA/Webb 1978)

Segment	Mean (%)	Standard deviation (%)
Thigh link / Femur length	93.3	0.9
Shank link / Tibia length	107.8	1.8
Upper arm link / Humerus length	89.4	1.6
Forearm link / Ulna length	98.7	2.7
Forearm link / Radius length	107.1	3.5

Anthropometric measurements are not taken from the centers of the anatomic joints but instead usually relate to externally discernible landmarks, such as bony protrusions on the skeleton—see Chap. 11. Hence, one task in developing a model depicting the human body is to establish the numerical relationships between standard anthropometric measures and link lengths. Simply expressing segment lengths as proportions of stature is usually incorrect because stature has very low correlations with most body measures but is highly correlated to only a few other height measures, as listed in Table 11.4 (Chap. 11). However, bone lengths can serve in calculations of link lengths by ratios or regression equations, as listed in Tables 4.4 and 4.5.

4.3.2 Body Volumes

Knowing the volume of whole the body or of a segment is often necessary for the calculation of inertial properties or for the design of close-fitting garments. Use of Archimedes' principle provides the body volume: immersing the body (or body segment) in a container filled with water, then measuring the volume of the dis-placed water yields the volume of the immersed body. The technique works well for obtaining volumes of limbs and, if changes due to respiration can be controlled, of the whole body. The immersion technique works with living persons and cadavers; yet, the used segmentations differ, as Fig. 4.5 shows.

One of the indirect methods for obtaining volume is to assume that geometric forms can represent body segments. Shapes with easily calculated volumes include sphere, cylinder, and truncated cone. Another method is to use information about the cross-section contours, which can be obtained by imaging scans (such as by x-rays via computed tomography, CT, or magnetic resonance imaging, MRI) or by

Table 4.4 Regression equations for estimating link lengths (in cm) from bone lengths (adapted from NASA/Webb 1978)

Empirical equation	Standard error of of estimate	Correlation coefficient
Thigh link length = 132.8253 + 0.8172 tibia length	16.57	0.73
Thigh link length = 92.0397 + 0.8699 fibula length	10.34 10.34	0.87 0.87
Shank link length = 8.2184 + 1.0904 fibula length	5.95 5.95	0.97 0.97
Shank link length = 1.0776 tibia length	nda nda	nda nda
Arm link length = 66.2621 + 0.8665 ulna length	9.90 9.90	0.94 0.94
Arm link length = 58.0752 + 0.9646 radius length	8.92 8.92	0.94 0.94
Forearm link length = 1.0709 radius length	nda nda	nda nda
Forearm link length = 0.9870 ulna length	nda nda	nda nda

nda: no data available

Table 4.5 Estimated link lengths (in cm) for the 1985 U.S. population (adapted from NASA/ Webb 1978)

	Female			Male		
	5th	50th	95th	5th	50th	95th
Thigh link	36.9	39.5	42.1	40.4	43.2	46.1
Shank link	34.7	37.4	40.0	38.9	42.1	45.3
Upper arm link	26.1	27.8	29.5	28.6	30.5	32.3
Forearm link (ulna)	22.7	24.1	25.5	25.6	27.1	28.7
Forearm link (radius)	22.7	24.1	25.5	25.9	27.5	29.2

Fig. 4.5 Defining body
segments on living persons
and cadaver bodies (adapted
from NASA/Webb 1978)

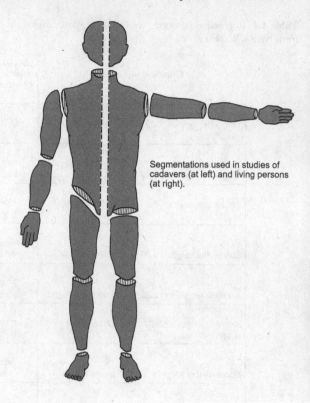

Segmentations used in studies of
cadavers (at left) and living persons
(at right).

dissection. If these cross-section contours are taken at sufficiently close separations
so that the changes between cross-sections can be assumed linear with distance, the
volume V is the sum of the cross-section areas A_i multiplied with their distances d_i
from each other:

$$V = \sum (A_i d_i)$$

Often, the distance d between cross-section contours is kept constant and
adjacent cross-sectional areas are averaged:

$$V = \sum \bar{A}_i d = \sum [(A_{i-1} + A_i)d/2]$$

Other approximations rely on the assumption that the cross-sections resemble
regular geometric figures. For example, if the body cross-section is elliptical, then
its area results from

$$A_i = \pi a_i b_i$$

where a_i is the semi-major axis and b_i s the semi-minor axis of a section i; the
volume is the sum of cross-sections and their distances, as above.

4.4   Inertial Properties

Knowledge of mass properties of the total body and its portions is important for establishing static and dynamic properties of the human body. Into the 1960s, measurements on cadavers provided basic benchmark data. More recent combinations of classic and 3D measurements taken on the body's surface, and of deep body scans (such as by CT or MRI scans) provide detailed information about mass distribution characteristics and the composition of the body's interior: presence and extent of adipose and of connective tissues, especially muscles, and of skeletal bone with the locations of body joints. Regression equations and other statistical predictors describe the volumes, masses, and moments of inertia of the whole body and its segments, including composite volumes, centers of volume, intersegment cut centroids, and principal inertial axes. Computerized mathematical models serve to describe the geometry and the internal composition of the body, using information derived from tomography, magnetic resonance imaging, and dual energy x-ray absorptiometry.

Newton's second law establishes weight W as a force that depends on body mass m and the gravitational acceleration g: $W = m\,g$. Weight is easily determined with a variety of scales; when measured in air, there is a slight (but usually unimportant) error due to buoyancy. Table 4.6 lists equations to estimate body segment weight from total body weight W, based on cadaver measurements.

The human body is not homogeneous throughout; its density varies depending on tissue cavities, water content, fat tissue, and bone components. Still, in many cases it is sufficient to assume that either the body segment in question or even the whole body is of constant (average) density.

Density D is the mass per unit volume V:

$$D = m/V = W/(g\,V)$$

Mass m can be calculated from volume V and density D:

$$m = D\,V = W/g.$$

Specific density D_s compares D to the density of water D_w:

$$D_s = D/D_w.$$

Information about the relations of body segment weight to total body weight, compiled in Table 4.7, provides some insight into mass and density distributions.

128

Chapter 4   Anthromechanics

Table 4.6 Prediction equations to estimate segment mass (in kg) from total body mass (weight W, in kg) (adapted from NASA/Webb 1978)

Segment	Empirical equation	Standard error of estimate	Correlation coefficient
Head	$0.0306W + 2.46$	0.43	0.626
Head and neck	$0.0534W + 2.33$	0.60	0.726
Neck	$0.0146W + 0.60$	0.21	0.666
Head, neck and torso	$0.5940W - 2.20$	2.01	0.949
Neck and torso	$0.5582W - 4.26$	1.72	0.959
Total arm	$0.0505W + 0.01$	0.35	0.829
Upper arm	$0.0274W - 0.01$	0.19	0.826
Forearm and hand	$0.0233W - 0.01$	0.20	0.762
Forearm	$0.0189W - 0.16$	0.15	0.783
Hand	$0.0055W + 0.07$	0.07	0.605
Total leg	$0.1582W + 0.05$	1.02	0.847
Thigh	$0.1159W - 1.02$	0.71	0.859
Shank and foot	$0.0452W + 0.82$	0.41	0.750
Shank	$0.0375W + 0.38$	0.33	0.763
Foot	$0.0069W + 0.47$	0.11	0.552

Detailed density data obtained from weight and volume measurements of dissected cadavers are in Tables 4.8 and 4.9. However, cadavers loose fluids and undergo chemical changes so these tissues are not exactly the same as in living subjects. Three-dimensional scanning, stereophotometry, and immersion and weighing techniques serve to determine density.

4.4.1   Lean Body Mass

Another useful concept to distinguish between the compositions of different bodies is that of Lean Body Mass (or Lean Body Weight). Total body weight W is the sum

Table 4.7 Relative weights of body segments (adapted from NASA/Webb 1978)

Groups of segments: % of total body weight	Single segment: % of segment group
Head and neck: 8.4	Head: 73.8
	Neck: 26.2

Groups of segments: % of total body weight	Single segment: % of segment group
Torso: 50.0	Thorax: 43.8
	Lumbar: 29.4
	Pelvis: 26.8

Groups of segments: % of total body weight	Single segment: % of segment group
Total leg: 15.7	Thigh: 63.7
	Shank: 27.4
	Foot: 8.9

Groups of segments: % of total body weight	Single segment: % of segment group
Total arm: 5.1	Upper arm: 54.9
	Forearm: 33.3
	Hand: 11.8

of lean body weight and fat weight. Whereas the fat component varies throughout the body and among individuals, the typical underlying assumption is that basic structural body components such as skin, muscle and bone have about the same percentage composition from individual to individual.

There are several techniques to determine body fat. Accurate procedures include: magnetic resonance imagining (MRI); computerized tomography (CT) scans; weighing of the body submerged in water (the skinnier the body, the less its buoyancy); or air displacement plethysmography. Generally, these techniques require expansive equipment. Skinfold measures are much less complicated (and less expensive) to perform: the fold thickness of loose skin with its underlying fat is measured with a caliper. However, skinfold measurement is not particularly reliable since different skinfolds may be grasped and compressed in different manners.

Table 4.8 Segment mass ratios in percent derived from cadaver studies (adapted from Roebuck et al. 1975)

	Harless 1860	Braune & Fischer 1889	Fischer 1906	Dempster 1955[1]	Clauser McConville & Young 1969	Average
	2 specimens	*3 specimens*	*1 specimen*	*8 specimens*	*13 specimens*	
Head	7.6	7.0	8.8	8.1	7.3	7.8
Trunk	44.2	46.1	45.2	49.7	50.7	47.2
Total arm	5.7	6.2	5.4	5.0	4.9	5.4
Upper arm	3.2	3.3	2.8	2.8	2.6	2.9
Forearm and hand	2.6	2.9	2.6	2.2	2.3	2.5
Forearm	1.7	2.1	-	1.6	1.6	1.8
Hand	0.9	0.8	-	0.6	0.7	0.8
Total leg	18.4	17.2	17.6	16.1	16.1	17.1
Thigh	11.9	10.7	11.0	9.9	10.3	10.8
Shank and foot	6.6	6.5	6.6	6.1	5.8	6.3
Shank	4.6	4.8	4.5	4.6	4.3	4.6
Foot	2.0	1.7	2.1	1.4	1.5	1.7
Total body[2]	100.0	100.0	100.0	100.0	100.0	100.0

[1] Dempster's values adjusted by Clauser, McConville & Young 1969.
[2] Calculated from head + trunk + 2(total leg + total arm).

Table 4.9 Body densities derived from cadaver studies (adapted from McConville et al. 1980)

	Harless 1860	Dempster* 1955		Clauser, McConville & Young 1969	
	Mean	Mean	SD	Mean	SD
Head	-	-	-	1.071	-
Head + neck	11.1	11.1	0.012	-	-
Neck + torso	-	-	-	1.023	0.032
Thorax	-	0.92	0.056	-	-
Abdomino-pelvic	-	1.01	0.014	-	-
Torso + limbs	-	1.07	0.016	-	-
Thigh	1.07	1.05	0.008	1.045	
Lower leg	1.10	-	-	-	-
Shank	-	1.09	0.015	1.085	0.014
Foot	1.09	1.10	0.056	1.085	0.014
Total arm	-	1.07	0.027	1.058	0.025
Upper arm	1.08	-	-	-	-
Forearm	1.10	1.13	0.037	1.099	0.018
Hand	1.11	1.16	0.110	1.108	0.010
Total Body	-	-	-	1,042	0,018

*Dempster's values adjusted by Clauser, McConville & Young 1969.

4.4.2   Locating the Center of Mass

For computational analyses of gait and other motions, the body mass is often considered concentrated at one point in the body, the center of mass* CM. The actual location of the mass center of a living person varies because respiration affects the mass distribution, as do shifting body fluids, muscular contractions, and food or fluid ingestion or excretion. Of course, there are major changes in the location of the center of mass with different body positions and, in particular, with body movements.

For the body at rest, various methods exist to determine the location of the center of mass. Most rely on the principle of finding the one location where a single support would keep the body balanced. One simple technique to determine CM is to

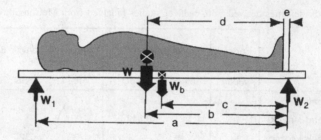

Fig. 4.6 Determining the center of mass of the body placed on a known board and on two scales

place the body on a platform which is supported by two scales at precisely known support points. The two forces at the support points counteract the body weight. Figure 4.6 shows this procedure for a supine body. Here, W_1 is the force at scale 1; a is the distance between scales 1 and 2; W is the weight of the body at its center of mass CM; W_b is the weight of the board at its CM, and c its distance from W_2; d is the distance between the CM of the body and the soles of the feet, and e their distance from W_2; and W_2 is the force at scale 2.

This allows the calculation of d, the distance from the CM to the soles of the feet. The sum of the vertical components of force and the sum of the moments about W_2 are both zero, since the body is at rest (not accelerated):

$$W_1 - W - W_b + W_2 = 0$$
$$a W_1 - b W - c W_b = 0$$

Geometrically,

$$b = d + e$$

Rearranging and inserting known force and distance values provide the value for d:

$$d = (a W_1 - c W_b)/(W_1 + W_2 - W_b) - e.$$

Note that the actual weight W of the subject does not appear in the equation for d.

Table 4.10 contains the results of several studies to determine mass center locations; three of the studies took place well over a hundred years ago.

Table 4.10 Locations of the centers of mass ratios of body segments, measured (in %) from their proximal ends (adapted from Roebuck et al. 1975)

	Harless 1860	Braune & Fischer 1889	Fischer 1906	Dempster 1955	Clauser, McConville & Young 1969
	2 specimen	*3 specimen*	*1 specimen*	*8 specimen*	*13 specimen*
Head *	36.2*	—	—	43.3*	46.6*
Trunk ^{nc}	44.88^{nc}	—	—	—	38.0^{nc}
Total arm	—	—	44.6	—	41.3
Upper arm	—	47.0%	45.0	43.6	51.3
Forearm and hand	—	47.2	46.2	67.7	62.6
Forearm ^{nc}	42.0^{nc}	42.1^{nc}	—	43.0^{nc}	39.0^{nc}
Hand ^{nc}	39.7^{nc}	—	—	49.4^{nc}	18.0^{nc}
Total leg ^{nc}	—	—	41.2^{nc}	43.3^{nc}	38.2^{nc}
Thigh ^{nc}	48.9^{nc}	44.0^{nc}	43.6^{nc}	43.3^{nc}	37.2^{nc}
Shank and foot	—	52.4	53.7	43.7	47.5
Shank	43.3	42.0	43.3	43.3	37.1
Foot	44.4	44.4	—	42.9	44.9
Total body **	58.6**	—	—	—	58.8**

*Percent of head length, measured from the crown down.
^{nc} The values on these lines are not comparable because the investigators used differing definitions for segment lengths.
**Percent of stature, measured from the floor up.

4.4.3   Moments of Inertia

To assess the mechanics of linear motions, information about the mass of involved body segments is necessary. Similarly, for the modeling of rotations, one also needs to know the moment of inertia* $I = \Sigma\ m_i r_i^2$. It describes a body's rotational inertia, which determines the torque needed to achieve angular acceleration about a rotational axis.

There are three principal approaches to determine the moment of inertia, I, of the body or of a segment. The first technique is to use geometric models whose selected shapes (often spheres or cylinders) simulate the actual body form; their densities must be constant and known. The second approach uses regression equations based on body weight, segment weight or segment volume. The third method employs the mechanical construct of the radius of gyration, $K_{\%L}$, expressed as a proportion of segment (or link) length, L, such as listed in Table 4.11. The radius can be calculated by multiplying length L (Table 4.5) and $K_{\%L}$ (Table 4.11). The resulting product is squared, then multiplied by the appropriate segment mass (Tables 4.6, 4.7, 4.8) to obtain an estimate of the moment of inertia:

$$I = m(L\,K_{\%L})^2$$

Table 4.11 Radius of gyration, $K_{\%L}$, in per cent of segment length, L. The principal axes are x, forward; y, to the right; and z, down (adapted from NASA/Webb 1978)

Name	Link L	Axis	K in %
Head	Head Length	x	31.6
		y	30.9
		z	34.2
Torso	Torso length: trochanterion height to suprasternale height	x	43.0
		y	35.2
		z	20.8
Thigh	Fibular height to trochanterion height	x	27.9
		y	28.4
		z	12.2
Shank	Fibular height	x	28.2
		y	28.2
		z	7.6
Foot	Foot length	x	26.1
		y	24.9
		z	12.2
Upper arm	Radiale to acromion length	x	26.1
		y	25.4
		z	10.4
Forearm	Stylion to radiale length	x	29.6
		y	29.2
		z	10.8
Hand	Hand breadth	x	50.4
		y	45.6
		z	26.6

4.5   Kinematic Chain Models

A commonly used model of the human body is built on and around a simplified skeleton that consists of straight links between adjacent joints. Figure 4.7 shows a such "stick person" model.

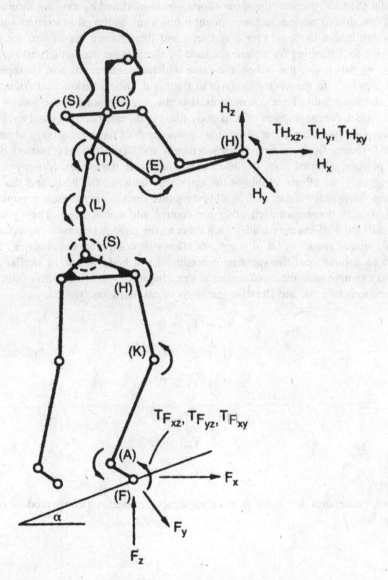

Fig. 4.7 Free-body diagram of the link-joint model of the human body, indicating the chain of forces (F) or moments (T) transmitted from the hand through arm, trunk, and leg to the foot on the floor

Forces exerted with the hand (H_x, H_y, or H_z) to an outside object or the moments generated in the hand (T_H) are transmitted along the links. First, the force exerted with the right hand, modified by the existing mechanical advantages, must be transmitted across the right elbow (E). (Also, at the elbow additional force or moment must be generated to support the mass of the forearm. However, for the purposes of this discussion, the model will be considered massless.) Similarly, the shoulder (S) must transmit the same efforts, again modified by existing mechanical conditions. In this manner, all subsequent joints transmit the effort exerted with the hands throughout the trunk, hips, and legs and finally from the foot to the floor. Here, one foot standing on a plane inclined by the angle α counteracts all the hand efforts. At this point, the orthogonal force and moment vectors can be separated again, similarly to the vector analysis at the hand. In a massless and frictionless body, the same sum of vectors must exist at the feet as started at the hand.

Of course, the assumption of zero body mass is unrealistic. This can be rectified by incorporating information about mass properties* of the human body segments, detailed above. Body motions (displacements and accelerations), instead of the static posture assumed here, would further complicate the model. Moreover, considering only the efforts visible at the interfaces between the body and the environment disregards the fact that, at all body joints, antagonistic muscle groups exist which usually counteract each other for control and stabilization. Their possibly high individual efforts may nullify each other to the outside observer. Nevertheless, for this simple model, a set of equations allows its computational analysis. These equations follow from the standard procedure for a body at rest (a similar equilibrium example was discussed earlier in this chapter). Setting the sum of forces in all directions to zero, and likewise the sums of all moments, yields

$$H_x + F_x = 0$$

$$H_y + F_y = 0$$

$$H_z + F_z = 0$$

$$T_{Hxz} + T_{Fx} = 0$$

$$T_{Hy} + T_{Fyz} = 0$$

$$T_{Hx} + T_{Fxy} = 0$$

Such procedures are basic to even rather complex mechanical models of the human body.

Notes

The text contains markers, *, to indicate specific references and comments, which follow.

Anthromechanics: King (1984), Chaffin and Andersson (1984), Kroemer et al. (1988), Chaffin et al. (1999), Winter (2004), Marras (2008), Marras and Radwin (2006), Chaffin (2008), Kroemer (2008), Kumar (2008), Oezkaya et al. (2016).

The proper unit for force measurement is the Newton (N): Some older tables of forces use the units "kg" or "kg$_f$" or "kp" and "lb" or "lb$_f$"; they all can be translated into Newton because the kilogram-"force"-unit equals 9.81 N; the pound-"force"-unit converts to 4.44 N.

Free-body diagram: At its boundaries, a free-body diagram must account for all forces and moments by which the body interacts with connected bodies and the environment, such as upper arm or gravity: see mechanics texts for details of that procedure; Oezkaya and Nordin (1999); Chaffin et al. (2006), Oezkaya et al. (2016) present worked examples.

Joint reaction force: It is routine to initially insert joint reactions into free-body analysis equations; experimental measurements, model assumptions or mathematically solving the equations will make it clear whether these joint forces or moments actually exist; and if so, what their magnitudes are and in which directions they act.

Center of mass CM: In physics, that point where the weighted relative position of the distributed mass of a rigid body sums to zero. A force applied at this point causes a linear acceleration but no angular acceleration.

Moment of inertia: In physics, the quantity of angular mass (rotational inertia) of a rigid body that determines the torque needed to achieve angular acceleration about a rotational axis.

Information about mass properties: Harless (1860), Braune and Fischer (1889), Fischer (1906), Dempster (1955), Clauser et al. (1969), Chandler et al. (1975), Roebuck et al. (1975), Herron et al. (1976), McConville et al. (1980), Young et al. (1983), Kaleps et al. (1984), King (1984), Zehner (1984); Fung (1990) and Kroemer (2008) provide historically interesting reviews and data sets. Hay (1973), Roebuck et al. (1975), NASA/Webb (1978), Chaffin and Andersson (1984), NASA (1989), Mungiole and Martin (1990), Roebuck (1995), Durlin et al. (2005), Chaffin et al. (2006) compiled data.

Summary

Biomechanical modeling of the human body, or of its parts, in general requires simplification of actual (physiologic, anatomic, anthropometric) characteristics to fit the methods and techniques derived from mechanics. While this establishes limitations regarding the completeness, reliability, and validity of anthromechanical procedures, it also allows research and conclusions with unique insights which would not have been possible if using traditional (physiologic, anatomic, anthropometric) approaches. However, one has to be keenly aware of the limitations imposed by the underlying simplifications and assumptions.

An important application of anthromechanics is in the calculation of forces that can be developed by muscles as moments about body joints. In reversing this procedure, one can assess the load on muscles, bones, and tissues generated by external loads on the body in various positions.

Body segment dimensions, their volumes and mass properties can be calculated from anthropometric data.

Kinematic chain models of linked body segments allow the prediction of total-body capability (such as lifting) from the consideration of body segment capabilities.

Although much information is available, anthromechanics is still a developing scientific and engineering field with a wide variety of focus points, research methods, and measurement techniques for producing new theoretical and practical results. While using the substantial data and knowledge base that exist, the practicing engineer must studiously follow the progress reported in the scientific and engineering literature in order to stay abreast of developments so that the most appropriate information can be applied to the design and management of human/machine systems.

Glossary

Acceleration Second time-derivative of displacement.

Anthromechanics Biomechanics (see there) applied to the human.

Biomechanics The study of the mechanics of a (human or animal) body, especially of the forces exerted by muscles and gravity on the skeletal structure; also, the mechanics of a part or function of a living body such as of the heart or of locomotion.

Center of mass In physics, the location in a body where the mass its considered concentrated.

Co-contraction Simultaneous contraction of two or more muscles.

Contraction (of muscle) Literally, "pulling together" the z lines delineating the length of a sarcomere, caused by the sliding action of actin and myosin filaments. Contraction develops muscle tension only if the shortening is resisted. During an isometric "contraction" no change in sarcomere length appears; in an eccentric "contraction" the sarcomere is actually lengthened. To avoid such contradiction in terms, it is often better to use the terms activation, effort, or exertion, see Chap. 2.

Dynamics A subdivision of mechanics that deals with forces and bodies in motion.

Ergonomics The application of scientific principles, methods and data drawn from a variety of disciplines to the design of engineered systems in which people play significant roles.

Force In physics, a vector that can accelerate a mass. As per Newton's second law, the product of mass and acceleration; the proper unit is the Newton, with $1\text{ N} = 1\text{ kg m s}^{-2}$. On Earth, one kg applies a (weight) force of 9.81 N (1 lb exerts 4.44 N) to its support due to gravity. Muscular force often is described as tension (pressure or stress) multiplied by transmitting cross-sectional area.

Kinematics A subdivision of dynamics that deals with the motions of bodies, but not the causing forces.

Kinetics A subdivision of dynamics that deals with forces applied to masses.

Lever arm One component of the formula for moment (or torque): "force × lever arm". Also called mechanical advantage, it represents the distance from a fulcrum at which the force vector acts.

Mechanics The branch of physics that deals with forces applied to bodies and their ensuing motions.

Moment The product of force and its lever arm when trying to rotate or bend an object; the stress in material generated by two opposing forces that try to bend the material about an axis perpendicular to its long axis; see also Torque.

Moment of inertia In physics, measure (in kg m^2) of the resistance of an object to a change in its <u>rotation</u> rate. Also called mass moment of inertia, rotational inertia, angular mass.

Muscle force The product of tension within a muscle multiplied with the transmitting muscle cross-section.

Muscle tension The pull within a muscle expressed as force divided by transmitting cross-section.

Physics The science of matter or energy and their interactions.

Physiology The branch of biology that deals with the physical and chemical functions of living organisms and their parts.

Radius of gyration The distance from the axis of rotation to a point where the total mass of the body may be concentrated, so that the moment of inertia about the axis remains the same.

Statics A subdivision of mechanics that deals with bodies at rest.

Tension Force divided by the cross-sectional area through which it is transmitted.

Torque A force applied at a lever arm that twists or rotates something about its central axis; see also Moment.

References

Braune W, Fischer O (1889) Determination of the moments of inertia of the human body and its limbs (in German). Translated 1963 as Human mechanics. AMRL-TDR-63-123. Aerospace Medical Research Laboratory, Wright-Patterson AFB

Chaffin DB (2008) Digital human modeling for workspace design (Chap. 2). In: Caldwell CM (ed) Reviews of human factors and ergonomics, vol 4. Human Factors and Ergonomics Society, Santa Monica

Chaffin DB, Andersson GBJ (1984) Occupational biomechanics. Wiley, New York

Chaffin DB, Andersson GBJ, Martin BF (2006) Occupational biomechanics, 4th edn. New York, Wiley

Chandler RF, Clauser CE, McConville JR, Reynolds HM, Young JW (1975) Investigation of inertial properties of the human body. AMRL-TR-74-137. Aerospace Medical Research Laboratory, Wright-Patterson AFB

Clauser CE, McConville JT, Young JW (1969) Weight, volume, and center of mass of segments of the human body. AMRL-TR-69-70. Aerospace Medical Research Laboratory, Wright-Patterson AFB

Dempster WT (1955) The anthropometry of body action. Dyn Anthr Ann N Y Acad Sci 63 (4):559–585

Durkin JL, Dowling JJ, Scholtes L (2005) Using mass distribution information to model the human thigh for body segment parameter estimation. J Biomech Eng 127(3):455–464

Fischer O (1906) Theoretical fundamentals for mechanics of living bodies with special application to man as well as to some processes of motions in machines (in German). Translated (circa 1960) in Air Technical Intelligence Translation ASTIA No. ATI-153668. Aerospace Medical Research Laboratory, Wright-Patterson AFB

Fung YC (1990) Biomechanics. Springer, New York

Harless E (1860) The static moments of the component masses of the human body (in German). Trans Math-Phys R Bavar Acad Sci 8:1, 69–96. Cited by Clauser, McConville J, Young (1969) as an unpublished translation

Hay JG (1973) The center of gravity of the human body. In: Kinesiology III, pp 20–44. American Association for Health, Physical Education, and Recreation, Washington

Herron RE, Cuzzi JR, Hugg J (1976) Mass distribution of the human body using biostereometrics. AMRL Technical report 75-18, Aerospace Medical Research Laboratory, Wright-Patterson AFB

Kaleps I, Clauser CE, Young JW, Chandler RF, Zehner GF, McConville J (1984) Investigation into the mass distribution properties of the human body and its segments. Ergonomics 27 (12):1225–1237

King A (1984) A review of biomechanical models. J Biomech Eng 106:97–104

Kroemer KHE (2008) Anthropometry and biomechanics: anthromechanics (Chap. 2). In: Kumar S (ed) Biomechanics in ergonomics, 2nd edn. CRC Press, Boca Raton

Kroemer KHE, Snook SH, Meadows SK, Deutsch S (eds) (1988) Ergonomic models of anthropometry, human biomechanics and operator-equipment interfaces. National Academy of Sciences, Washington

Kumar S (ed) (2008) Biomechanics in ergonomics, 2nd edn. CRC Press, Boca Raton

Marras WS (2008) The working back: a systems view. Wiley, New York

Marras WS, Radwin RG (2006) Biomechanical modeling (Chap. 1). In: Dickerson RS (ed) Reviews of human factors and ergonomics, vol 1. Human Factors and Ergonomics Society, Santa Monica

McConville JT, Churchill T, Kaleps I, Clauser CE, Cuzzi J (1980) Anthropometric relationships of body and body segment moments of inertia. AFAMRL-TR-80-119. Aerospace Medical Research Laboratory, Wright-Patterson AFB

Mungiole M, Martin PE (1990) Estimating segment inertial properties: comparison of magnetic resonance imaging with existing methods. J Biomech 23(10):1039–1046

NASA (1989) Man-systems integration standards. Revision A., NASA-STD 3000. LBJ Space Center, Houston

NASA/Webb (1978) Anthropometric sourcebook. NASA Reference Publication 1024. LBJ Space Center, Houston

Oezkaya N, Nordin M (1999) Fundamentals of biomechanics, 2nd edn. Springer, New York

Oezkaya N, Leger D, Goldsheyder D, Nordin M (2016) Fundamentals of biomechanics, 4th edn. Springer, New York

Roebuck JA (1995) Anthropometric methods. Human Factors and Ergonomics Society, Santa Monica

Roebuck JA, Kroemer KHE, Thomson WG (1975) Engineering anthropometry methods. Wiley, New York

Winter DA (2004) Biomechanics and motor control of human movement, 3rd edn. Wiley, New York

Young JW, Chandler RF, Snow CC, Robinette KM, Zehner GF, Lofberg MS (1983) Anthropometric and mass distribution characteristics of the adult female. FAAAM- 83-16. Federal Aviation Administration, Washington

Zehner GF (1984) Analytical relationships between body dimensions and mass distribution characteristics of living populations. In: Mital A (ed) Trends in ergonomics/human factors I. Elsevier, Amsterdam

Chapter 5
Respiration

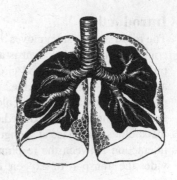

Overview

The respiratory system provides oxygen for energy metabolism and dissipates metabolic byproducts, especially carbon dioxide. In the lungs, oxygen is absorbed into the blood. Blood circulation transports oxygen (and nutrients) throughout the body, particularly to the working muscles. Blood circulation also moves metabolic byproducts to the skin and the lungs. At the skin, both water and heat are dissipated, whereas in the lungs, carbon dioxide as well as water and heat are dispelled while more oxygen is absorbed.

This chapter does not address the topic of internal (cellular) respiration, which is the utilization of oxygen and the production of carbon dioxide and other metabolites by cells—this is discussed in Chap. 7.

The Model

The human respiratory system provides a two-way gas exchange in the lungs: absorbing oxygen from the inhaled air and infusing it into the blood stream; in the opposite direction, extracting carbon dioxide from the blood and dispelling it into the air to be exhaled. Both functions require diffusion of gas through the lung membranes that separate the air space from the blood in arteries and veins.

The lungs are located within the ribcage. Pumping action of muscles in the thorax generates air flow that fills and empties the lungs. The airways between lungs and throat and nose are not merely pathways, but they also serve to condition the air inhaled from the human's surroundings.

© Springer Nature Switzerland AG 2020
K. H. E. Kroemer et al., *Engineering Physiology*,
https://doi.org/10.1007/978-3-030-40627-1_5

Introduction

The respiratory system moves air into and out of the lungs. Here, part of the oxygen contained in the inhaled air is absorbed and transferred to the blood flowing in the pulmonary arteries. (Blood circulation is further discussed in Chap. 6) The lungs also remove carbon dioxide, water, and heat (all byproducts of the metabolic processes, see Chap. 7) from blood that arrives in the pulmonary veins and transfers them into the air to be exhaled. The absorption of oxygen and the expulsion of the metabolites take place at sponge-like surfaces in the lungs. Obviously, there is close interaction between the respiratory system and the circulatory system which provides the transport of oxygen, carbon dioxide, water and heat—see Fig. 5.1.

Fig. 5.1 Model of interrelated functions of the respiratory and circulatory systems

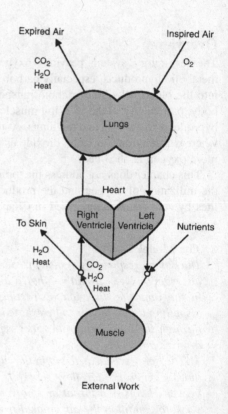

5.1 Architecture

Figure 5.2 sketches the respiratory system. Air enters the body through the nose and mouth and passes through the throat (pharynx), voice box (larynx), and the windpipe (trachea) into the so-called bronchial tree, a series of branching divisions of the bronchus. In each lung, the airways bisect repeatedly into ever smaller ducts, bronchi into bronchioles, until they terminate in the saclike alveoli of the lungs.

The adult respiratory tract has twenty-three successive paired branchings. The first sixteen branches merely conduct the air with little or no gas exchange. The next three branches are respiratory bronchioles (with diameters of about 1 mm) with some gas exchange. The final four generations of branches form the alveolar ducts which terminate in the alveoli. These are small air sacs with diameters of about 400 μm that have thin membranes between capillaries and other subcellular structures allowing the exchange of oxygen (O_2) and carbon dioxide (CO_2) and water between blood and air. Figure 5.3 shows the scheme of the respiratory tree. Altogether, between 200 and 600 million alveoli provide a grown person with 70–100 m^2 of exchange surface.

Air flow is the result of the pumping action of the thorax. The diaphragm, which separates the chest cavity from the abdomen, descends about 10 cm when the abdominal muscles relax (which brings about a protrusion of the abdomen). Furthermore, inspiratory muscles connecting the ribs contract and, by their

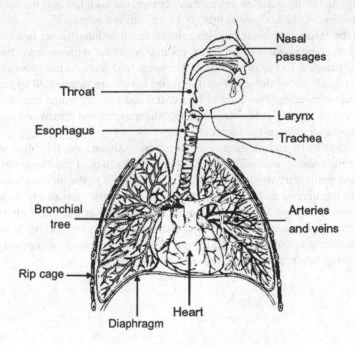

Fig. 5.2 Schematic of main structures of the respiratory system (adapted from Asimov 1963)

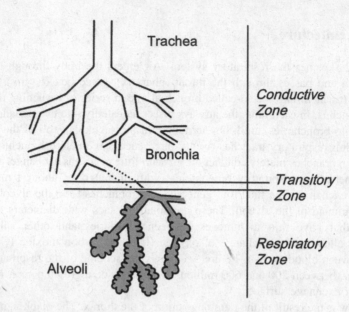

Fig. 5.3 Model of the respiratory tree

anatomical and mechanical arrangements, raise the ribs. Hence, the dimensions of the rib cage and of its included thoracic cavity increase both towards the outside and in the direction of the abdomen: this sucks air into the lungs.

When the inspiratory muscles relax, elastic recoil in lung tissue, in the thoracic wall, and the abdomen restores their resting positions without involvement of expiratory muscles: this expels air from the lungs. However, if heavy work creates ventilation needs above those at rest, the recoil effects are augmented by actions of the respiratory muscles. The internal intercostal and the abdominal muscles reduce the chest and lung cavity; the external intercostal muscles and downward movement of the diaphragm enlarge the thoracic cavity.

Thus, inspiratory and expiratory muscles are activated reciprocally, and both overcome the resistance provided by the elastic properties of the chest wall and of airways and pulmonary tissue. Most of that resistance is in the airways because the air flow in the trachea and the main bronchi is turbulent, particularly so at heavy exercise which requires high flow velocities. However, in the finest air tubes air flow is slow and laminar. Altogether, the energy required for breathing is relatively small, amounting to only about 2% of the total oxygen uptake of the body at rest and increasing to not more than 10% at heavy exercise.

5.2   Functions

The primary task of the respiratory system is *external respiration*: to absorb oxygen from air and provide it to the body, and to dispel carbon dioxide, heat, and water to the air. (Chap. 7 deals with *internal respiration*, the utilization of oxygen and the production of carbon dioxide and other metabolites by the cells, hence also called *cellular* respiration.)

The specific task of breathing is to move air into, through, and out of the lungs. This includes conditioning the inspired air: adjusting the temperature and moisture content of the inward flowing air. The temperature regulation is so efficient that the inspired air is at about body core temperature (37 °C) when it reaches the end of the pharynx, whether inspiration is through the mouth or through the nose.

The passageways clean the incoming air from particles of foreign matter. Cleansing mostly takes place at mucus-covered surfaces of the nasal passages, which trap particles larger than about 10 μm by small hairs and moist membranes. Smaller particles settle in the walls of the trachea, the bronchi, and the bronchioles. An occasional sneeze or cough (with air movements at approximately the speed of sound in the deeper parts of the respiratory system) helps to expel foreign particles. Thus, the deep lungs are kept nearly sterile.

In a normal climate, about 10% of the total heat loss of the body, whether at rest or work, occurs in the respiratory tract. This percentage increases to about 25% at outside temperatures of −30 °C. In a cold environment, heating and humidifying the inspired air cools the mucosa; during expiration, some of the water condenses out of the warm humid air—hence the "runny nose" in the cold.

5.3   Respiratory Volumes

The volume of air exchanged in the lungs depends largely on the activation of respiratory muscles. When they remain relaxed, there is of course still air left in the lungs; a forced maximal expiration reduces this volume of air in the lungs to the so-called *residual capacity*, see Fig. 5.4. A maximal inspiration adds the volume called *vital capacity*. Both volumes together are the *total lung capacity*. During rest or submaximal work, only the so-called *tidal volume* is moved, leaving both an inspiratory and an expiratory reserve volume within the vital capacity.

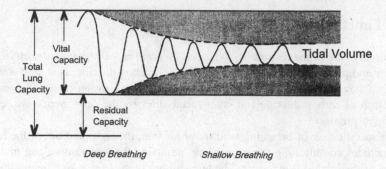

Fig. 5.4   Respiratory volumes

Dead volume, or dead space, is the volume—about 150 mL—of the conductive zone of the human airways where inhaled air does not come in contact with alveoli and, therefore, does not contribute to gas exchange. Accordingly, with a normal inspiration (tidal volume) of about 500 mL in a healthy adult, only 350 mL reach the alveoli.

Vital capacity and other respiratory volumes are usually measured with a spirometer. The results depend on the age, training, gender, body size, and body position of the subject. Total lung volume for highly trained tall young males is between 7 and 8 L and their vital capacity is up to 6 L. On average, women have lung volumes that are about 10% smaller than those of men. Untrained persons have volumes of about 60–80% of their athletic peers.

Pulmonary ventilation is the movement of gas in and out of the lungs. Its volume is calculated by multiplying the frequency of breathing by the expired tidal volume. This is called the (respiratory, expired) *minute volume*. At rest, an adult person breathes around 12 times per minute. During light exercise, primarily the tidal volume and not the rate of breathing increase. With heavier work the respiratory frequency quickly rises, up to about 45 breaths per minute in adults. Children breathe much faster at maximal effort; 5-year olds do so about 70 times per minute and 12-year olds about 55 times per minute. With these large variations, breathing frequency, although easy to measure, is not a very reliable indicator of the demands of work performed.

Alveolar ventilation, the volume of fresh air that enters the alveoli each minute, is a suitable measure of an individual's respiratory effectiveness. Assuming that 350 mL normally reach the alveoli with each breath, typical alveolar ventilation is around 4.2 L/min with 12 breaths per minute. In a maximal effort, alveolar ventilation can be as high as 100 L/min; at the opposite extreme, a person can stay alive with as little as 1.2 L/min.

5.4 Measuring Respiration

Following the demands of work, the human respiratory system is able to increase its moved air volumes and absorbed oxygen by large multiples compared to those at rest. The minute volume can be boosted by a factor of 20, increasing from about 5 to about 100 L or more per minute. Though not exactly linearly related to minute volume, the oxygen consumption shows a similar increase.

Observing the changes in pulmonary functions allows assessing a person's strain while performing physical work: breathing rate, tidal volume and minute volume are of particular interest. Of these, breathing rate is easily recorded with a plethysmograph (which measures changes in volume of the lungs) or an air flow meter (which measures the mass of air flowing through a tube). However, in general these measurements provide less useful information to the ergonomist than do recordings of circulatory events (especially heart rate, see Chap. 6) and of metabolic functions (especially oxygen consumption, see Chap. 7).

Summary

The respiratory system moves air into and out of the lungs where oxygen contained in the inhaled air is absorbed and transferred to the blood flowing in the pulmonary arteries. Further, the lungs remove carbon dioxide, water and heat from blood that arrives in the pulmonary veins and transfer them into the ambient air to be exhaled.

Normally, only a portion (the tidal volume) of the available volume (vital capacity) of the respiratory passages is utilized for respiratory exchange.

An adult at rest has a breathing frequency of fewer than 20 beaths per minute. With light work, the frequency remains near resting level, but the tidal volume gets larger. In heavy work, both breathing rate and tidal volume increase.

In response to physically demanding work, the respiratory system can easily increase its moved air volume by a factor of 20 over resting conditions, accompanied by a similar increase in oxygen intake.

Glossary

Ergonomics The application of scientific principles, methods and data drawn from a variety of disciplines to the design of engineered systems in which people play significant roles.

Metabolism (Greek, change) The set of chemical reactions that sustain life in organisms: 1. conversion of food/fuel to energy for running cellular processes; 2. conversion of food/fuel to building blocks for proteins, lipids, nucleic acids, and some carbohydrates; 3. elimination of nitrogenous wastes. Metabolic reactions can be catabolic (breaking down compounds) or anabolic (building up (synthesis) of compounds. Usually, catabolism releases energy, and anabolism consumes energy.

Metabolite A substance formed in or necessary for metabolism—see there.

Plethysmograph Instrument to record variations in the size (volume) of parts of the body, such as of the chest circumference with breathing.

Respiration The process of breathing, inhaling and exhaling, which moves air into and out of the lungs and provides oxygen for energy metabolism and dissipates metabolic byproducts. (Chap. 7 covers "internal" or "cellular respiration", the utilization of oxygen and the production of carbon dioxide and other metabolites by cells.)

Spirometer Instrument to measure the flow of air into and out of the lungs.

References

There are many texts on human physiology discussing respiration; among them

Asimov I (1963) The human body. Signet, New York

Astrand PO, Rodahl K, Dahl HA, Stromme SB (2003) Textbook of work physiology. In: Physiological bases of exercise, 4th edn. Human Kinetics, Champaign

Hall JE (2016) Guyton and Hall Textbook of medical physiology, 13th edn. Saunders, Philadelphia

Kenney WL, Wilmore JH, Costill DL (2020) Physiology of sport and exercise, 7th edn. Human Kinetics, Champaign

Chapter 6
Circulation

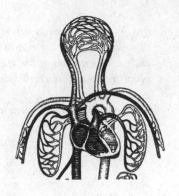

Overview

Two transport systems transfer materials between body cells and tissues: the blood circulation and, as an accessory, the lymphatic circuit. These transport networks move nutritional materials from the digestive tract to cells for catabolism, synthesis, and deposit. Blood takes oxygen from the lungs to the consuming cells; it also carries carbon dioxide to the lungs to be expelled, lactic acid to the liver and kidneys for processing, and water to the skin and lungs for heat dissipation. The blood stream also is part of the body control system by carrying hormones from endocrine glands to receptive cells.

In the blood, red cells carry oxygen, required for the heart to beat, the brain to function, and the muscles to work. White cells defend against invasion by pathogens. Platelets and proteins in plasma form clots that can prevent hemorrhages. Stem cells in the bone marrow incessantly renew blood: red cells every few months, platelets and most white cells every few days. The lymphatic system primarily helps remove excess (interstitial) fluids from body tissues and participates in the body's immune system response.

The Model

The human blood circulation is a closed fluid loop powered by two pumps in series: the right and left sections of the heart. After passing through the left pump, blood flows within ever branching and narrower arteries until it slowly moves through a delicate network of fine vessels in an organ, such as muscle. Here, as needed, oxygen, hormones and other materials are exchanged with the surrounding tissues; byproducts of metabolic processes

© Springer Nature Switzerland AG 2020
K. H. E. Kroemer et al., *Engineering Physiology*,
https://doi.org/10.1007/978-3-030-40627-1_6

are gathered; heat is taken up or dispersed. This accomplished, the blood then drains from the organ into small vessels which combine to a vein that ends in the atrium of the right half of the heart. After passing this pump, the blood reaches fine capillaries in the lungs where it releases metabolites and gathers oxygen.

The circulatory system of the blood, with the lymphatic system as an accessory, is able to serve many different target locations throughout the body, with often-changing demands and workloads.

Introduction

Blood circulation is essential for providing nutrients and oxygen to muscles and other organs; and for removing metabolic by-products. The flow of blood is accomplished by the heart, a "pump" whose beat (contraction rate) is often taken as the indicator of the work load of the total body. The lymphatic circuit, separate from the blood vessels, collects excess fluid from body tissues and returns it the blood; in this respect, the lymphatic system is an accessory to the blood flow system.

6.1   Body Fluids

Water is the largest weight component of the body: in women about 50% of body weight, close to 60% in men. Slim individuals have a higher percentage of total water than obese persons since adipose tissue contains very little water. The relationship between water and lean (fat-free) body mass is about 72% in healthy adults. Of the total body water (say, 40 L in an adult), about 70% (25 L) is *intracellular fluid*, within body cells. The other 30% surround the cells: this *extracellular fluid* has an ionic composition similar to ocean water. The concentrations of various electrolytes are different in the intra- and extracellular fluids. The cell membrane, which separates those two fluids, acts as a barrier for positively charged ions (*cations*).

Of the extracellular fluid, some is contained within blood vessels (*intravascular fluid)* while the rest is between the blood vessels and the cell membrane (*interstitial fluid*); the cells are "bathed" in interstitial fluid. Substance exchange between the intravascular and interstitial fluids takes place primarily through the walls of the capillaries; however, plasma and blood corpuscles cannot pass through these walls.

6.1.1 Blood

Approximately 10% of the total fluid volume of the human body consists of blood. The volume of blood is variable, however, and depends on age, gender, and training. Volumes of 4–4.5 L of blood for women and 5–6 L for men are normal. About two-thirds of the total volume are usually located in the venous system, while the other third is in the arterial vessels. When a human is at rest, the total volume of blood circulates through the body in about one minute.

Blood can carry heat of about 3.6 J/kg/°C; this is called the specific heat of blood. This heat capacity* is below that of water (4.18 J/kg/°C), but is sufficient to transports heat to or away from tissues.

Blood consists of two components: the fluid is called plasma, the suspended cells (and cell fragments) are called formed elements. By volume, blood is half water and nearly as much, 45%, red blood cells. Plasma solids (chiefly proteins) constitute about 4% of blood volume, and the remaining 1% consists of white blood cells and platelets.

The ratio of the volume of red blood cells to the total blood volume is called hematocrit. Red cells look like flattened donuts without holes (bi-concave disks), have a large, flexible surface area, and carry hemoglobin, which binds to gases, especially oxygen and carbon dioxide.

Some white cells (*lymphocytes, monocytes*) have their origin in lymph system tissues. Bone marrow produces red blood cells (*erythrocytes*), white blood cells (*leukocytes, granulocytes*), and platelets (*thrombocytes*); 1 mm^3 (1 mL) of blood contains about 300,000 cells and platelets.

Loss of less than 10% of the total blood volume is not critical for a healthy person. If the loss approaches 20%, blood pressure diminishes and changes in pulse and breathing set in. At a blood loss of 40% or more, death is imminent if blood and fluids are not infused.

6.1.1.1 Blood Groups

Depending on the content of certain antigens and antibodies, blood is classified into four groups: A, B, AB, 0. These classifications reflect incompatibility reactions in blood transfusions. Another blood group classification relies on the rhesus (Rh) factor that relates to an obstetrical problem which may occur when an Rh-negative woman is pregnant with an Rh-positive baby.

6.1.1.2 Functions of Blood Components

The plasma carries dissolved materials to the cells, particularly oxygen and nutritive materials (including monosaccharides, neutral fats, amino acids from catabolism—see Chap. 7) as well as enzymes, salts, vitamins and hormones. It removes waste products, particularly dissolved carbon dioxide and also heat.

The red blood cells transport gases. Oxygen attaches to their hemoglobin, an iron-containing protein molecule which makes the blood cell appear red. Each molecule of hemoglobin contains four atoms of iron which combine loosely and reversibly with four molecules of oxygen. Hemoglobin also has a strong affinity to carbon monoxide (CO), which takes up spaces otherwise filled by oxygen; this explains the high toxicity of CO gas exposure. Carbon dioxide (CO_2) molecules can bind to amino acids of the hemoglobin protein; since these are different binding sites, hemoglobin molecules can react simultaneously with oxygen and carbon dioxide.

Each gram of hemoglobin can combine with at most 1.34 mL of oxygen. With 150 g hemoglobin per liter of blood, a fully O_2-saturated liter of blood can carry 0.20 L of oxygen. In addition, a small amount of oxygen, about 0.003 L, is dissolved in the plasma.

6.1.2   Lymph

Lymph is a colorless fluid containing white blood cells. It "bathes" tissues and drains through the lymphatic system into the blood stream. The lymphatic vessels are separate from the blood vessels yet acts as an accessory to the blood flow system. One of its tasks is to collect excess interstitial and extracelluar fluid from body tissues and return it the bloodstream.

Lymphatic capillaries exist in most tissue spaces, close to blood capillaries. The lymphatic capillaries are thin-walled and permeable so that even large particles and protein molecules can pass directly into them. Hence, the fluid in the lymphatic system is really an overflow from tissue spaces; lymph is much the same as interstitial fluid. The lymphatic capillaries have closed ends and one–way flap valves, which serve to direct the flow toward larger ducts, called lymphatics. At intervals, they pass through lymph nodes: these filter out foreign materials, including invading micro-organisms. Finally, lymphatic vessels lead to the neck where they empty into the blood circulation at the juncture of the left internal jugular and left subclavian veins.

The flow of lymph within its tubing system depends on the interstitial fluid pressure: the higher the pressure, the larger the lymph flow. Another factor affecting lymph flow is the so-called lymphatic pump: excess lymph stretches the lymph vessel which then automatically contracts. This contraction pushes the lymph past its one-way valves. The contractions occur periodically, one at every six to ten seconds. Motion of tissues surrounding the lymph vessels, such as by the contraction of skeletal muscle surrounding a vessel, can also pump lymph.

These lymph-pumping mechanisms generate a partial vacuum in the tissues so that they can collect excess fluid. Lymph flow is highly variable, on the average 1–2 mL/min; usually, this is enough to drain excess fluid and especially excess protein that otherwise would accumulate in the tissue spaces. Swelling of the feet and lower legs in the course of prolonged sitting with little motion (for example, during long airplane rides) is a common example of the collection of fluids, particularly in the lymphatic system.

6.2   The Circulatory System of the Blood

Working muscles and other organs require generous blood flow to obtain oxygen and nutrients, to remove metabolites, and for hormonal control. These demanding organs are located throughout the body, and their supply needs frequently change. Hence, varying and substantial demands are placed on the transport system, the circulatory system of the blood.

6.2.1   Architecture of the Circulatory System

It is convenient, and anatomically and functionally correct, to model the human blood circulation as a closed loop with two fluid pumps in series: the right and left sections of the heart. Propelled by the pumping action of the heart, blood flows within ever branching and narrower arteries until it moves slowly through a delicate network of fine vessels, the "vascular bed" (see below) of an organ. Here, oxygen and other carried goods are exchanged with the surrounding tissues. This accomplished, the blood drains from the organ into small vessels that combine to form veins, ending in the atrium of the other side of the heart.

The left side of the heart powers the *systemic* subsystem. It carries oxygen-rich blood to the capillary beds, which transverse organs such as the skeletal muscles. There, it releases oxygen and energy carriers such as glucose and glycogen, while it picks up metabolic by-products, especially carbon dioxide, water and heat. The vascular bed drains into the venules of the *pulmonary* subsystem, powered by the right side of the heart. After passing through that pump, the blood reaches the capillary beds of the lungs, where it releases metabolic byproducts and picks up oxygen, to be carried to the next consumer of fuel.

Blood pumped by the heart flows through a series of vessels, arteries, arterioles, capillaries; and then through venules, and veins before returning to the heart. Arteries transport blood away from the heart and branch into smaller vessels, arterioles. Arterioles distribute blood throughout capillary beds, the sites of exchange with the body tissues (see below). Having passed through the capillary bed, blood is lead back to small vessels, venules, which connect to the larger veins. Veins eventually guide blood back into the heart.

The arterial system is a relatively high-pressure system, so arteries have thick walls that appear round in cross section. The venous system is a lower-pressure system, containing veins that have larger lumens and thinner walls.

The total circulatory system consists of a large number of parallel or serial, often interconnected sections which supply individual organs. It interacts closely with the respiratory systems, as shown earlier in Fig. 5.1 of Chap. 5.

6.3   The Heart as Pump

The heart is, in essence, a hollow muscle which produces, via contraction and with the aid of valves, blood flow through it. Each of the halves of the heart has an antechamber (*atrium*) which opens into the chamber (*ventricle*), the pump proper. The blood flows from the main vein into the atrium; it then moves through an open valve into the ventricle. Musculature surrounding the ventricle compresses it after its valves close; this is the *systole*. When the internal pressure is equal to the pressure in the aorta, the exit valve opens so that the ventricle can expel the pressurized blood into the aorta, which is the main artery. The left side of the heart pushes blood into the systemic system, the right heart side pushes blood into the pulmonary system. Then follows the *diastole*, a phase of the heart relaxing; then the cycle of systole to diastole repeats. The events in the heart halves are similar, but the right side generates pressure in the pulmonary arteries that is only about 1/5 the pressure that the left heart produces in its systemic arteries.

The mechanisms for excitation and contraction of the heart muscle are similar to those of skeletal muscle; however, specialized cardiac cells (*sinoatrial nodes*) in the atrium serve as "pacemakers" which do not need external nervous impulses to function. These specialized cells determine the frequency of heart contractions by propagating stimuli to other cells in the ventricle, especially the Purkinje fibers, which make the ventricular muscles contract.

The heart's own intrinsic control system operates, without external influences, at rest with 50–70 beats per minute (depending on the individual). Changes in heart action stem from the central nervous system via the autonomic nervous system (see Chap. 3). Stimulation to augment heart action comes through the sympathetic system, mostly by increasing the heart rate, the strength of cardiac contraction, and the blood flow through the coronary blood vessels supplying the heart muscle. The parasympathetic system causes a decrease of heart activities, particularly a reduced heart rate through weakened contraction of the atrial muscle, slowed conduction of impulses (lengthening the delay between atrial and ventricular contraction), and decreased blood flow through the coronary blood vessels. The parasympathetic system is dominant during rest periods. (The coordinated actions of the sympathetic and parasympathetic nervous systems are another example of control by two opposing systems, also exhibited by the agonist/antagonist set-up of skeletal muscle.)

The electrokardiogram (EKG, also spelled electrocardiogram and abbreviated ECG) records myocardial action potentials. Letters mark certain events in the EKG: P names the wave associated with electrical stimulation of the atrium while the Q, R, S, and T waves identify ventricular events. The EKG mostly serves for clinical diagnoses of heart irregularities; however, with appropriate apparatus it can be used for counting the heart rate. Figure 6.1 shows the electrical, pressure, and sound events during a contraction-relaxation cycle of the heart.

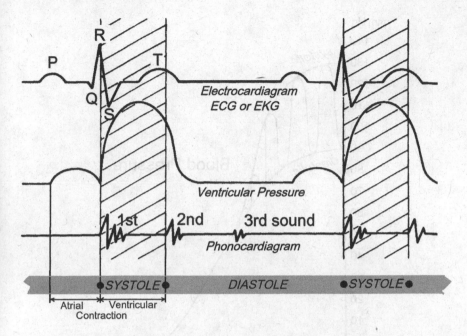

Fig. 6.1 Electrocardiogram, pressure fluctuation, and phonogram of the heart

As already mentioned, the blood flow from the ventricle depends on the strength of contraction during systole: since the volume of blood cannot move as quickly through the aorta as the heart expels into the aorta, continuing contraction of the heart increases blood pressure even further. Part of the excess volume is kept in the aorta and its large branches, which together act as an elastic pressure vessel (German *windkessel*). After the aortic valve closes at the beginning of the relaxation (diastole) of the heart, the elastic properties of the aortic walls propel the stored blood into the ever-branching arterial system, where the elastic blood vessels smooth out the pressure waves of blood volume. By the time blood reaches the smallest branchings of the arterial tree, the blood flow is even and slow—see Fig. 6.2.

At rest, about half the volume in the ventricle is ejected (*stroke volume*) while the other half remains in the heart (*residual volume*). Under exercise load, the heart ejects a larger portion of the contained volume. When the heart strains to supply much blood, such as during very strenuous physical work with small muscle groups or during maintained isometric contractions, the heart rate can become very high (such as 180 beats per minute).

At a heart rate of 75 beats per minute, the diastole takes less than 0.5 s and the systole just over 0.3 s. At a heart rate of 150 beats per minute, the periods are close to 0.2 s each: an increase in heart rate occurs mainly by shortening the duration of the diastole. At rest, blood pressure is typically around 70 mmHg during diastole and 120 mmHg during systole, depending on age and other factors. These values may double with heavy exercise.

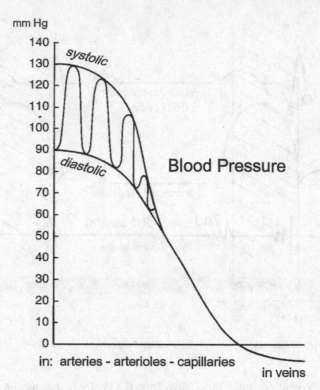

Fig. 6.2 Scheme of the smoothing and reduction of blood pressure along the circulatory pathways

6.3.1   Cardiac Output

The output of the heart mainly depends on two factors: the pressure in the blood generated by each contraction, and the frequency of contractions (*heart rate*, *pulse rate*). Both together determine the cardiac minute volume. Since the available blood volume in the body does not vary, the rate of blood flow determines the capacity to do work. The cardiac output of an adult at rest is around 5 L per minute: for example, a stroke volume of 70 mL with a heart rate of 70 beats per minute. When performing strenuous exercise, this volume can increase five-fold to about 25 L/min; a well-trained athlete may reach up to 35 L/min. Table 6.1 shows the needs of organs for blood supply at rest and during heavy exercise.

The ability of the heart to adjust its minute output volume to the requirements of the activity depends on two factors: on the effectiveness of the heart as a pump, and on the ease with which blood can flow through the circulatory system and return to the heart. A healthy heart can pump much more blood through the body than usually needed. Hence, cardiac output in healthy persons is more likely to be limited by the transporting capability of the vascular portions of the circulatory system than by the

Table 6.1 Blood supply for organs during rest and work

Consumer	at rest*	at heavy work*
Muscle	15%	75%
Heart	5%	5%
Digestive tract	20%	3%
Liver	15%	3%
Kidneys	20%	3%
Skin	5%	5%
Bone	5%	1%
Fatty tissues	10%	1%
Brain	5%	5%

*Total cardiac output is about 5L/min at rest, about 25L/min at heavy work.

heart itself. In the vascular system, the arterial section (upstream of the metabolizing organ) has relatively strong, elastic walls which act as a pressure vessel transmitting pressure waves far into the body, though with much loss of pressure along the way. At the arterioles of the consumer organ, the blood pressure is reduced to approximately one-third its value at the heart's aorta. Figure 6.2 shows the smoothing of pressure waves and reduction of pressure along the blood pathway schematically. As blood seeps through the consuming organ (such as a muscle) via capillaries, blood flow becomes slow and blood pressure drops. The remaining pressure differential (positive on the arterial side, negative on the venous side) helps to maintain blood transport through the intricate network of capillaries called the capillary bed.

6.4 The Capillary Bed

The anatomical patterns of the blood vessels that penetrate the tissues of the lungs and of other organs in need of blood supply are quite similar—see Fig. 6.3. In the direction of the blood flow, the blood vessels branch prolifically into ever smaller vessels: from arteries into arterioles into capillaries (altogether, there are about 10^9 capillaries in the human body). Thus, their total available vascular volume increases even as each vessel's cross-sectional area decreases. (For example: Oxygen enters the blood stream in the pulmonary capillaries with a perfusion area of about 90 m^2 and is delivered in the systemic capillaries through a surface more than double as large.) In the capillary bed, this increase in area makes the blood flow slow enough to allow diffusion through the thin cell walls. On the exit side of the capillary bed,

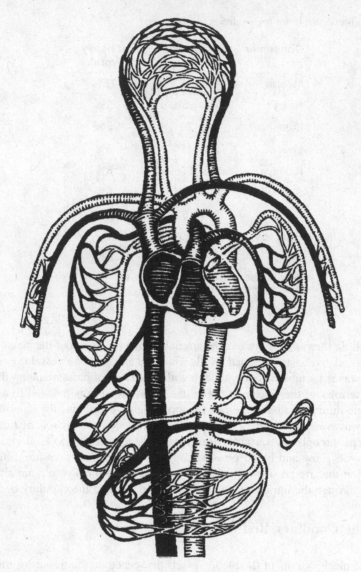

Fig. 6.3 Sketch of the circulatory system (adapted from Asimov 1963)

the small venules combine to form veins, fewer but larger vessels that lead blood back to the heart.

The diagram of a capillary bed in Fig. 6.4 shows blood entering through an arteriole, encircled by smooth muscles. These muscles control the diameter (*lumen*) of the blood vessel by contracting or relaxing as stimulated by the sympathetic nervous system and in response to local accumulation of metabolites. Next, the blood flows into a metarteriole, which has fewer enclosing muscle fibers. Then, blood passes

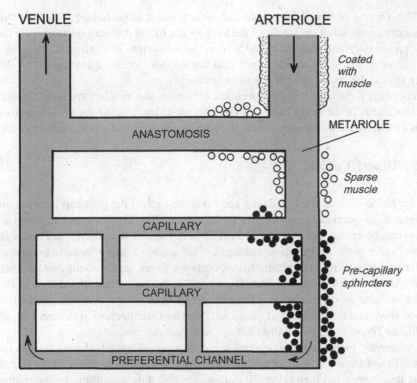

Fig. 6.4 Schematic diagram of the capillary bed

through the pre-capillary sphincters, which control the openings to the capillaries. The metarteriolic and the pre-capillary muscles primarily respond to local tissue conditions. Finally, blood reaches the venules on the exit side of the vascular bed.

Contraction or relaxation of the smooth muscles around a blood vessel change its lumen, hence affecting the flow resistance and the blood pressure. If high volume of blood flow is needed, the muscles allow pathways to remain open; constriction of the capillary bed reduces local blood flow, channeling it to other organs with higher demand for blood. If a short-cut of circulation is needed, blood may use an anastomosis, a shunt that directly connects arteriole to venule.

When blood passes through a larger vessel relatively quickly, there is little opportunity for providing nourishment through vessel cells or for removing waste. However, flow through capillaries is slow because of their small diameter; the smallest (about 5 μm) are just wide enough to let an erythrocyte squeeze through. Slow flow enables diffusion of the cell tissues which provides for exchange of gases, metabolic constituents, and heat.

Vasoconstriction and vasodilation in the arterioles are the primary controls of the distribution of blood flow and pressure. Yet, striated muscle (such as during a sustained strong isometric effort—see Chap. 2) can also compress blood vessels, which may be hinder or shut off blood flow.

The venous portion of the systemic system has a large lumen with little flow resistance; only about one-tenth of the total loss in blood pressure occurs here. (This low pressure system may be called a *capacitance* system in contrast to the arterial *resistance* system.) Valves are built into the venous system, allowing blood flow only in one direction: toward the right ventricle.

Pulmonary circulation and systemic circulation use similar structures, but the capillary beds of the lungs have less constriction or shunting of the vessels that surround the thin-walled air sacs (*alveoli*) where the exchange of gases and heat occurs.

6.5   Blood Vessels

Of the blood vessels, the capillaries and (to some extent) the post-capillary venules provide semi-permeable membranes to the surrounding tissue so that nutrients and gases can be exchanged. All other blood vessels, such as the arteries, arterioles and veins, serve only as transport channels. The walls of these vessels consist (in different compositions) of elastic fibers, collagen fibers, and smooth muscle fibers. The walls are thick in the big arteries and thin in the big veins. Blood flow at each point in these blood vessels is primarily determined by the pressure head of the blood wave and the diameter of the vessel. Size and architecture of different vessels influence blood flow, blood distribution, and vascular resistance.

As already mentioned, the arteries serve as pressure vessels during the ejection of the blood from the heart. Their elastic tissues stretch under the systolic impact, store this energy and then release it during diastole, thus smoothing an intermittent flow to a more continuous stream. Still, the ejection of the blood from the left ventricle causes a pressure wave to travel along the blood vessels at speeds of 10–20 times the velocity of the blood in the aorta (which is about 0.5 m/s when the body is at rest). Since the pressure waves distort blood vessels even at some distance from the heart, measuring how frequently these pulses occur (such as with a plethysmographic instrument attached to a finger tip) reveals the heart rate. Far away from the heart, in the capillaries, venules, and veins, there is no appreciable pressure variation associated with the heart's systolic and diastolic phases.

The flow resistance in the large arteries and veins is small, since the diameter of these vessels is large and the flow velocity is high. In contrast, the peripheral resistance in the arterioles, metarterioles, and capillaries is substantial; that causes a significant reduction in blood pressure, even if the velocity of the (turbulent) blood flow is still high. The diameter of the vessels, generally below 0.1 mm, can be further reduced by smooth muscle fibers that wrap around the vessels—see above. Contraction of these muscles constricts the vessels, which recoil to their normal size when the muscles relax. Thus, contracting these smooth muscles around the blood vessels (and possibly the pressure generated by contraction of nearby striated muscle) can change the flow characteristics and achieve reduction or complete shut-off of circulation to organs where blood supply is not so urgent, hence allowing better supply to those organs in greater metabolic need. To avoid a decrease in arterial blood pressure, any increase in vessel diameter by vasodilation (opening blood vessels beyond their lumen at regular vasomotor tone) must be compensated by vessel constriction in other areas and/or by an increase in cardiac output.

The blood-collecting venules have an external coating of connective tissue and smooth muscle rings, which is only intermittently spaced near the arterial bed but develops into a complete layer in their distal parts, toward the larger veins. Thus, alternately contracting and relaxing muscles act as a pump which moves venous blood towards the right heart, aided by valves in the veins of extremities that prevent backflow of the blood; this is of importance because the blood pressure in the veins is very low, approaching zero just as blood reaches the heart.

The venous system usually carries about three-quarters of the total blood volume, while arteries contain about 20% and capillaries about 5% of the blood volume at any given time.

6.6   Hemodynamics

As in the physics of classical fluid dynamics, the important factors in the dynamics of the blood flow (hemodynamics) are the capacity of the pump (the heart) to do the work; the physical properties of the fluid (blood) to be pumped, particularly its viscosity ("internal friction"); and the properties of the transport pipes (blood vessels) in regard to the required flow rates (volume per unit time) and flow velocities.

The difference in internal pressure between the start and the end of the flow pathway is the main determiner of flow rate and flow velocity; this pressure differential results from the resistance to flow. These factors are related by the equation for the pressure gradient:

$$\Delta p = Q \times R$$

with Δp the pressure gradient ($p_{start} - p_{end}$), Q the flow rate, and R the peripheral resistance.

One key principle of fluid dynamics is that at any point in a fluid, the pressure is equal in all directions. This means that fluid pressure can be measured as the pressure against the lateral walls of the containing vessel. Thus, the usual measure of "blood pressure" (commonly assessed with a cuff around the left upper arm) reflects this lateral pressure, but the elastic properties of the containing blood vessel and other intervening tissues modify that reading.

The flow resistance of a deformable blood vessel is quite different from the formula used for rigid manufactured pipes (such as carry municipal water), but still essentially depends on the diameter of the vessel, the length of the vessel, and on the viscosity of the blood. Blood viscosity is about 4 to 5 times that of water, and is determined by the protein content of the plasma as well as the number, dimensions, and shapes of the blood cells; the more hematocrit, the higher the resistance to flow.

The flow of blood is often not smoothly stream-lined (laminar) but turbulent. Turbulence occurs when the outer layers of the blood stream are in physical contact with the inner walls of the blood vessels and, therefore, move more slowly than the more central sections of the blood flow. Turbulence increases the energy loss within the moving fluid, therefore (at a given pressure gradient) the average flow is slower than in a laminar stream.

The static pressure in a column of fluid depends on the height of that column (Pascal's law). However, the hydrostatic blood pressure in the feet of a standing person is not as large as expected from principles of physics since the valves in the veins of the extremities modify the pressure. For example, in a person standing upright, the arterial pressure in the feet may be only about 100 mmHg higher than in the head. Nevertheless, blood, lymph and other primarily extracellular fluids can pool in the lower extremities, particularly when one stands (or sits) still for a long enough time, resulting in swelling the volume of the lower extremities (*edema*).

6.7   Regulation of Circulation

If organs such as muscles need increased blood flow, flow regulation takes place primarily at two sites: at the heart and in the organ to be supplied.

The heart regulates blood flow by adjusting its cardiac output. The blood pressure in the aorta, however, depends not only on cardiac output, but also on peripheral resistance, elasticity of the main arteries, viscosity of the blood, and total blood volume. The local flow is mainly determined by the pressure head and the diameter of the vessel through which it passes. The smooth muscles encompassing the arterioles and veins continuously receive nerve impulses that keep the internal diameter (lumen) of the vessels more or less constricted. This local vasomotor tone is controlled by vasoconstricting fibers driven from the medulla in the brain stem (see Chap. 3); alterations of the vessel's muscle tone keep the systemic arterial blood pressure on a level suitable for the actual requirements of all vital organs. (Changes in heart function and in circulation are initiated at brain levels above the medullary centers, probably at the cerebral cortex; see Chap. 3.)

If at a particular site metabolite concentrations rise above an acceptable level, this local condition directly causes nearby metarteriolic and sphincter muscles to relax, allowing more blood flow. When skeletal muscles work hard and therefore require more blood flow, signals from the brain's motorcortex can activate vasodilation of precapillary vessels in the muscles and simultaneously trigger a vasoconstriction of the vessels supplying the abdominal organs. This leads to a remarkable and very quick redistribution of the blood supply which favors skeletal muscles over the digestive system: the "muscles-over-digestion" principle.

However, even in heavy exercise, the systemic blood flow is controlled in such a way that the arterial blood pressure provides adequate blood supply to brain, heart, and other vital organs. For this purpose, neural vasoconstrictive commands can override local dilatory control. For example, the temperature-regulating center in the hypothalamus can affect vasodilation in the skin if this is needed to maintain a suitable body temperature, even if this means a reduction of blood flow to the working muscles: the "skin-over-muscles" principle (see Chap. 8).

Thus, on the arterial side, circulation at the organ/consumer level is regulated both by local control and by impulses from the central nervous system, the latter having overriding power. Vasodilation in organs needing increased blood flow, together with vasoconstriction where blood is not so necessary, regulate local blood supply. At the same time, the heart increases its output by higher heart beat frequency and higher blood pressure.

At the venous side of circulation, constriction of veins, combined with pumping actions of dynamically working muscles and forced respiratory movements, facilitate return of blood to the heart. This makes increased cardiac output possible, because the heart cannot pump more blood than it receives.

Heart rate generally is coupled with oxygen consumption; hence heart rate increases with energy production of dynamically working muscles in a linear fashion from moderate to heavy work (see the section on "Indirect Calorimetry" in Chap. 8). However, the heart rate at a given oxygen intake is higher when the work is performed with the arms than with the legs. This reflects the use of different muscles and muscle masses with different lever arms to perform the work. Smaller muscles doing the same external work as larger muscles experience more strain and therefore require more oxygen.

Static (isometric) muscle contraction increases the heart rate, apparently because the body tries to bring blood to the tensed muscles. However, it is difficult to compare this effect in terms of efficiency (as in "heart beats per effort") with the increase in heart rate during dynamic efforts because, in the isometric contraction, a muscle does no "work" (in the physical sense: there is force but no displacement) whereas in the dynamic case, work is done.

Physical work done in a hot environment causes a higher heart rate than in moderate climate, as explained in Chap. 9. Finally, emotions, nervousness, apprehension, and fear can affect the heart rate, especially at rest and during light work.

Given the vital role of the heart in maintaining circulation and the fact that many persons suffer from malfunctions, it is fortunate that diagnoses, medical treatments and repairs of deficiencies in proper functioning have made great advances. Clogged arteries, malfunctioning heart valves, and problems in pacing can often be relieved, even completely resolved, by surgical procedures and implants*.

6.8   Measurement Opportunities

With current technology it is impractical to measure the changes in blood supply at the working muscle, although this would be of special interest because the local supply determines largely whether the muscle can perform its job. Yet, without using invasive techniques, available instrumentation makes it easy to assess functions of the heart, even when the person is at rest or work: heart (pulse) rate can be recorded by electrical and volume (pressure) changes, and the sound can be heard easily even without a stethoscope. With heart beat frequency so closely related to body effort and its observation so convenient, measuring heart rate is a very important tool for medical and ergonomic purposes.

Notes

The text contains markers, *, to indicate specific references and comments, which follow.

Heat capacity: ITIS (2016)

Surgical procedures and implants in the USA:

- coronary stents: >644,000 patients, >965,000 stents (in 2009)—Auerbach et al. (2012);
- cardiac pacemakers: ~200,000 (in 2016)—van der Zee and Doshi (2016);
- artificial heart valve: >182,000 (estimate for 2019)—iData Research (2019);
- cardiac defibrillators: >200,000 (in 2009)—Mond and Proclemer (2011).

Summary

In the human body, two transport systems transfer materials between body cells and tissues: the blood and the lymphatic systems. These sytems move nutritional materials from the digestive tract to cells for catabolism, synthesis, and deposit; and they transport oxygen and metabolic byproducts.

The blood circulation system is a closed loop with two fluid pumps in series: the right and left sides of the heart. After passing through one of the pumps, the blood flows within branching and narrowing arteries until it slowly moves through a delicate network of fine vessels in an organ. Here, as needed, oxygen and other materials as well as heat are exchanged with the surrounding tissues. This accomplished, the blood drains into small vessels that combine to a vein that ends in the atrium of the next half of the heart. The circulatory system of the blood, with the lymphatic system as an accessory, is able of adapting to serve many different target locations throughout the body, with often-changing demands.

Blood plasma carries dissolved materials to the cells, particularly oxygen and nutritive materials as well as enzymes, salts, vitamins and hormones. It removes waste products, particularly dissolved carbon dioxide and heat. The red blood cells perform the oxygen transport. Both oxygen and carbon dioxide attach to hemoglobin, an iron-containing protein molecule in the red blood cell.

Working muscles and other organs require generous blood flow to supply them with oxygen and nutrients, to remove metabolites, and for control through the hormonal system. These consumers are located throughout the body, and their supply needs change. Hence, very different and substantial demands are placed on the transport system, the circulatory system of the blood.

Flow through the arterial and subsequent venous subsystems is determined by the organs depending on their need of blood and by local conditions of smooth muscle contraction. The volume of blood pumped per minute is about 5 L at rest and can increase up to sevenfold with strenuous work. This increase is brought about mostly by changes in heart rate and in blood pressure.

Heart rate (pulse rate) is the number of ventricular contractions per minute; it creates pressure waves in the arteries. Its range is usually 50 to 70 beats per minute at rest and increases about threefold with strenuous exercise.

Stroke volume is the volume of blood ejected from the left heart into the main artery during each ventricular contraction. It is usually 60 to 100 mL at rest and may increase threefold with hard work.

Cardiac output, also called (cardiac) minute volume, equals stroke volume multiplied with heart rate. This volume of blood injected into the main artery per minute can increase five- to sevenfold over resting values.

Blood pressure is the internal pressure in the arteries near the heart; at rest it is about 70 mmHg during diastole and 120 during systole. These values may double with heavy exercise.

The flow in the circulatory system is accomplished by the pumping heart, whose beat rate is often taken as the indicator of the work load of the total body.

Glossary

Absorption Here, passing through cell walls.

Anastomosis A shunt (shortcut) from arteriole to venule.

Arteriole Terminal branch of an artery, especially a small artery joining a larger artery to a capillary.

Artery A muscular, elastic tube that carries blood away from the heart.

Blood groups Classification of blood (into four groups: A, B, AB, 0) according to the content of certain antigens and antibodies.

Blood pressure The internal pressure in the arteries near the heart, often measured using a pressure cuff on the upper arm.

Cardiac minute volume Heart stroke volume multiplied by heart rate: the volume of blood injected into the main artery per minute (also called cardiac ouput).

Cardiac output Heart stroke volume multiplied by heart rate: the volume of blood injected into the main artery per minute (also called cardiac minute volume).

Catabolism Metabolic breakdown of complex molecules into simpler ones, often with release of energy.

Diastole Relaxation of the heart ventricle.

Diffusion Here, permeation of cell walls.

ECG Electrocardiogram, see there.

Edema Abnormal accumulation of (primarily) extracellular fluids, resulting in swelling.

EKG Electrokardiogram, German spelling of electrocardiogram; see there.

Electrocardiogram (ECG, also spelled electrokardiogram, EKG) The electrical events associated with the heart beat.

Erythrocytes Red blood cells.

Granulocytes White blood cells.

Heart rate Number of ventricular contractions per minute (also called pulse rate).

Hematocrit The percentage of red cell volume in the total blood volume.

Hemoglobin An iron-containing protein molecule in the red blood cell.

Leukocytes White blood cells.

Lumen The open cross-section (internal diameter) of a blood vessel.

Plasma (of blood) The fluid portion of blood.

Platelets Thrombocytes.

Plethysmograph Instrument to record variations in the size of parts of the body, such as of the finger circumference with blood pressure pulses.

Pressure Force per surface area.

Pulmonary system The portion of the blood circulation supplied by the right side of the heart.

Pulse rate Number of ventricular contractions per minute (also called heart rate).

Red blood cells Erythrocytes.

Sinoatrial nodes Specialized cardiac cells that act as the natural "pacemakers" of the heart.

Specific heat The amount of heat required to raise the temperature of a mass unit by one degree; as compared to the amount of heat required to raise the temperature of a gram of water by one degree.

Stroke volume The volume of blood ejected from the heart into the main artery during each ventricular contraction.

Systemic system The portion of the blood circulation supplied by the left side of the heart.

Systole Compression of the heart ventricle.

Thrombocytes Platelets.

Vasoconstriction Reducing the lumen of a blood vessel below its lumen at regular vasomotor tone.

Vasodilation Opening of a blood vessel beyond its lumen at regular vasomotor tone.

Vein A large membranous tube that carries blood to the heart.

Venule A small vein joining a capillary to a larger vein.

Viscosity Resistance to flow, "internal friction" of a fluid.

White blood cells Leukocytes, granulocytes.

References and Further Reading

Asimov I (1963) The human body. Signet, New York

Auerbach DI, Maeda JL, Steiner C (2012) Hospital stays with cardiac stents, 2009. HCUP Statistical Brief # 128. Agency for Healthcare Research and Quality, Rockville. https://www. hcup-us.ahrq.gov/reports/statbriefs/sb128.jsp. Accessed 29 Oct 2019

iData Research (2019) Burnaby. https://idataresearch.com/over-182000-heart-valve-replacements-per-year-in-the-united-states/. Accessed 26 Oct 2019

ITIS Foundation (2016) Tissue properties. https://itis.swiss/assets/Downloads/TissueDb/ Files20161012/materialemreftable2016.pdf. Accessed 8 Nov 2016

Mond HG, Proclemer A (2011) The 11th world survey of cardiac pacing and implantable cardioverter-defibrillators: calendar year 2009–a World Society of Arrhythmia's project. Pacing Clin Electrophysiol 34(8):1013–1027. https://doi.org/10.1111/j.1540-8159.2011.03150. x. Epub 2011 June 27. Accessed 26 Oct 2019

van der Zee A, Doshi SK (2016) Permanent leadless cardiac pacing. Newsletter dated 23 May 2016. American College of Cardiology, Washington

There are many texts on human physiology and circulation for further reading, among them

Astrand PO, Rodahl K, Dahl HA, Stromme SB (2003) Textbook of work physiology. Physiological bases of exercise, 4th ed. Human Kinetics, Champaign

Hall JE (2016) Guyton and hall textbook of medical physiology, 13th ed. Saunders, Philadelphia

Kenney WL, Wilmore JH, Costill DL (2020) Physiology of sport and exercise, 7th edn. Human Kinetics, Champaign

Chapter 7
Metabolism

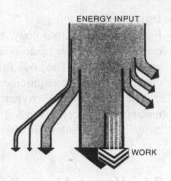

ENERGY INPUT

WORK

Overview

Overall, the healthy human body maintains a balance between energy input and energy output. Nutrients are the input from which the body's metabolic processes liberate chemically stored energy. By converting energy from the supply of nutrients from food, the body keeps itself running, produces heat, and generates physical work. Work is measured in terms of physical energy transmitted to objects outside the body.

The Model
A Hybrid Electric/Combustive Automobile

uses alternatively an electric motor powered by a chemical battery, which requires periodic re-charging; or an internal combustion engine, powered by explosions of a fuel–air mixture. The combustion engine must be cooled and waste products must be removed.

In comparison: *The "Human Energy Machine"*

runs on energy drawn from an "ATP battery", which derives its energy from the breakdown of ATP to ADP. Continual rebuilding of ATP from ADP is necessary, which requires energy. That energy comes from metabolic "combustive" breakdown of glucose, glycogen and triglycerides in the presence of oxygen.

  In the "human machine" muscles are both cylinders and pistons: bones and joints are the gears. Metabolic byproducts, including heat, need removal.

© Springer Nature Switzerland AG 2020
K. H. E. Kroemer et al., *Engineering Physiology*,
https://doi.org/10.1007/978-3-030-40627-1_7

174

Introduction

For engineering physiology, human metabolism is important primarily as it concerns how energy is made available to be used by muscles, that is, to do useful and purposeful work. The body first assimilates chemical energy and then liberates this energy to meet the requirements of physical work*.

The utilization of oxygen and the production of carbon dioxide and other metabolites by cells is called internal or cellular respiration. The topic of "respiration" as in breathing is discussed in Chap. 5.

7.1   Human Metabolism and Work

The first law of thermodynamics states that all real-world processes involve transformations of energy which conserve the total amount of energy. Since energy can neither be created nor destroyed, the energy consumed by an animal or human has a definite course, as sketched in Fig. 7.1. The digestive tract is not completely efficient and does not absorb all energy content of the ingested food; the remainder passes through and is purged as feces (F). In some animals, especially cattle, noteworthy quantities of energy may also be eliminated as methane gas.

Furthermore, some byproducts of the metabolic processes in the body are energetically useless, such as urea and water, which get removed in the urine (U); some of the water evaporatesin the lungs and on the skin. Subtracting these from the initial consumption results in the assimilated (digested, absorbed) energy, the net energy input (I).

The body uses the net input energy (I) in different ways. Some digested materials serve as building blocks (B) for synthesizing more complex components, especially during growth and as replacements cells. Other energy gets stored in cells for later use (S). Secretions, such as saliva and sweat, contain some energy. However, the largest

Fig. 7.1 Flow of energy from intake to expenditure (adapted from Speakman 1997)

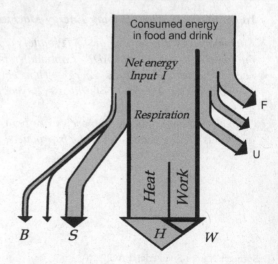

portion of the energy input is used for "internal (cellular) respiration": the process of consuming oxygen and generating CO_2 and heat (H) when the body performs a biochemical activity which requires energy. That energy comes from the breakage of bonds in the high-energy molecule ATP. The freed energy makes muscle contraction possible, such needed to perform work on an object outside the body.

In the human body, the balance between net energy input (I) and outputs may be represented, somewhat simplified, by the equation

$$I = M = H + W + S$$

where M is the metabolic energy generated: this divides into the heat energy H (generated by the body as part of muscular effort as well as basic maintenance of body function), which the body must then dispel to the outside environment; the work W (or exercise) performed; and the energy storage S in the body (labeled positive if it increases, negative if it decreases).

> The measuring units for energy (work) are Joules (J) or calories (cal) with $4.19 \, J = 1$ cal. (Exactly: $1 \, J = 1 \, Nm = 0.239$ cal $= 10^7$ ergs $= 0.948 \times 10^{-3}$ BTU $= 0.7376$ ft lb.) Often one uses the metric kilojoule ($\times 1 \, kJ = 1000 \, J$) or the kilocalorie (1 Cal = 1 kcal = 1000 cal) to measure the energy content of foodstuffs. The units for power (time rate of energy) are Watt (1 W = 1 J/s) or kcal/h (1 kcal/h = 1.163 W).

Assuming that there is no change in energy storage and that no net heat is gained from the environment or lost to it (such as by radiation or convection by hot or cold surroundings; see Chaps. 8 and 9 for details about actual heat exchanges), the energy balance equation simplifies to

$$I = H + W$$

Human energy efficiency (work efficiency) is the ratio between work performed and energy input:

$$e(\text{in percent}) = 100 \, W/I = 100 \, W/M$$

In everyday activities, only about 5% or less of the energy input converts into work, which is the energy usefully transmitted to outside objects; highly trained athletes may attain, under favorable circumstances 25%, perhaps more*. Obviously, the body converts by far most of its energy input into heat, usually at the end of a long chain of internal metabolic processes.

Skeletal muscles (see Chaps. 2 and 8) are able to convert chemical energy into physical energy. From a resting state, a muscle can increase its energy generation up to 50-fold. Such enormous variation in metabolic rate not only requires quickly adapting supplies of energy and oxygen to the muscle but also generates large amounts of waste products (mostly heat, carbon dioxide, and water) which need removal. Thus, while performing physical work, the body's ability to maintain an internal equilibrium (homeostasis) largely depends on coordinated functioning of the circulatory and respiratory systems: they supply the involved muscles with energy carriers and oxygen and remove wastes and heat. Figure 7.2 indicates the interactions among energy inputs, metabolism, and outputs.

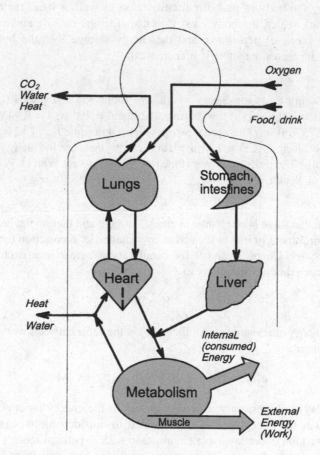

Fig. 7.2 Interactions among energy inputs, metabolism, and outputs

7.2 Energy Liberation in the Body

Food and drink, entering the body through the mouth, introduce their energy into the body. The main energy carriers are carbohydrates, fats, and proteins. After passing through the stomach, the small intestines absorb the nutrients into blood and lymph. Then follows a chemical conversion that assimilates the absorbed nutrients for energy storage as tissue-building fat, and as glycogen or as glucose for imminent use. However, if cellular activity requires immediate supply, the chemical bonds in fats, glycogen or glucose are too slow in disbanding their energy; yet, existing stores of adenosine triphosphate (ATP) provide instantly and easily accessible energy while it degrades to adenosine diphosphate (ADP). However, the energy stored in the body in form of ATP is limited; converting ADP back into ATP is necessary to rebuild the energy store. That conversion of ADP into ATP requires energy which is taken from the body's main energy stockpile: fats, glycogen, and glucose.

At the muscle, the chemically stored energy in ATP is released at the mito-chondria; it enables muscular contractile work which generates kinetic (mechanical) energy.

7.2.1 Energetic Reactions

Energy transformation in living organisms involves particularly two chemical reactions:

- Anabolism, the formation of bonds constructing more complex molecules from smaller chemical units, requires energy input: these reactions are called ender-gonic (or endothermic).
- Catabolism, the breakage of molecular bonds breaking down more complex molecules into smaller chemical units, liberates energy: such reactions are called exergonic (or exothermic).

Depending on the molecular combinations, bond breakages yield different amounts of released energies. Often, reactions do not go simply from the most complex to the most broken-down state, but progress in steps with intermediate and temporarily incomplete stages.

7.3 Digestion

Figure 7.3 shows the path of digestion: Energy enters the body through the mouth in food or drink (ingestion). Chewing breaks the food into small particles and saliva starts its chemical breakdown. Saliva in the mouth is 99.5% water and contains salt, enzymes and other chemicals including the enzyme mucus. Crushing and mixing the food particles with saliva (mastication) makes them stick together in a size (bolus) convenient for swallowing and lubricates it for passage (deglutition) down to the stomach. The enzyme lysozyme destroys bacteria which otherwise would attack the mucous membranes and teeth.

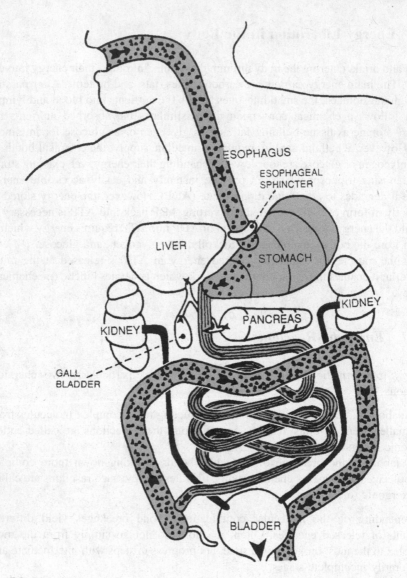

Fig. 7.3 Digestion

During swallowing, breathing stops for a couple of seconds; the epiglottis closes so that the bolus does not get into the windpipe (trachea) but can slide down the gullet (esophagus). The slide takes up to 8 s for a bolus of formerly solid food but just about a single second for liquids.

Inside the stomach, churning movements, 2 to 4 per minute, help gastric juice to break up the bolus until it becomes fully liquefied (chyme). At this stage of digestion, alcohol is quickly absorbed into the bloodstream. The enzyme pepsin and hydrochloric acid initiate digestion by breaking down proteins in the stomach, but there is little effect on the molecules of fats and carbohydrates. Fatty food stays in the stomach for up to six hours; protein-rich materials pass through more quickly, while carbohydrates leave within two hours.

Most of the chemical digestion of foods takes place in the duodenum, which is the first approximately 25 cm of the small intestine (which at 3 cm has a smaller diameter than the large intestine but at 6 m length is about four times as long). Circular muscles in the intestinal wall mix its contents, while longitudinal muscles contract in waves and propel the contents downstream. The inner surface of the small intestine has many finger-like projections (villi) which increase the contact area. The surfaces embed capillaries with blood and lymph, which absorb digested foods. The pancreas adds digestive enzymes and hormones. The liver and gall bladder add bile, a salt-rich fluid that helps to emulsify and absorb fats. During the three to five hours the foodstuffs stay in the small intestine, about 90% of all nutrients get extracted.

The final removal of available nutrients takes place in the large intestine, which has a diameter of about 5 cm and a length of approximately 1.5 m. The first section, the cecum, mostly does chemical digestion; the following colon absorbs water and electrolytes. Again, circular muscles mix and longitudinal muscles propel its contents. Finally, the intestine shapes solid wastes and undigested food components into a soft paste (feces) for egestion (defecation). The blood transports nitrogenous wastes to the kidneys, which excrete them into the urine.

Thus, digestion begins in the stomach but mostly occurs in the small intestines. Digestion changes foodstuff chemically by breaking large complex molecules into components suitable to be transported through membranes of body cells and then absorbed. The large variety of chemical reactions that ensue on the digested and absorbed foodstuffs is called assimilation: a re-assembly into molecules that can be easily degraded to release their energy content, or stored as energy reserves, or used as raw materials for body growth and repair. All together, it takes 5–12 h after eating to extract the nutrients and energy from food.

7.3.1   Energy Content of Nutrients

Our food consists of various mixtures of organic compounds (foodstuffs) and water, salts and minerals, vitamins and other items, and of fibrous organic material (primarily cellulose). The fibrous bulk (roughage) improves mechanical digestion by stretching the walls of the intestines and helping the passage of contents, but does not release energy.

A bomb calorimeter can be used to measure the energy value of foodstuffs: it burns food material electrically so that it is completely reduced to carbon dioxide (CO_2), water (H_2O), and nitrogen oxides. The developed heat is the measure of the energy content. However, in the human body, some of that energy is "lost" during internal conversion processes. The nutritionally usable energy contents per gram are, on average*:

- Fat, 37 kJ (9 kcal)
- Alcohol, 29 kJ (7 kcal)
- Carbohydrate, 17 kJ (4 kcal)
- Protein, 17 kJ (4 kcal)

The energy content of our daily food and drink depends on the mixture of the basic foodstuffs therein; examples (predominantly from North America) are listed in Table 7.1.

Table 7.1 Approximate energy content of foods and drinks in North America

	kJ/100g or kJ/100mL	kcal /100g or kcal/100mL
Meat, Poultry, Fish		
Bacon, fried	2440	590
Chicken, fried	1100–1200	260–285
Chicken, skinless, roasted	800	190
Halibut	710	170
Hamburger beef, cooked	1100	265
Hot dog, beef	1140	270
Red snapper	390	95
Roast beef, lean	700	170
Salmon	760	185
Shrimp	380	95
Steak, cooked	880	210
Tuna, canned in oil	880	210
Tuna, canned in water	460	110
Turkey, white meat	655	155
Bread	925–1275	220–280
Butter	3025	720
Honey	1260	300
Jam	1150	275
Olive oil	3570	850
Sugar	1610	385
"Fast Foods"		
French fries	1255	300
Hamburger, "Deluxe, Double, Big" and the like	2520	600

(continued)

Table 7.1 (continued)

Hamburger, "Single"	1260	300
Pizza, "Super, Supreme" and the like	1650	390
Pizza, cheese, regular	970	230
Sandwich	800–1390	210–330
Snacks		
Brownies	1680	400
Candy, hard	1630	390
Cookies, chocolate chip	1925	460
Cookies, oatmeal with raisins	1965	470
Cupcakes, chocolate icing	1510	360
Peanuts, roasted	2460	585
Potato chips	2400	575
Pretzels	1655	400
Beverages		
Beer, light	125–170	30–40
Beer, regular	170–210	40–50
Coffee	8	2
Hot chocolate, hot cacao	360	85
Juice, orange, apple	190	45
Liquor, bourbon, scotch, gin, vodka and others, Alcohol 40 Vol% (80 proof) 45 Vol% (90 proof)	980 1085	225 260
Milk, low fat	205	50
Milk, whole	265	65
Milkshake	545	130
"Soft drinks", such as Cola, Sprint, Fanta and others	170	40
Tea	0	0
Wine, red or white	325	80

1000 J = 1 kJ   1000 cal = 1 kcal = 1 Cal

7.3.2   Digestion of Carbs, Fats, and Proteins

The energy in our food exists essentially in the molecular bonds of carbon, hydrogen, and oxygen; and of nitrogen in proteins.

Carbohydrates come in small to rather large molecules; most consist of only the three chemical elements carbon, oxygen, and hydrogen. (The ratio of H to O usually is 2 to 1, just as in water: hence the name carbohydrate, meaning watery carbon). Carbohydrates exist as simple sugars (monosaccharides), double sugars (disaccharides) and in form of a large number of monosaccharides joined into chains (polysaccharides). The most common natural polysaccharides are plant starch, glycogen, and cellulose.

Carbohydrate digestion starts by breaking the bonds between saccharides so that the compounds reduce to simple sugars. The monosaccharides produced in the digestion process are principally glucose (about 80%); fructose and galactose also result and then eventually are converted to glucose. Glucose is a monosaccharide which the blood transports to all body tissues. For storage, the body keeps it in form of glycogen (a multibranched polysaccharide of glucose) in the liver and also in muscles. When needed, the glycogen in the liver is converted to glucose and then transported to active tissues, where it is metabolized.

Fat is a triglyceride, a lipid molecule of one glycerol nucleus joined to three fatty acid radicals. Unsaturated fat has double bonds between adjacent carbon atoms in one or more of the fatty acid chains; hence, the compound is not saturated with all the hydrogen atoms it could accommodate. Saturated fat has no double bonds between the carbon atoms and is fully saturated with hydrogen atoms. The more unsaturated a fat is, the more it is a liquid (oil). Most plant fats are polyunsaturated while most animal fats are saturated and hence solid.

Digestion of fat takes place in the small intestine; chemically, it is the breakage of bonds linking the glycerol residue to the three fatty acid residues. Glycerol and fatty acid molecules are small enough to cross cell membranes and hence can be absorbed. The bloodstream can transport only water-soluble materials such as glycerol. Many fatty acids are water repellent: these are absorbed into the lymph vessel system (see Chap. 6). Most fat is stored subcutaneously and in the viscera, while only a small portion is stored within muscles.

Fats can provide a large portion of the energy needed to sustain prolonged but moderately intense exercise. For cellular metabolism it must first be reduced from its complex triglyceride form to its basic components, glycerol and free fatty acids (FFAs). Only FFAs are used to form ATP.

Proteins are chains of amino acids joined together by peptide bonds. Many such different bonds exist, and thus proteins come in a large variety of types and sizes. Digestion, which occurs partially in the stomach but mostly in the small intestine, breaks the bonded protein into amino acids, which are then absorbed into the bloodstream. The blood carries them to the liver, which disperses cells throughout the body to be rebuilt into new proteins. Other amino acids become enzymes, which are organic catalysts that control chemical reactions between other molecules without being consumed themselves. Still other amino acids become hemoglobin (the oxygen carrier in the blood), or hormones, or collagen. Hence the body has many important uses for proteins beyond using them as a source of energy; energy liberation from protein in the body is less efficient than the energy yield from carbohydrates and fats.

7.3.3   *Absorption and Assimilation*

The stomach absorbs primarily water, salts, and certain drugs, among them alcohol. As already mentioned, most of the absorption and assimilation of foodstuffs takes place in the small intestine. There, digested foodstuffs enter the blood capillaries or (especially fatty acids) the lymphatic system. The blood capillaries drain eventually into the hepatic portal vein. This vein also receives inputs from the stomach, pancreas, gall bladder, and spleen. Lymph flows from the intestinal walls through the thoracic duct to the left subclavian vein, where the lymph enters the blood stream and becomes part of the blood plasma. The liver receives blood from the portal vein and from the hepatic artery.

Liver cells remove digestion products from the blood and store or metabolize them. Further, the liver generates glycogen $(C_6H_{10}O_5)_x$, the storage form of carbohydrate, from glucose $(C_6H_{12}O_6)$. The liver also synthesizes neutral fat (*triglyrides*) from glucose (hence, one can "get fat" without eating any), fatty acids, and amino acids derived from proteins. Thus, the liver controls much of the fat utilization of the body. Fat serves as the body's main energy storage.

7.4   Energy Release

After digestion, absorption and assimilation, the principal energy carriers are in the body in form of

- glucose and its storage form, the closely related glycogen (polysaccharide of glucose),
- triglycerides (neutral fat), and
- amino acids in protein.

Of these, glucose is the most easily used energy carrier. Some glucose is present at the mitochondria of muscles, more is easily available from the blood stream. Its catabolism can occur with the use of oxygen (aerobically), or without oxygen (anaerobically).

Chemically, oxidation is a loss of electrons from an atom or a molecule (conversely, reduction is a gain of electrons). In such reactions, electrons appear in the form of hydrogen atoms and the oxidized compound is de-hydrogenated. In human metabolism, organic fuels (glucose, fats, and occasionally amino acids) constitute the major electron donors, while oxygen is the final electron acceptor (oxidant) of the fuel.

7.4.1   Aerobic Metabolism of Glucose

Oxidation of glucose follows the stoichiometric formula

$$C_6H_{12}O_6 + 6O_2 = 6CO_2 + 6H_2O + (\text{up to } 38) \text{ ATP}$$

This means that one molecule of glucose combines with six molecules of oxygen, imported to the process from the outside; this generates six molecules each of carbon dioxide and water, while energy (up to 2891 kJ/mol, equivalent of up to 38 ATP) is released. Equal volumes of water, carbohydrate, and oxygen take part in this conversion: therefore, measuring the amount of oxygen uptake, or the quantities of water and carbon dioxide produced, allows accurate assessment of the actual energy use in the body, discussed in Chap. 8.

7.4.2   Aerobic Metabolism of Glucose

Another method of oxidation is breaking glucose molecules into several fragments and letting these fragments oxidize each other. This means energy yield under anaerobic conditions, since no external oxygen is imported; the process is called glycolysis (*lysis* means breakdown). The energy yield is 2 molecules of ATP per 1 molecule of glucose.

Glucose catabolism (and fat catabolism as well) takes place in sequential biochemical reactions, which produce intermediary metabolites. The first phase is anaerobic: the 6-carbon compound glucose breaks into two 3-carbon molecular fragments, each of which naturally becomes a 3-carbon compound pyruvic acid molecule. This can become lactic acid if not oxidized: lactic acid is a normal metabolic intermediate of some importance for "muscle fatigue", mentioned earlier in Chap. 3 and discussed in more detail in Chap. 8. This process releases some energy; in stoichiometric terms,

$$C_6H_{12}O_6 + 2 \text{ ATP} = 2 \text{ Lactic Acids} + 4 \text{ ATP}$$

The second phase starts out to be anaerobic: pyruvic acid splits into carbon dioxide and hydrogen in a series of self-renewing reactions known as the Krebs cycle (also called citric acid cycle or tricarboxylic cycle) shown in Fig. 7.4. Here, hydrogen atoms separate in pairs from the intermediary metabolites, the first of which is pyruvic acid. (The removal of hydrogen atoms from the intermediary metabolites, called dehydrogenation, is particular to the Krebs cycle). As oxygen becomes available, hydrogen reacts with it to form water. Now, glucose completely metabolizes and produces six CO_2 molecules, six H_2O molecules, and releases energy, as in the aerobic glucose oxidation equation above, to release up to 38 ATP.

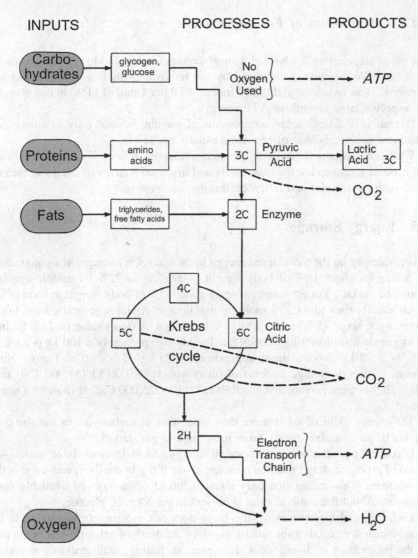

Fig. 7.4 Breakdown of foodstuffs, Krebs cycle. Number of carbon and hydrogen atoms in squares

7.4.3 Metabolism of Carbohydrate

The oxidation of carbohydrates employs three processes: aerobic glycolysis, Krebs cycle, and electron transport chain, as shown in Fig. 7.4. The oxidation of glucose generates up to 38 molecules of ATP from 1 molecule of muscle glucose.

7.4.4   Metabolism of Fat and Protein

The use of fat relies on lipolysis of its triglycerides (neutral fat) to free fatty acids: in turn, the FFAs become metabolized by the removal of hydrogen and associated electrons. Their oxidation yields 129 mol of ATP per 1 mol of FFA. In this way, fat can supply a large amount of ATP energy.

Oxidation of amino acids, components of protein, is used only in emergency situations; so, metabolism of protein is usually negligible.

While fat accounts for most of the energy reserves in the body, already existing glucose and glycogen are the most easily and first used sources of energy at the cell level, both by muscles and in the central nervous system.

7.5   Energy Storage

Fat provides by far the most stored energy in the body. On average, in a young man, fat makes for about 16% of body weight, increasing to 22% by middle age and more if he "is fat". Young women average about 22% of body weight in form of fat, which usually rises to 34% or more by middle age. Athletes generally have lower percentages, about 15% in men and 20% in women. A (low) value of 15% fat in a 60 kg person results in 9 kg of body fat. Twenty-five percent of a 100 kg person, or 33% of a 75 kg person, mean approximately 25 kg of body fat. These values amount to energy storage in form of fat of about 340,000 kJ (81,000 Cal) in a skinny, lightweight person, and about 941,000 kJ (225,000 Cal) or more in a heavy person.

The energy yield of fat is more than twice that of carbohydrates but the conversion of fat to usable energy takes more time to get started.

Glycogen provides much less stored energy. Most humans have about 400 grams of glycogen stored near the muscles, about 100 g in the liver, and some in the bloodstream. This means that only about 9,200 kJ of energy are available from glycogen. About the same amount is present in the form of glucose.

Under normal circumstances, the body does not use protein amino acids for energy since their catabolism usually involves the death of cells, protein being part of the protoplasm of living cells. However, in fasting, malnutrition, starvation, certain illnesses, and during all-out physical efforts catabolism of proteins does occur.

7.6   Energy for Muscle Work

Muscle mitochondria store "quick-release" energy in form of the molecular compound adenosine triphosphate, ATP. This intracellular vehicle of chemical energy transfers its energy into mechanical energy by donating its high-energy phosphate group to fuel the process of muscle contraction.

Hydrolysis can break the ATP phosphate bonds easily and quickly to provide the energy needed for muscle contraction:

$$ATP + H_2O = ADP + energy(released)$$

However, high energy demands consume the small stores of available ATP within seconds; therefore, ATP needs constantly to be replenished. Creatine phosphate, CP, (also called phosphocreatine, PCr) achieves this by transferring a phosphate molecule to adenosine diphosphate, ADP. However, this reaction requires energy:

$$ADP + CP + energy(required\ input) = ATP + H_2O$$

The cycle of converting ATP into ADP (which releases energy) and then re-converting ADP into ATP (which requires energy) is anaerobic. The ATP-ADP-ATP conversion acts immediately but is not fully efficient because the amount of energy gained during the breakdown is less than the energy required to re-build ATP from ADP. (This is consistent with the second law of thermodynamics which states that entropy in a system increases, and thus perpetual motion is impossible.)

Because ATP provides quick energy for only a few seconds, its re-synthesis from ADP is necessary for continuous operation. The required energy derives from the breakdown of glucose, glycogen, and fat (and possibly protein): breaking their complex molecules to simpler ones, ultimately to CO_2 and H_2O, provides the ultimate energy source for the rebuilding of ATP from ADP. Figure 7.5 shows a schematic overview of the energy conversion from ingestion to output.

As discussed in Chaps. 2 and 3, a signal traveling from a nerve causes a muscle twitch (contraction). A twitch consists of four periods, as described in Table 7.2. The latent period, typically lasting about 10 ms, shows no reaction yet of the muscle fiber to the motor neuron stimulus. Shortening takes place usually within 40 ms for a fast-twitch fiber. Released heat energy causes the cross-bridges between actin and myosin to undergo a thermal vibration, which creates a "ratcheting" action that causes the heads of the actin rods to slide towards each other along the myosin filaments. At the end of this period, the muscle element has reached its shortest length and can develop tension. During the relaxation period, also commonly about 40 ms in a fast-twitch fiber, the bridges stop oscillating, the bonds between myosin and actin break, and the muscle is pulled back to its original length either by the action of antagonistic muscle or by an external load. During the recovery period, again taking about 40 ms, the metabolism of the muscle is aerobic, with glucose and stored glycogen being directly oxidized for the final regeneration of ATP and phosphocreatine.

Fig. 7.5 Schematic overview
of the energy flow from
ingestion, digestion and
assimilation and release via
metabolism to output as either
heat or work

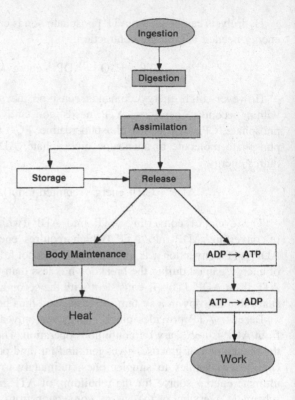

Table 7.2 Energy metabolism in a single twitch of a fast-twitch fiber

Period, duration	Energy metabolism	Muscle action (see Chapter 2)
Latent, about 10 ms	none	No reaction yet in the muscle fiber to the motor neuron stimulus
Shortening, about 40 ms	Energy for this process is freed from the ATP complex, mostly anaerobically	Cross-bridges between actin and myosin vibrate and "ratchet" the heads of the actin rods towards each other along the myosin
Relaxation, about 40 ms	ATP is re-synthesized to ADP	Cross-bridges stop oscillating, bonds between myosin and actin break, the muscle returns to its original length
Recovery, about 40 ms	Aerobic metabolism oxidizes glucose and glycogen; final regeneration of ATP and phosphocreatine	None

7.6.1 The First Few Seconds of Muscular Effort

At the very beginning of intensive muscle work, breaking the phosphate bonds of ATP releases the "quick energy" for muscular contraction. However, the contracting muscle consumes its local supply of ATP in about two seconds.

7.6.2 The First Ten Seconds of Muscular Effort

The next source of immediate energy is creatine phosphate (CP). It anaerobically transfers a phosphate molecule to the newly created molecule of ADP and thus turns it back into ATP. Skeletal muscle contains enough CP to synthesize ATP for 8 to 12 s of intensive muscle activity.

7.6.3 Effort Lasting Longer Than Ten Seconds

After about 10 s of ATP–ADP–ATP reactions, the supply of CP is exhausted. However, the muscle can continue to do moderately high muscular work by producing ATP through glycolysis, the breakdown of glucose and, if not enough is available, of glycogen through glycogenolysis. The energy yield of these processes is 2 mol of ATP per 1 mol of glucose, 3 mol of ATP for 1 mol of glycogen.

7.6.4 Muscular Work Lasting Minutes to Hours

If the physical activity has to continue, it must be at a level where the oxygen supply suffices to keep the energy conversion processes going. Hence, extended muscular work is necessarily at a lower level of energy than a quick burst of maximal effort can generate. To sustain enduring work, the oxygen supply at the mitochondria must allow continual aerobic energy conversion. Without oxygen, a molecule of glucose yields two molecules of ATP; with oxygen, the glucose energy yield is up to 38 molecules of ATP.

When muscular work lasts from several minutes to hours, this provides time for pyruvic acid, a product of glycolysis or glycogenolysis, to enter the mitochondria. There, its conversion in the Krebs cycle produces additional energy for ATP resynthesis.

Fat is rich in energy. The metabolic use of fat relies on lipolysis of triglycerides into free fatty acids: in turn, the FFAs become metabolized by the removal of hydrogen and associated electrons. Their oxidation yields 129 molecules of ATP per 1 molecule of FFA. In this way, fat supplies a large amount of ATP energy. Use of fat for energy takes time to start, yet it is highly effective and entirely aerobic.

Oxidation of amino acids, components of protein, is used only in emergency situations; so, metabolism of protein is usually negligible.

7.7   Aerobic and Anaerobic Work

As just discussed, at the start of intense muscular effort, energy liberation can proceed without use of oxygen: during the initial 10 seconds, ATP and CP provide energy for very high performance; then, anaerobic glycolysis supplies energy for a few minutes of strenuous muscular work. However, for longer lasting work, even when of lower intensity, only aerobic energy conversion can provide the necessary energy.

In aerobic conditions, energy yield is sufficient to keep up fairly high energy expenditure as long as ATP is reformed as quickly as it is used up. The conversion of glucose and glycogen is simple, but the utilization of fats (glycerol and fatty acids) requires a more complex process: they convert to intermediary metabolites and enter the Krebs cycle. Their final energy yield is approximately 9,805 kJ per mole, more than threefold that from glucose.

Still, if very heavy expenditure is required over long time, such as in a marathon run, the interacting metabolic system and the oxygen-supplying circulatory system might become overtaxed. A runner who "hits the wall" (figuratively) most likely has used up the body's glycogen supply and also gone into "oxygen debt"—more in Chap. 8.

However, in our every-day activities, we regulate our energy (work) output to match our body's ability to develop energy with the available supply of oxygen. If needed, we simply take a break during which our body resynthesizes accumulated metabolic byproducts and returns the metabolic, circulatory and respiratory systems to their normal states.

Overall, work of a sustainable intensity is aerobic, even if many of the single intermediate steps in the metabolic reactions are in fact anaerobic. For example, glucose breakdown is first anaerobic, followed by an aerobic phase: oxygen is required for the complete metabolism of glucose. Glycogen stores near muscles deplete much more quickly when the muscles must work anaerobically than when able to work aerobically. The combustion of all fat derivatives is strictly aerobic.

Usually, at rest and during moderate work, the oxygen supply is more than sufficient and, hence, the energy metabolism is essentially aerobic. This leads to high ATP and low ADP concentration. To meet intermediate work (energy) demands, the breakdown of glucose speeds up. Above a critical intensity of labor, the oxygen-transporting system cannot provide enough oxygen to the cells and pyruvic acid transforms into lactic acid instead of going through the Krebs Cycle. During continued intermediate work, the lactate that has been developed may reconvert to glycogen in the liver; this can take place even at the muscle, if aerobic conditions exist. With increasingly higher work intensity, more phases of the metabolic processes become anaerobic, which may eventually require cessation of the muscular work.

Table 7.3 Anaerobic and aerobic energy liberation during maximal efforts

Energy released	Duration of the greatest possible effort				
	10 s	1 min to 10 min		1h to 2h	
By anaerobic processes in kJ *in %*	100 *85*	170 *65 to 70*	150 *10 to 15*	80 *2*	65 *1*
By aerobic processes in kJ *in %*	20 *15*	80 *30 to 35*	1,000 *85 to 90*	5,500 *98*	10,000 *99*
Total in kJ	120	250	1,150	5,580	10,065
Primary energy source	ATP splitting	CP	Glucose, glycolysis	Glycogen and fatty acids, Krebs Cycle	
Type of process	Anaerobic	Mixed	Mixed	Aerobic	

Table 7.3 shows schematically the contributions of aerobic and anaerobic energy liberation during short and long maximal efforts. During maximal work of short duration, up to one minute or so, the energy available depends on ATP splitting. If hard work lasts up to about 10 min, the fuel conversion is complex: at the start, anaerobic utilization of ATP and creatine phosphate predominates. Then, anaerobic conversion of glucose to lactate takes over increasingly. In prolonged heavy work, lasting an hour or more, the maximal work output depends on the oxidation of glycogen and fatty acids.

7.8   Energy Use and Body Weight

The simplified equation I = H + W + S, at the beginning of this chapter, describes the balance between energy input, energy output (in the form of heat and work), and energy storage. If the input exceeds the output, energy storage in form of body mass ("weight") increases; conversely, mass decreases if the input is smaller than the output. A change of one kilogram in body weight is the result of an increase or a reduction of 29,000 to 34,000 kJ (7,000 to 8,000 kcal) in energy input.

Apparently, the body tries to maintain a given energy storage. This means that normally a person's body weight (which mostly derives from the weight of water, bones and tissues, and from the mass of fat as stored energy) stays at some present level. Changing that "set point" requires definite changes in health, in nutritional habits (reduction or increase of energy in food and drink supply), and in energy expenditures (such as vigorous and repeated exercising, or reduction of physical activities). As long as the body keeps this set point, body weight remains constant.

For example, if one reduces food intake slightly, the body tries to extract enough energy from the remaining intake to preserve the existing body weight. A continued starvation diet lowers fat storage, and hence, body weight. However, returning to the previous eating habits after having achieved a lowered body weight allows the body to re-attain its previous weight, unless the set point has been lowered. Changing the set point usually requires that one stays an extended time at that new level of nutrition and exercise.

Notes

The text contains markers, *, to indicate specific references and comments, which follow.

Assimilates chemical energy and liberates energy to do physical work: There are many textbooks that explain biological and chemical physiology in general; other books focus on special fields such as medical physiology, for example by Hall (2016); on work and exercise, for instance by Astrand et al. (2003); or on sports, such as by Kenney et al. (2020).

Energy usefully transmitted to outside objects: Astrand et al. (2003); Wilmore et al. (2008).

Nutritionally usable energy: FAO (2003).

Summary

The metabolic breakdown of foodstuffs releases energy. Most of that energy transforms into heat, and only a small portion can be transmitted as "work" (mechanical energy) to an outside object.

When the body performs physical work, the muscular efforts require energy. This energy is provided by a biochemical process called internal (or cellular) respiration, which consumes oxygen and generates carbon dioxide, water, and heat. The energy becomes available from the breakage of bonds in the high-energy molecule ATP.

ATP is stored at the mitochondria of muscles. When a muscle requires energy, the existing ATP is degraded to ADP; this liberates energy anaerobically. The conversion acts immediately but is not fully efficient because the amount of energy gained by the breakdown of ATP is less than the energy required to re-build ATP from ADP.

The ATP supply in the muscle is so small that it can sustain only a few seconds of muscular effort. For the next approximately 10 s of activity, skeletal muscle uses CP to synthesize ATP. Then, with ATP and CP stores at the muscle depleted, moderately high muscular work can continue via production of ATP through breakdown of glucose (glycolysis) and, if needed, of glycogen (glycogenolysis).

When moderate muscular work extends to minutes or hours, cellular aerobic respiration of carbohydrates and fats can produce ATP.

The oxidation of carbohydrates employs three processes: aerobic glycolysis, Krebs cycle, and electron transport chain. The oxidation of carbohydrates generates 37 to 39 mol of ATP from 1 mol of muscle glycogen; if glucose is used, the maximal gain is 38 ATP moles.

The use of fat relies on lipolysis of its triglycerides into free fatty acids (FFAs): in turn, they become metabolized by the removal of hydrogen and associated electrons. Their oxidation yields 129 molecules of ATP per 1 molecule of FFA. In this way, fat supplies a large amount of ATP energy.

Oxidation of amino acids (components of protein) is used only in emergency situations; thus, metabolism of protein is usually negligible.

Glossary

Absorption Here, passing through cell walls.

Anabolism Formation of bonds, which requires energy input: endergonic (or endothermic) reactions.

Assimilation Transformation of a digested nutriment into a part of the organism.

Catabolism Breakage of molecular bonds, which liberates energy: exergonic (or exothermic) reactions.

Digestion The act of changing food chemically into components suitable for assimilation (see there) into the body.

Endergonic (same as endothermic) reaction Chemical reaction requiring energy input.

Endothermic See endergonic.

Entropy A thermodynamic property that is the measure of a system's thermal energy per unit of temperature that is unavailable for doing useful work.

Ergonomics The application of scientific principles, methods and data drawn from a variety of disciplines to the design of engineered systems in which people play significant roles.

Exergonic (same as exothermic) reaction Chemical reaction yielding energy, usually as heat.

Exothermic See exergonic.

Laws of thermodynamics Three commonly accepted rules to describe how physical quantities (temperature, energy, and entropy) behave. The first law states that energy cannot be created or destroyed in an isolated system and hence remains constant ("Law of Conservation of Energy"); the second law states that the entropy of any isolated system increases; the third law states that the entropy of a system approaches a constant value as the temperature approaches absolute zero.

Metabolism Chemical processes in the living body; in a narrower sense, energy-yielding processes.

Mole Abbreviated mol; one mole equals 6.022×10^{23} molecules (Avogadro's number).

Respiration As in breathing is treated in Chap. 5; "internal" or "cellular respiration" is discussed here in this chapter.

References

Astrand PO, Rodahl K (1977/1986) Textbook of work physiology. 2nd and 3rd eds. McGraw-Hill, New York

Astrand PO, Rodahl K, Dahl HA, Stromme SB (2003) Textbook of work physiology. Physiological bases of exercise. 4th ed. Human Kinetics, Champaign

FAO (Food and Agriculture Organization of the United Nations) (2003) Food energy—methods of analysis and conversion factors. FAO Food and nutrition paper 77. FAO, Rome

Hall JE (2016) Guyton and Hall Textbook of medical physiology. 13th ed. Philadelphia, Saunders

Kenney WL, Wilmore JH, Costill DL (2020) Physiology of sport and exercise, 7th edn. Human Kinetics, Champaign

Speakman JL (1997) Doubly labelled water: theory and practice. Chapman & Hall, London

Wilmore JH, Costill D, Kenney WL (2008) Physiology of sport and exercise, 4th edn. Human Kinetics, Champaign

Chapter 8
Work and Exercise

Overview

Performing the same intensity of physical work and exercise strains individuals differently, in accordance with their physique and training. In order to match the physical requirements of a task with a person's capacity for this effort, one needs to know both the individual's energetic capacity, and how much of this capacity a given job demands. This chapter first considers the assessment of a person's capacity, then addresses task demands.

> **The Model**
> *For matching the work to the person it is necessary to know both the individual's energetic capacity and how much of this capacity a given task demands,*
>
> *Having a person perform calibrated exercise allows to measure the person's related capacities. Conversely, measuring the physiological reactions of "standard" persons as they perform their work task allows categorization of the physical demands of the task.*

Introduction

The "heaviness" of physical work (that is, the demands on the body) or of exercise (such as at sports) is measured in units of energy, the same as discussed in Chap. 7.

> The measuring units for energy (work) are Joules (J) or calories (cal) with 4.19 J = 1 cal. (Exactly: 1 J = 1 Nm = 0.239 cal = 10^7ergs = 0.948 × 10^{-3} BTU = 0.7376 ft lb.) Often one uses the metric kilojoule (1 kJ = 1000 J) or the kilocalorie (1 Cal = 1 kcal = 1000 cal) to measure the energy content of foodstuffs. The units for power (time rate of energy) are Watt (1 W = 1 J/s) or kcal/hr (1 kcal/hr = 1.163 W).

© Springer Nature Switzerland AG 2020
K. H. E. Kroemer et al., *Engineering Physiology*,
https://doi.org/10.1007/978-3-030-40627-1_8

Different persons have differing capabilities to produce physical efforts. Their individual abilities can be measured and then compared to the situational demands imposed. To be successful in performing a task, a person's capability must meet or exceed the task demands.

8.1   Capacity for Physical Exercise and Work

The capability of the human body to perform work or exercise depends on its ability to generate varying energy levels over various time periods. These capabilities are determined by the individual's capacity for energy output (especially physique, health, skill); by the muscular and neuromuscular function characteristics (such as muscle strength and coordination of motion); by psychological factors (such as motivation, not a topic of this text); and by the thermal environment (see Chap. 9). Figure 8.1 illustrates these relations.

Among the currently used procedures to assess internal metabolic capacities, four techniques are most commonly used:

1. diet and weight observation;
2. direct calorimetry;
3. indirect calorimetry; and
4. subjective rating of perceived effort.

8.1.1   Diet and Weight Observation

To maintain the body's energy balance (homeostasis), all energy that enters the body in form of nutrients, solid or fluid, must in some manner leave the body: most leaves in terms of energy internally consumed to maintain the body (finally converted to heat) or as external work or exercise performed—see Fig. 8.1. This balance assumption means that there is neither energy storage (when the person gets fatter, more voluminous and heavier) nor use of stored energies (the person getting slimmer). Accordingly, diet and weight observation provides a means for assessment of internal energetic processes, but requires rather lengthy observation periods, usually weeks or months.

If such balance of energy can be enforced, then measurement of the energy content in food and drink intake provides information about the energy output related to external work. However, strict and complete control of energy intake and output may be feasible in a laboratory setting but is rather difficult in an everyday setting where the subject is expected to simply (and comprehensively) record the intake.

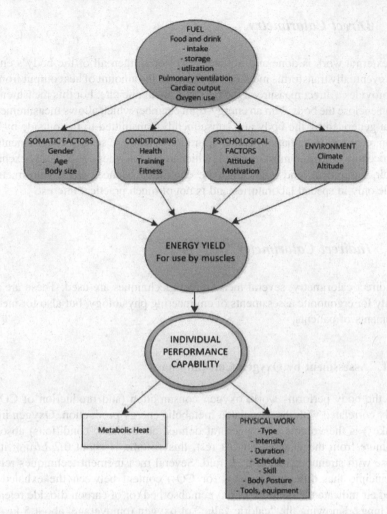

Fig. 8.1 Main determiners of individual physical work capacity. Adapted from Kroemer et al. (2003), Kroemer Elbert et al. (2018)

Techniques to determine body fat, or conversely lean body mass, include imaging and displacement techniques as well as skinfold thickness measurements (see Chap. 4). Another approach is to calculate the Body Mass Index (BMI) from weight and stature*; however, this is not much of an improvement over comparing waist and hip circumferences.

8.1.2   Direct Calorimetry

If no external work is done and no energy is stored, then all of the body's energy intake eventually transforms into heat. In this case, the amount of heat output from the body provides a direct measure of the person's metabolic rate. For this measurement, one can enclose the body with an energy-tight chamber which allows measurement of all heat generated by the body and subsequently transmitted to the outside by conduction, convection, radiation, and evaporation. There are several requirements for this procedure: the room must be small (which limits activities), and energy exchange with air, equipment, and walls must be controlled. Hence, direct calorimetry is feasible only in special laboratories and is not of much practical interest.

8.1.3   Indirect Calorimetry

For indirect calorimetry, several measurement techniques are used. These are used not only for ergonomic assessments of engineering physiology, but also for medical assessments of patients.

8.1.3.1   Assessment by Oxygen Consumption

When the body performs work, oxygen consumption (and production of CO_2) is directly correlated with the associated metabolic energy production. Oxygen intake ("uptake") is the volume of oxygen (at defined atmospheric conditions) absorbed per minute from the inspired air. At rest, this volume is about 0.2 L/min; it can increase with strenuous exercise 30-fold. Several measurement techniques rely on the principle that differences in O_2 (or CO_2) content between the exhaled and inhaled air indicate the quantity of oxygen absorbed (or of carbon dioxide released) in the lungs. Knowing the "caloric value" of oxygen (on average, about 5 kcal per liter of O_2), the volume of oxygen consumed for an activity allows calculating the energy that the body converts. (Brief descriptions of techniques to measure oxygen uptake are in Appendix A to this chapter.)

The respiratory exchange quotient (RQ, also called the Respiratory Exchange Rate, RER) compares the volumes of carbon dioxide exhaled to oxygen consumed. Metabolizing one gram of carbohydrate requires 0.83 L of oxygen and releases the same volume of carbon dioxide (see Chap. 7). Hence, for carbohydrates, the RQ is one (unit). The energy released is 18 kJ per gram of O_2, 21.2 kJ per liter O_2. Table 8.1 shows the RQs for carbohydrate, fat and protein conversion. Thus, measuring the volumes of CO_2 and O_2 during work indicates which energy carrier is metabolized, and how much energy is consumed. (Note that the average values listed in Table 8.1 assume the construct of a "normal" adult on a "normal" diet doing "normal" work; this can be a misleading simplification.)

Table 8.1 Oxygen needed, RQ, and energy released in nutrient metabolism

		Carbo-hydrate	Fat	Protein	Average*
O_2 consumed, L/g		0.83	2.02	0.79	na
RQ , RER		1.00	0.71	0.80	na
Energy Yield	kJ/g	18	40	19	na
	kJ/LO$_2$	21.2	19.7	18.9	**21**
	kcal/LO$_2$	5.05	4.69	4.49	**5**

*Assumes the construct of a "normal" adult on a "normal" diet doing "normal" work.

8.1.3.2   Assessment by Heart Rate

Metabolic processes and their support systems interact closely: for proper functioning, the working muscles or other metabolizing organs must be supplied with nutrients and oxygen, and metabolic byproducts must be removed. Blood circulation, powered by the heart, provides the transport system. Therefore, heart rate (a primary indicator of circulatory functions) and oxygen consumption (representing the metabolic conversion) have a linear and reliable relationship in the range between light and heavy work, as schematically shown in Fig. 8.2. Given this relation, heart rate (HR, also called pulse rate) often can simply be substituted for measurement of metabolic processes, particularly of O_2 uptake. This is a very attractive shortcut since pulse counts are easier and less cumbersome to perform than oxygen measurements.

The simplest techniques of counting the pulse are to palpate an artery, often in the wrist or perhaps in the neck, or to listen to the sound of the beating heart. All the measurer has to do is count the number of pulses over a given period of time (such as 15 s) and from this calculate an average heart rate per minute. More refined techniques utilize various plethysmographic methods, which rely on tissue deformations due to changes in filling of the imbedded blood vessels, for example in a finger. Other approaches rely on electric signals associated with the pumping actions of the heart, usually sensed by electrodes placed on the chest (see Chap. 6, Fig. 6.1).

Measurement of heart rate has another major advantage over oxygen consumption as indicator of metabolic processes: heart rate responds more quickly to changes in work demands. Hence, heart rate more easily reflects quickly occurring responses in body functions due to changes in work requirements.

The reliability of measurement of heart rate to reflect energy expenditure is limited primarily by the intra- and inter-individual relationships between circulatory

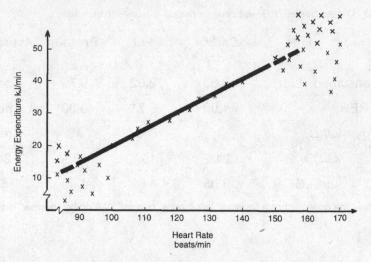

Fig. 8.2 Heart rate in relation to energy expenditure (oxygen uptake)

and metabolic functions. Statistically speaking, the regression line shown in Fig. 8.2, which associates heart rate with oxygen uptake (energy production), differs in slope and intersect (at-rest starting point) from person to person and from task to task, and may change with training or dis-conditioning of a person.

In addition, the scatter of the data around the regression line (statistically indicated by the coefficient of correlation) is also variable, primarily at the extremes of very light and very strenuous exercise. Very light exercise barely elevates the heart rate, which can be influenced easily by psychological events (such as excitement, startling noise, or fear) that may be completely unrelated to the task itself. At very heavy work, the O_2-to-HR relation may also lose predictive ability (that is, statistical value): for example, cardiovascular capacities may be exhausted before metabolic or muscular limits are reached. Further, the presence of heat load (see Chap. 9) also influences the relationship between O_2 and heart rate, adding one more confounding factor to the analysis.

8.1.4 Assessment by Subjective Rating of Perceived Effort

Humans are able to perceive and communicate the strain that a given work task generates in their bodies, and they can judge this effort. Assessing and rating the relationship between the physical stimulus (the work performed) and its perceived sensation have been used as long as people have expressed their preference for one type of work over another—long before qualitative techniques of measuring were available.

Since the 1960s, formal techniques for "rating the perceived exertion" (RPE)* associated with different kinds of efforts have been available. A common procedure is to appraise the effort on a nominal scale from "light" to "hard." Such a verbally anchored scale also provides a means to "measure" the strain subjectively perceived while performing standardized work.

Examples of rating scales appear in Appendix B to this chapter.

8.2   Standardized Tests

Most current medical and physiological assessments of human work capacities rely on measuring oxygen consumption and heart rate. To assess and compare persons' capacities, several tests with standardized external work are in common use. They typically employ bicycle ergometers, treadmills, or steps located in a laboratory on which the subject performs controlled exercises.

8.2.1   Bicycle, Treadmill and Step Tests

Exercising on bicycles, treadmills, and steps stresses certain body parts and body functions. For example, bicycling requires predominantly use of the leg muscles. Since the legs constitute (both in their mass and their musculature) large components of the human body, their extensive exercising in a bicycle test also strains pulmonary, circulatory, and metabolic functions of the body. However, a person who is particularly strong in the upper body but not well trained in the use of legs would show different strain reactions in bicycle ergometry than, say, a trained bicycle racer.

The treadmill also stresses primarily lower body capabilities, but (in contrast to sitting on a bicycle) the whole body weight must be supported and propelled by the feet and legs. If the treadmill is inclined, the body must also be lifted. Hence, this test strains the body in a somewhat more complete manner than bicycling but still does not involve trunk and arms. Furthermore, it requires somewhat more bulky equipment than a stationary bicycle.

Another standard test technique is to have the subject step up onto a raised platform and then step down again, repeated for the duration of the test. This "step test" technique requires only simple equipment; it stresses body functions in a fashion somewhat similar to running on a treadmill, however by primarily making the subject elevate the total body weight instead of moving it forward. A person heavier in weight and with shorter legs than another subject would show larger energy consumption. As in the other tests, muscular capabilities of the upper body are not assessed at all.

8.2.2   Challenges

Test selection often relies on availability of equipment and ease of use rather than on theoretical considerations. Obviously, the bicycle, treadmill, and steps test techniques do not reproduce the requirements of regular work. They exercise almost exclusively the lower extremities, not trunk or upper body functions. Consequently, various improvements on test equipment and procedures have been proposed; for example, a ladder-mill on which one climbs using both arms and legs, or an enhanced bicycle ergometer on which the test subject simultaneously operates hand-operated cranks.

In spite of their shortcomings, the current test techniques (especially those using a bicycle or treadmill) have become standard procedures to measure a person's capabilities for exercising at known intensity, judged by oxygen consumption and/ or heart rate and/or by subjective ratings. This has provided a rich database against which relative comparisons may be drawn.

The capability of the respiratory system to absorb oxygen into the blood and the related ability of the circulatory system to deliver this oxygen to the consuming organs (usual muscles) is the most commonly used assessment of a person's capacity for physical work. These interrelated capabilities are measured as the maximal volume of oxygen which a person can consume during intense exercise. The "VO_2max" ranges from 2.5 to 4.5 L/min in well-trained women, and from 3 to 6 L/min in athletic men, at sea level. Instead of using such absolute numbers, relative values also consider body size; these ratios result from dividing the maximal volume of oxygen (in milliliters) by body mass (in kilograms). The "relative VO_2max" is likely to be near 20 mL/min/kg in untrained persons and may go up to 80 mL/min/kg in a champion athlete*.

Persons so "calibrated" may then be sent to work in agriculture, on a construction site, or in a space ship; their reactions allow an assessment of how straining their work tasks are. This is the rationale by which tables were established that describe energy expenditures or heart rates associated with certain jobs or occupations (see below). Unfortunately, there is little or no assurance that the published field data describing the demands ("heaviness") of jobs indeed stem from measurements on "calibrated" subjects under "controlled" conditions. Furthermore, tasks and conditions of work may vary widely even if called the same. Consequently, there is wide diversity in published data.

In many jobs, exertion levels do not remain constant but vary over time. To take that variation into consideration, one can assess the operator's effort by frequent sampling or by continuous measuring over a sufficiently long time, so that peaks and lows in energy intake, conversion, and output "average out". For example, equations in Chap. 7 simply excluded energy storage. However, imbalances between energy intake and output of hundreds of kilojoules over a day are common; accordingly, there can in reality be significant changes in energy storage.

Most of the energy store in a healthy adult is in the form of adipose tissue, amounting to about 400,000 kJ—see Chap. 7. A substantial change in storage should result in a change in body weight. (Data on energy consumption are often reported in relation to body weight.) However, body water disturbs the seemingly simple relationship between weight and energy storage, because water is a large and rather labile component of body weight but contributes nothing to the energy stores.

A further challenge is to assure that the measuring process does not influence the measurement results. For example, putting on and carrying apparatus to monitor oxygen consumption often hinder the subject's motions and breathing, and even just being observed can lead to task performance that differs from the normal procedures. In these cases, energetic and circulatory processes may diverge from those at regular work, confounding the results.

To assess and use data measured during physical work, a reliable "baseline" is needed. A subject's basal metabolism is a solid reference for energetic processes. However, it is quite difficult for a subject to achieve a true, reliable basal condition with its stringent requirements (see below). Therefore, assessment of basal metabolism is often replaced by measurements on a resting subject, where conditions and measurements are more variable. Uncertainty about the baseline is of particular concern for light efforts that elevate body functions only slightly. (This may explain some of the percent-wise large variations found in different texts concerning the energy requirements of light work.).

8.3   Energy Requirements at Work

While the foregoing discussion primarily concerned the body's energetic capabilities, the following text concentrates on the demands that different activities impose on the human body.

8.3.1   Procedures to Catalogue Metabolic Requirements

Using "calibrated subjects" (as discussed above) allows the measurement of the metabolic demands of work tasks and the grouping of the results into convenient categories. For the measurements one assumes, as usual, no change in energy storage in the body. Either basal or resting metabolism establishes a baseline.

8.3.1.1   Basal Metabolism

A minimal amount of energy is necessary to keep the body functioning even if it does no additional activities. This basic metabolism is measured under strict conditions: usually after fasting for 12 h, with protein intake restriction for at least two days; and during complete physical rest in a neutral ambient temperature. Under these conditions, the basal metabolic values depend primarily on age, gender, height, and weight. (Height and weight are occasionally replaced by body surface

area in ergonomic databases.) In healthy adults there is little inter-individual variation, hence a commonly accepted value is 4.2 kJ (1 kcal) per kg per hour; this is
equivalent to 4.9 kJ/min for a person of 70 kg (corresponding to a daily caloric
expenditure of nearly 1700 kcal to fuel metabolic function).

8.3.1.2   Resting Metabolism

The highly controlled conditions needed to measure basal metabolism are rather
difficult to accomplish. Thus, one usually measures the metabolism (and the heart
rate) before the working day, with the subject at rest as much as possible.
Depending on the given conditions, resting metabolism is around 10–15% higher
than basal metabolism.

8.3.1.3   Work Metabolism

The increase in metabolism from resting to working is called work metabolism: this
represents the amount of energy needed to perform the work.

As Fig. 8.3 illustrates, at the start of physical work the actual oxygen uptake lags
behind the demand. During the first 45–90 s of working at an intensity near 50% of
the maximum possible, the energy metabolism is almost completely anaerobic
because the oxygen supply needs time to meet the demand. Thus, during the first
minutes of physical work, a discrepancy develops between required oxygen and

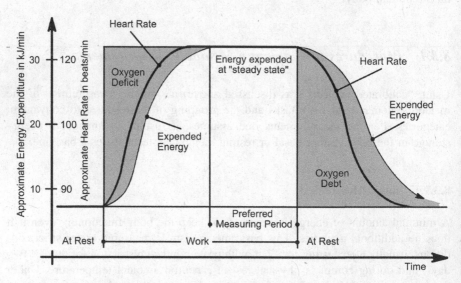

Fig. 8.3 Schematic illustration of energy liberation, energy expenditure, and heart rate before,
during, and after "steady state" work

available oxygen. After its slow onset, oxygen intake rises rapidly and finally approaches the level at which it fulfills the oxygen demands of the work. The initial oxygen deficit must be repaid at some time, usually during the following work if it is not overly demanding, or during a rest period.

When the demands of work stay below approximately 50% of the worker's capacities for maximal oxygen uptake, then heart rate and cardiac output increase so that they can finally achieve the required supply level. When they maintain this supply level, then the body is at "steady state"*.

When work ends, the oxygen demand falls back to its resting level, quickly at first and then leveling off. During this recovery time, depleted ATP stores are refilled, and myoglobin and hemoglobin are refurbished with oxygen. However, elevated tissue temperatures and epinephrine concentrations, augmented cardiac and respiratory functions, reconversion of lactate to glycogen, and other phenomena mean that, as a rule, the oxygen debt repaid is approximately twice as large as the oxygen deficit incurred*.

Given the close coupling between the circulatory and metabolic systems, heart rate reacts similarly; however, heart rate increases faster than oxygen uptake at the start of work, and falls back more quickly to its resting level after work. This "quick response" makes pulse rate measurements attractive to assess energy expenditure.

If the workload overtaxes the worker's metabolic (circulatory, respiratory, muscular) capacities, the person will have to stop the effort. However, if oxygen uptake, heart rate, cardiac output, ventilation, circulation, and other functions can supply the on-going demands of the workload, then "steady state" is achieved, as sketched in Fig. 8.4. Work that exceeds the demands of "steady state" requires more rest, as sketched in Figs. 8.4 and 8.5. Obviously, a well-trained person can attain this equilibrium between demand and supply at a relatively higher workload, whereas an unfit person can attain a steady state only at a lower demand level. Measuring maximal oxygen uptake indicates individual capacities for hard exercise and labor.

If the energetic work demands exceed about half the person's maximal O_2 uptake capacity, anaerobic energy-yielding metabolic processes play increasing roles. Onset of the feeling of "fatigue" (see below) usually coincides with depletion of glycogen deposits in the working muscles, drop in blood glucose, and increase in blood lactate. However, the associated physiological and motivational processes are still not well characterized: highly motivated persons (such as in sports competitions) may continue to perform such demanding effort for some time in spite of the physiological obstacles.

Vigorous activity may cause incapacitating oxygen deficit, unacceptably high lactate content of the blood, and excessive circulatory strain; so a "steady state" balance between demands and supply cannot be achieved and the body must stop working. Rest periods can counteract impending exhaustion, as sketched in Fig. 8.5, with recovery steepest at the beginning of a break. Given the same ratio of "total resting time" to "total working time," frequent, short rest periods have more recovery value than fewer, longer rest periods.

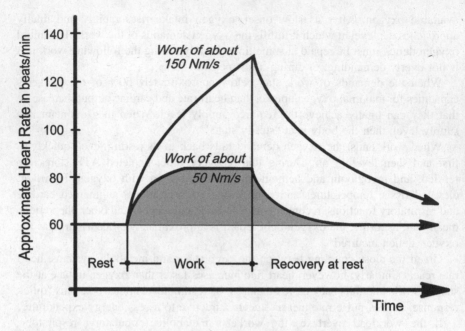

Fig. 8.4 Examples of heart rate at exhausting and during "steady state" work

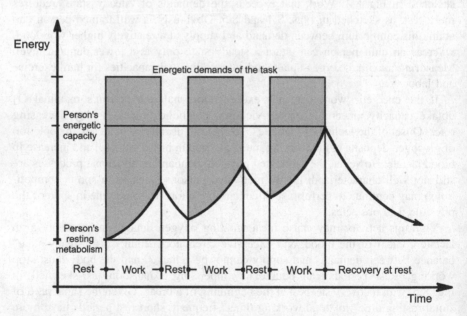

Fig. 8.5 Typical metabolic reactions to attempt work that exceeds one's capacity even with interspersed rest periods

Taken during steady state work, measurements of oxygen intake and heart rate dependably indicate the metabolic and circulatory demand-and-supply functions in the body. Oxygen and HR measurements conducted during the first few minutes of work do not reflect the long-time work demands on the body because of the slow responses of oxygen uptake and heart activity. These lagging responses to sudden changes in work demand make it doubtful that O_2 consumption and HR can reliably reflect the effects of physical work that contain rapid increases and decreases in effort. Measurements taken after the cessation of the work could be indicators of the severity of the preceding work but are difficult to interpret, partly because it is usually uncertain how long the recovery actually takes*.

8.3.2   Techniques to Estimate Energy Requirements

Besides actual measurements, two other means are often employed to obtain information on the energetic requirement of enduring physical effort: using tabulated data and calculating energy requirements.

8.3.2.1   Using Tabulated Data

Physiologists have measured energy expenditures in many jobs and occupations and compiled the data: Table 8.2 is an excerpt reflecting conditions in the 1940–1960s. Of course, professions and jobs change over time. Yet, some basic task elements remain the same: Table 8.3 presents information on energy expended in common body positions and activities. Table 8.4 contains data on energy consumption in various recreational sports. The data in Tables 8.3 and 8.4 apply to prototypical healthy young men of 75 kg, with estimates for women generally 10% to 15% lower. Of course, it is necessary to be cautious when using these tables: first, the actual conditions may not be typical; second, the tabulated data may be listed in different units. For example, Table 8.2 contains information on *total energy cost per day,* Table 8.3 describes *work metabolism per minute,* and Table 8.4 lists *hourly* rates for each *kg of body weight.*

8.3.2.2   Calculating

Summarized tables (such as Table 8.2) contain measurements that were taken when many jobs or activities differed from what they are today: housewives now seldom wash and wring loads of laundry by hand, secretaries no longer pound on mechanical typewriters, lumbermen don't usually cut down trees with hand saw and ax, few farmers walk behind a plow pulled by oxen or horses. Furthermore, these older data may have been "averaged" over unknown subjects or attained with only a few subjects.

Table 8.2 Total energy expenditures in various professions, in kcal per day (adapted from Astrand and Rodahl 1977)

	Energy expenditure, kcal/day		
Men	Minimum	Mean	Maximum
Coal miners	2970	3660	4560
Elderly industrial workers	2180	2820	3710
Elderly Swiss peasants	2210	3530	5000
Farmers	2450	3550	4670
Forestry workers	2860	3670	4600
Laboratory technicians	2240	2840	3820
Steelworkers	2600	3280	3960
University students	2270	2930	4410

Women			
Elderly housewives	1490	1990	2410
Middle-aged housewives	1760	2090	2320
Elderly Swiss peasants	2200	2890	3860
Factory workers	1970	2320	2980
Laboratory technicians	1340	2130	2540
University students	2090	2290	2500

The physical demands are likely to be different now from what they were when measured in the 1960s and 1970s.

A more analytic approach avoids related problems of using summary data: one can compose the total energetic cost of job or task activities by adding up the energetic costs of distinct (and clearly defined) work elements which, combined, make up this activity. Basic elemental costs do not change over time. If one knows the time spent in a given activity element and its metabolic cost per time unit, one can simply calculate the energy requirements of this element by multiplying its unit metabolic cost with its duration time. Repeating this for all other work elements, and summing all results, provides an estimate of the total energetic job effort.

Table 8.3 Energy consumption (to be added to basal metabolism) at various activities (adapted from Astrand and Rodahl 1977; Guyton 1979; Rohmert and Rutenfranz 1983; Stegemann 1984)

	kJ/min
Lying down, sitting, standing	
Resting while lying down	0.2
Resting when sitting	0.4
Sitting with light work	2.5
Standing still and relaxed	2.0
Standing with light work	4.0
Walking without load	
Walking on horizontal smooth surface at 2 km/h	7.6
Walking on horizontal smooth surface at 3 km/h	10.8
Walking on horizontal smooth surface at 4 km/h	14.1
Walking on horizontal smooth surface at 5 km/h	18.0
Walking on horizontal smooth surface at 6 km/h	23.9
Walking on horizontal smooth surface at 7 km/h	31.9
Walking on grass at 4 km/h	14.9
Walking in pine forest, smooth surface, at 4 km/h	18 to 20
Walking on plowed heavy soil at 4 km/h	28.4
Walking and carrying on smooth solid horizontal ground	
1 kg on the back at 4 km/h	15.1
30 kg on the back at 4 km/h	23.4
50 kg on the back at 4 km/h	31.0
100 kg on the back at 3 km/h	63.0
Walking downhill on smooth solid ground at 5 km/h	
5° decline	8.1

(continued)

Table 8.3 (continued)

10° decline	9.9
20° decline	13.1
30° decline	17.1

Walking uphill on smooth solid ground at 2.5 km/h	
10° incline, gaining altitude at 7.2 m/min	
No load	20.6
20 kg on back	25.6
50 kg on back	38.6
16° incline, gaining altitude at 12 m/min	
No load	34.9
20 kg on back	44.1
50 kg on back	67.2
25° incline, gaining altitude at 19.5 m/min	
No load	55.9
20 kg on back	72.2
50 kg on back	113.8

Climbing stairs or ladder	
Climbing stairs, 35° incline, steps 17.2 cm high	
100 steps per minute, gaining altitude at 17.2 m/min, no load	57.5
Climbing ladder, 70° incline, rungs 17 cm apart	
66 steps per minute, gaining altitude at 11.2 m/min, no load	33.6

Table 8.4 Total energy expenditure per kg body weight at various sports

	Energy expenditure kJ/kg, per hour
Badminton	53
Bicycling, 9 km/h	15
Bicycling, 16 km/h	27
Bicycling, 21 km/h	40
Cross-country skiing, 9 km/h	38
Cross-country skiing, 15 km/h	80
Ice skating, 21 km/h	41
Jogging, 9 km/h	40
Running, 12 km/h	45
Running, 16 km/h	68
Swimming, breaststroke, 3 km/h	45
Walking, 4 km/h	13
Walking, 7 km/h	25

For example:
For a person resting (sleeping) eight hours per day, at an energetic cost of 5.1 kJ/min, the total energy cost is 2448 kJ (5.1 kJ/min × 60 min/h × 8 h). If the person then does six hours of light work while sitting, at 7.4 kJ/min, this adds another 2664 kJ to the energy expenditure. With an additional six hours of light work done standing, at 8.9 kJ/min, and further with four hours of walking at 11.0 kJ/min, the total expenditure during the full 24-h day comes to 10,956 kJ.

When using tables of metabolic requirements of job elements, one has to check whether or not they include the basal or resting rates: Table 8.3 does not contain the baseline values, but Tables 8.2 and 8.4 do. In developed countries, typical daily expenditures range from about 6,000–20,000 kJ/day, with observed median values of about 10,000 kJ (2,400 Cal) for women and about 14,000 kJ (3,300 Cal) for men.

8.3.2.3   Light or Heavy Jobs?

With a linear relationship between heart rate and energy uptake, one can generally use heart rate to establish the "heaviness" of work. Of course, descriptive labels as "light" or "easy" or "heavy" reflect individual judgments that rely very much on the current socio-economic concept of what is permissible, acceptable, comfortable, or hard. Depending on the circumstances, there may be a diversity of opinions about how physically demanding a given job is. Table 8.5 lists ratings of job severity according to energetic and circulatory demands. This is a unisex table: some persons might find the work lighter, and others might perceive the effort to be heavier, than labeled.

Light work is associated with rather small energy expenditure (about 10 kJ/min including the basal rate) and a heart rate of approximately 90 beats/min (bpm). At this level of work, the oxygen available in the blood and from glycogen at the muscle is sufficient to meet the needs of the working muscles. During *medium* work, with about 20 kJ/min and 100 bpm, the oxygen requirement at the working muscles is still being met, and lactic acid initially developed is re-synthesized to glycogen during the activity. During *heavy* work, with about 30 kJ/min and 120 bpm, enough oxygen is still available if the person is physically fit and trained for such work. However, excess lactic acid generated during the initial minutes of the work remains until the end of the work period, only to be reduced back to normal levels after cessation of the work.

Table 8.5 Classification of work (performed over an entire work shift) from "light" to "extremely heavy" according to energy expenditure and heart rate

	Total energy expenditure		Heart rate
Classification	kJ/min	kcal/min	beats/min
Light work	10	2.5	90 or less
Medium work	20	5	100
Heavy work	30	7.5	120
Very heavy work	40	10	140
Extremely heavy work	50	12.5	160 or more

In light, medium and even in heavy work, metabolic and other physiological functions can attain a steady-state condition throughout the work period, provided the person is capable and trained. This is not the case with *very heavy* work, where energy expenditures are in the neighborhood of 40 kJ, and heart rate is around 140 bpm. Here, the original oxygen deficit increases throughout the duration of work, making intermittent rest periods necessary or even forcing the person to stop working altogether. At *extremely heavy work*, with energy expenditures attaining 50 kJ/min, associated with heart rates of 160 bpm or higher, oxygen deficit and accumulation of byproducts of the biochemical processes are of such magnitudes that frequent rest periods are needed; even highly trained and capable persons may be unable to perform this job throughout a full work shift.

An individual's maximal oxygen uptake provides a personalized measure of effort: work that requires 30–40% of the person's maximal uptake can be done for 8 h; but if the job requires 50% or more, exhaustion is likely to occur.

8.4   Overall Changes in Body Functions in Response to Work Loads

As stated at the beginning of this chapter, there is close interdependency among respiration, circulation, and metabolic functions. The least capable of these functions sets the limit for an individual's physical performance. Of course, other specific capabilities may also cap the possible output: for example, muscular strength frequently limits performance. Furthermore, changes in health, fitness, skill or motivation affect performance abilities.

Table 8.6 lists the main changes in physiological functions, from rest to maximal work, for energy and oxygen consumption, heart actions, and respiration. Any and all of these variables are useful for assessing a person's response to a given work

Table 8.6 Changes in physiological functions from rest to maximal effort

Energy consumption	From 1 to 20 kcal/min	× 20
Oxygen uptake	From 0.2 to 4 L	× 20
Cardiac action	Heart rate: from 60 to 180 beats/min	× 3
	Stroke volume: from 50 to 150 mL	× 3
	Cardiac output = minute volume × heart rate: from 5 to 35 L/min	× 7
	Blood pressure, systolic: from 90 to 270 mmHg	× 3
Respiration	Breathing rate: from 10 to 50 breaths/min	× 5
	Minute volume = tidal volume × breathing rate: from 5 to 100 L/min	× 20

load, or for judging the demands imposed by the task. However, measures of ventilation or metabolic responses are suitable only for assessing "dynamic" work. "Static" efforts, in which muscles are contracted and kept contracted, hinder or completely cut off the blood supply by compression of the capillary bed; so, relatively little additional energy is consumed. Therefore, techniques relying on respiratory or metabolic functions cannot properly assess static efforts even when these efforts may be fatiguing.

8.5   Fatigue

The term "fatigue" is used in this text as an operational description of a temporary state of reduced ability to continue muscular contraction or physical work. This phenomenon is best researched for maintained static (isometric) muscle contraction. As described in Chap. 2, when an effort exceeds about 15% of the maximal voluntary contraction (MVC) capability of a muscle, blood flow through the muscle becomes reduced due to the pressure build-up within the muscle; in a maximal effort, the muscle may compress its blood vessels completely, in spite of a reflexive increase in systolic blood pressure. Insufficient blood flow brings about an accumulation of potassium ions in the extracellular fluid and depletion of extracellular sodium. Combined with intracellular accumulation of phosphate (from the degradation of ATP), these biochemical events perturb the coupling between nervous excitation and muscle fiber contraction. Depletion of the energy carriers ATP or creatine phosphate, and the accumulation of lactate (which was long believed to be the main reason for fatigue) also occur. In addition, the increase of positive hydrogen ions, resulting from anaerobic metabolism, causes a drop in intramuscular pH, in turn inhibiting enzymatic reactions, notably those in the ATP breakdown.

Discomfort, pain, and de-coupling of central nervous system (CNS) control and muscle action signal the onset of physiological fatigue. However, persons with high motivation to overcome the feeling of fatigue or exhaustion can continue their physical effort for considerable time whereas others stop shortly after fatigue symptoms begin.

Fatigue can be avoided by reducing or stopping the effort that causes fatigue. During a rest period, accrued metabolic byproducts can be metabolized, the respiratory and circulatory systems recover, and the will to continue may be restored. As Figs. 8.3, 8.4 and 8.5 demonstrate, the recovery benefits are largest immediately after stopping work but the rate of recovery slows during the rest intermission. Therefore, as already stated, many short rest pauses have more value in avoiding or overcoming fatigue than fewer but longer stops. In the case of heavy physical work, for example, this means that one should not "push through" the task and then rest, after becoming more or less exhausted; rather, the manager should encourage laborers to take frequent but short breaks.

8.6 Human Engineering/Ergonomics

Calorimetric information is especially useful for assessing

- Capability: an individual's energetic functions in response to standardized activities: this is a measurement of individual metabolic work capacity.
- Task demand: the energetic requirements of a given activity as reflected by the energetic response of a "standard person", or of a specifically selected and trained individual, doing that work.

In the interest of standardization, test activities used to generate metabolic data should be kept neutral, not resembling actual job tasks. However, the current standard tests (bicycle, treadmill, steps) primarily involve the use of leg muscles, not of the upper body and arms; in contrast, in most everyday jobs, arm and upper body muscles are primarily used. Development of new standard tests which better align with real-world conditions would be helpful to better determine job-related capabilities and demands.

Much of the currently used equipment to measure metabolic reactions to work tasks is awkward to wear and requires that a physician or physiologist performs the testing, often in a laboratory setting rather than at a regular workplace. Improvements in test instrumentation tto become less cumbersome and invasive than current equipment would make testing easier and more realistic. Even if some of the present techniques should be maintained, it would be useful to determine categories of work tasks in which simple assessments, such as by heart rate or by subjective judgment, would suffice.

Managers and engineers establish the required task and how it shall be done, and often have control over the external environment. They must adjust the work to be performed (and the work environment) to suit the operator's physiologic capabilities—which includes the encouragement, and not just tolerance, of operators freely taking frequent rest periods during heavy work. Managers and engineers must also understand the physiologic (and psychologic) procedures to determine performance limits; how to interpret this information (which currently mostly stems from laboratory tests that do not resemble actual work tasks); and how to apply it to a real-world jobs. If job demands exceed human capabilities, it is necessary to re-design the work* by changing tasks, tools, and environment as appropriate and needed. In addition, training of the operator in procedure and fitness may also be advisable. This approach of matching requirements with abilities is shown in Fig. 8.6.

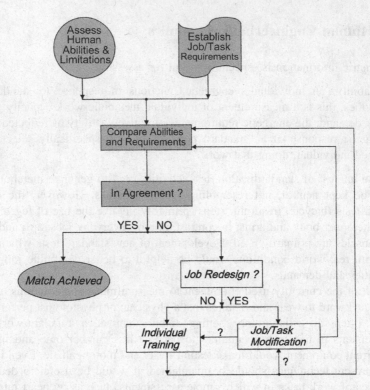

Fig. 8.6 Matching task demands with human abilities

Notes

The text contains markers, *, to indicate specific references and comments, which follow.

Body Mass Index (BMI) from weight and stature: Centers for Disease Control, http://www.cdc.gov/nchs

Techniques for "rating the perceived exertion" (RPE): Borg (1962, 1982, 2001, 2005), Pandolph (1983)

VO$_2$max in untrained persons and in athletes: Kenney et al. (2020)

The body is at "steady state": In reality, even in a so-called steady-state situation, heart rate, cardiac output, pulmonary ventilation can still increase gradually, which may be a consequence of increasing body temperature or due to an increasing use of fatty acids as substrate

The oxygen debt repaid is approximately twice as large as the oxygen deficit incurred: "The body must pay 100% interest on the oxygen borrowed from the anaerobic bank." Astrand and Rodahl (1986)

Measurements taken after the cessation of the work ... uncertain how long the recovery actually takes: Lehmann (1953, 1962), Brouha (1960)

Re-design the work by changing tasks, tools, and environment as appropriate and needed: Kroemer Elbert et al. (2018).

Summary

Among the techniques to assess metabolic processes, measurements of oxygen uptake and of heart rate are practical and reliable; equipment is commercially available. Subjective ratings of perceived effort require no hardware at all.

Common tests measure an individual's metabolic and circulatory capabilities in response to standardized exercise, which is usually performed on bicycle ergometers or treadmills. Reversing that process, observation of workers' metabolic and circulatory efforts while performing work allows categorizing work requirements.

Energy requirements depend on the activity level of the body: "basal" metabolism suffices to keep the body alive while no work effort is being done, "resting" metabolism keeps the body functioning while at rest, and "work" metabolism maintains the body at work.

At the start of work, increases in both oxygen consumption and heart rate lag behind increased work demands but heart rate responds more quickly. If the demand level is suitable, these body functions finally achieve a steady state. After cessation of the effort, oxygen intake and heart rate fall back to their resting levels, heart rate again more quickly than O_2 uptake.

Measurement of oxygen consumption or heart rate as indicators of the physical task requirements should be performed during the steady-state phase of the exercise. Task "heaviness" can be classified by energy and heart rate requirements.

In some cases, energy requirements can be taken from tabulated listings. Such lists apply to given conditions, for example pertaining to work details, the individual worker, or the climate. Often, one can calculate the energy requirement of a job by breaking it into elemental subtasks for which the (average) energy requirements are known: the total energy need is the sum of the energy needs for each subtask.

If the physical demands are too high for the individual, no steady rate can be achieved, and the work must be interrupted for rest and recovery. Such finding gives reason to redesign the task.

Appendix A: Techniques of Indirect Calorimetry

Measuring the oxygen consumed over a sufficiently long period of time is a practical way to assess the metabolic processes. (A properly trained physician or physiologist should perform this test.) As discussed earlier, one liter of oxygen consumed releases about 21 kJ (5 kcal) of energy in the metabolic processes. This assumes a normal diet, a healthy body oxidizing primarily carbohydrates and fats under conditions of light to moderate work, and suitable climatic conditions (The "normality" of the metabolic conditions can be judged, to some degree, by the respiratory exchange quotient, RQ, mentioned earlier.).

Classically, indirect calorimetry has been performed by collecting all exhaled air during the observation period in airtight (Douglas) bags. The volume of the exhaled air is then measured and analyzed for oxygen and carbon dioxide as needed for the determination of the RQ. From these data one can calculate the amount of oxygen, and hence energy, used during the collection period. This requires a rather complex air collecting system, including nose clip and a mouthpiece with intake and exhaust valves. This apparatus can become quite uncomfortable for the subject and hinders speaking, which limits this procedure mostly to the laboratory. A major improvement was done by diverting only a known percentage of the exhaled air into a small collection bag. This 1950s procedure is still in use since the subject must carry only a relatively small device, which does not greatly affect most daily activities.

Use of instantaneously reacting oxygen sensors brought significant progress because their placement into the air flow of the exhaled air allows breath-by-breath analysis. Since the volume of exhaled air can also be measured by suitable sensors, "open" face masks draw a stream of air across the face of the subject who simply inhales from this air flow and exhales into it. This allows free breathing, speaking, and drinking; it even cools the face (which might in fact improve the working capacity to a small extent). Another technical variation employs a hood-like enclosure of the head. The differences in oxygen measured with either equipment are usually less than 15%; therefore, for most field observations, the accuracy of bagless procedures is quite sufficient.

Another technique uses measurement of isotopes. It relies on doubly-labelled water (DLW), in which both the hydrogen and the oxygen have been partly or completely replaced with uncommon heavy isotopes of these elements, usually deuterium and oxygen-18. The depletion of these isotopes indicates the generation of water in the body which parallels the metabolizing process. A dose of water with these isotopes is injected into or drunk by the subject, animal or human. Samples of saliva, urine, or blood are taken to permit measurements of the isotope concentrations and their elimination rates. At least two such samples are needed: the first one after the isotopes have reached equilibrium in the body, and a second sample some time later. The measuring period maybe as short as 24 h in small animals and as long as 14 days in adult humans. The DLW isotope technique is particularly useful for measuring an average metabolic rate ("field metabolic rate") over days or weeks when direct or indirect calorimetry would be impractical.

Appendix B: Rating the Perceived Effort

Early in the 19th century, models of the relationships between a physical stimulus and one's perceptual sensation of that stimulus (its psychophysical correlate) were developed. In 1834, Weber suggested that the "just noticeable difference" Δ I depends on the absolute magnitude of the physical stimulus I:

$$\Delta I = a\,I$$

where a is constant.

In 1860, Fechner related the magnitude of the "perceived sensation" P to the magnitude of the stimulus I:

$$P = b + c \log I$$

where b and c are constants.

In the 1950s, Stevens and Ekman introduced ratio scales (with a zero point and equidistant scale values), to describe the relationships between the perceived intensity I and the physically measured intensity P of a stimulus:

$$P = d\,I^n$$

where d is a constant and n ranges from 0.5 to 4, depending on the modality.

Beginning in the 1960s, Borg and collaborators modified these relationships to take into account deviations from previous assumptions (such as zero point and equidistance), and to describe the perception of different kinds of physical efforts. Borg's "General Function" is:

$$P = e + f(I - g)^n$$

The constant e represents "the basic conceptual noise" (normally below 10% of I) and the constant f indicates the starting point of the curve; g is a conversion factor that depends on the type of effort.

Ratio scales indicate only proportions between percepts, but do not indicate absolute intensity levels. They neither allow intermodal comparisons nor comparisons between intensities perceived by different individuals. Borg has tried to overcome this problem by assuming that the subjective range and intensity level are about the same for each subject at the level of maximum intensity. In 1960, this led to the development of a "category scale" for the Rating of Perceived Exertion (RPE). The scale ranges from 6 to 20 (to match heart rates from 60 to 200 bpm) with equidistant steps. Every second number is verbally anchored:

Borg RPE Scale

- 6 (no exertion at all)
- 7—extremely light
- 8
- 9—very light
- 10
- 11—light
- 12
- 13—somewhat hard
- 14
- 15—hard (heavy)
- 16
- 17—very hard
- 18
- 19—extremely hard
- 20 (maximal exertion)

In 1980, Borg proposed the CR-10 Scale, a category scale with steps in constant ratios, said to retain the same correlation (of about 0.88) with heart rate as the RPE scale, particularly if large muscles are involved in the effort.

Borg CR-10 Scale

- 0—nothing at all
- 0.5—extremely weak, just noticeable
- 1—very weak
- 2—weak, light
- 3—moderate
- 4—somewhat strong
- 5—strong, heavy
- 6
- 7—very strong
- 8
- 9
- 10—extremely strong, almost maximal
- 11 or higher—The individual's absolute maximum, highest possible.

(Note: The terms "weak" and "strong" may be replaced by "light," or "hard" or "heavy," respectively).

Just as any other psychophysical measurements, Borg scales need rigidly controlled procedures to yield credible results. For example, the experiment's protocol may state to put the CR1-0 Scale in front of the person who will rate the perceived intensity of work, with these instructions:

You do not have to specify your feelings, but do select the number that most correctly reflects your perception of the job demands. If you feel no loading, you should answer zero, nothing at all. If you start to feel something just noticeable, your answer is 0.5, extremely weak. If you have an extremely strong impression, your answer will be 10. So, the more you feel, the higher the number you are choosing. Keep in mind that there are no wrong numbers; be honest, do not overestimate or underestimate your ratings.

Borg claimed in 2001 and 2005 that his scales have high reliability (meaning that a repeated test yields the same results as the initial test) and high validity because the test results correlate well with measures of heart rate. Wilson and Corlett discussed other rating scales in 2005.

Glossary

Aerobic In the presence of oxygen

Anaerobic In the absence of oxygen

Basal metabolism Energy that suffices to keep the body alive while no effort is done

Calorimeter · A sealed chamber used to measure the heat eliminated from or stored in a body

Calorimetry Measurement of the heat eliminated from or stored in a system

Direct calorimetry Measurement of heat actually produced by the body; the body is confined in a sealed chamber, called calorimeter (see there)

Energy In physics, the capacity of a physical system to do work

Ergometer A device, usually a stationary braked bicycle, used for measuring the amount of work done by the whole body or by arms, legs, or specific muscles

Ergometry The study of physical work activity, often done with testing equipment such a treadmills or bicycle ergometer (see there)

Ergonomics The application of scientific principles, methods and data drawn from a variety of disciplines to the design of engineered systems in which people play significant roles

Fatigue In this text, an operational description of a temporary state of reduced ability to continue muscular contraction or physical work

Indirect calorimetry Estimation of produced body energy by means of the differences of oxygen and carbon dioxide in the inspired and expired air

Lactate Lactic acid that is missing one proton; so it becomes its conjugate base, lactate

Metabolism Chemical processes in the living body; in a narrower sense, the energy-yielding processes

Oxygen deficit The discrepancy between required oxygen and available oxygen

Plethysmograph Instrument to record variations in the size (volume) of parts of the body, such as of the chest circumference with breathing or of the finger volume with blood pulsesPulse

Resting metabolism Energy that suffices to keep the body functioning while at restResting metabolism

Work metabolism Energy above the resting metabolism that suffices to keep the body functioning while at work.

References

Astrand PO, Rodahl K (1977,1986) Textbook of work physiology. 2nd and 3rd eds. McGraw-Hill, New York

Astrand PO, Rodahl K, Dahl HA, Stromme SB (2003) Textbook of work physiology. Physiological bases of exercise, 4th edn. Human Kinetics, Champaign

Borg G (1962) Physical Performance and Perceived Exertion. Gleerups, Lund

Borg G (1982) Psychophysical bases of perceived exertion. Med Sci Sports Exerc 14(5):377–381

Borg G (2001) Rating scales for perceived physical effort and exertion. In: Karwowski W (ed) International encyclopedia of ergonomics and human factors. Taylor and Francis, London

Borg G (2005) Scaling experiences during work: perceived exertion and difficulty. In: Stanton N, Hedge A, Brookhuis K, Salas E, Hendrick H (eds) Handbook of human factors and ergonomics methods. CRC Press, Boca Raton

Brouha L (1960) Physiology in industry. Pergamon Press, New York

Ekman G (1964) Is the power law a special case of Fechner's law? Percept Mot Ski 19:730

Fechner GT (1860) Sachen der psychophysik. Breitkopf und Hertel, Leipzig

Guyton AC (1979) Physiology of the human body, 5th edn. Saunders, Philadelphia

Kenney WL, Wilmore JH, Costill DL (2020) Physiology of sport and exercise, 7th edn. Human Kinetics, Champaign

Kroemer KHE, Kroemer HB, Kroemer-Elbert KE (2003) Ergonomics: How to design for ease andefficiency, 2nd amended edn. Prentice Hall, Englewood Cliffs, NJ

Kroemer Elbert KE, Kroemer HB, Kroemer Hoffman AD (2018) Ergonomics: how to design for ease and efficiency, 3rd edn. Academic Press, London

Lehmann G (1953, 1962) Praktische arbeitsphysiologie, 1st and 2nd edn. Thieme, Stuttgart

Pandolph KP (1983) Advances in the study and application of perceived exertion. Exerc Sport Sci Rev 11:117–158

Rohmert W, Rutenfranz J (eds) (1983) Praktische arbeitsphysiologie, 3rd edn. Thieme, Stuttgart

Stegemann J (1984) Leistungsphysiologie, 3rd edn. Thieme, Stuttgart

Stevens SS (1957) On the psychophysical law. Psychol Rev 64:151–181

Weber EH (1834) De pulse, resorptione, auditu et tactu. Kochler, Leipzig

Wilmore JH, Costill D, Kenney WL (2008) Physiology of sport and exercise, 4th edn. Human Kinetics, Champaign

Wilson JR, Corlett N (eds) (2005) Evaluation of human work, 3rd edn. Taylor and Francis, London

Chapter 9
Thermal Environment

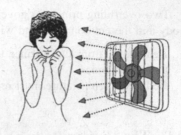

Overview

The body generates energy; some of this energy is used to perform work, while the rest must be dissipated as heat, mostly by convection and evaporation. The body may also receive heat from or lose heat to the environment by radiation, conduction and convection. The body strives to balance these exchanges. There are physiologic and physical means to affect energy transfer; engineering controls of the macro- and micro-climates are available to provide suitable ergonomic conditions.

The Model

The human body generates heat while also exchanging (gaining or losing) heat with the environment. The basic amount of heat generated stems from the body's life-sustaining functions; further heat is generated depending on the work which the body performs. The internal flow of heat is primarily controlled by blood circulation in the body masses between skin and core.

Since a rather constant core temperature must be maintained, suitable heat flow between the body and its surrounding environment must be achieved. Clothing affects the heat exchange with the surrounds. Within built structures, the ambient climate usually can be controlled by technical means.

Introduction

It is convenient to use the simple concepts of "core" and "shell" temperatures of the human body although, in reality, the body's temperature regulation system maintains diverse temperatures at various locations under different conditions. The "core" is thought to consist mainly of brain, heart, lungs, and abdominal organs; the "shell" is essentially the skin.

© Springer Nature Switzerland AG 2020
K. H. E. Kroemer et al., *Engineering Physiology*,
https://doi.org/10.1007/978-3-030-40627-1_9

Two overriding principles govern the main aspects of flow of heat between the body and the environment, and within the body:

1. Heat flows from the warmer to the colder matter, as per the second law of thermodynamics.
2. Body core temperature must remain close to 37 °C

9.1   The Human Body as a Thermo-Regulated System

There are small temperature changes in the human body throughout the day due to circadian rhythm and work schedules, discussed in Chap. 10. However, the main impact on the human thermal regulatory system results from the interaction between metabolic heat generated within the body and external energy gained in hot surroundings or lost in cool environments.

. The human body has an intricate regulation system to maintain its core temperature near 37 °C, and tries to keep temperatures in its outer layers well above freezing or, respectively, cooler than the 40s. Digressions beyond ±2 °C from the 37 °C core temperature impair body functions as well as physical and mental work capacities, and hence task performance.

If the temperature in a human cell exceeds 45 °C, heat coagulation of proteins takes place. If the temperature falls to freezing, ice crystals break the cell apart. Deviations of the core temperature of ±6 °C from 37 °C can be lethal.

9.1.1   The Energy Balance

When discussing the metabolic system in Chap. 7, the energy balance between body energy inputs and outputs was described as

$$I = M = H + W + S.$$

- I is the energy input via nutrition which is converted into the body's rate of metabolic energy production M;
- H is the rate of heat that must be released from the body;
- W is the rate of external work done;
- S is the rate of energy storage in the body.

Traditionally, energy rates are discussed in units of Watts or in Joules per second (1 W = 1 J/s), and per square meter of the involved body surface*.

If not all metabolic energy can be dispelled to the environment and/or energy is transferred from the environment to the body, the heat storage (and hence, energy storage) S in the body increases. If more than H gets lost from the body to the environment, S becomes smaller.

When there is no change in the quantity of energy storage S, the system is in balance. Furthermore assuming that the quantities I and W remain unchanged allows simplifying the equation that describes the energy exchange with the thermal environment. In this case, thermal balance is achieved when the energy

$$H = I - W$$

is dissipated to the environment while no energy is gained from the environs.

9.2   Energy Exchanges with the Environment

The body exchanges heat energy with the environment* through radiation, conduction, convection, and evaporation.

An object at temperature greater than absolute zero emits radiant energy. Heat exchange through radiation (R) depends primarily on the temperature difference between two opposing surfaces, for example between a window pane and a person's skin. (The exchange of heat by radiation does not depend on the temperature of the air between the two exchanging surfaces.) Heat always radiates from the warmer to the colder surface: from the skin to the cold window pane in the winter; in the summer, to the body from a sun-heated pane.

The body can lose or gain heat through radiation, as sketched in Figs. 9.1 and 9.2.

Energy exchanges through radiation follow the Stefan-Boltzmann law of radiative heat transfer: the total heat radiated from a surface is proportional to the fourth power of its absolute temperature—in kelvin* (K), see Fig. 9.3. In mathematical terms, this law allows the calculation of the amount of radiating energy Q_R (in W = J/s) gained (+) or lost (−) by the human body:

Fig. 9.1 Heat loss through radiation. (adapted from Kroemer and Kroemer 2005)

230 Chapter 9 Thermal Environment

Fig. 9.2 Heat gain through radiation (adapted from Kroemer and Kroemer 2005)

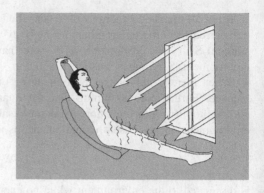

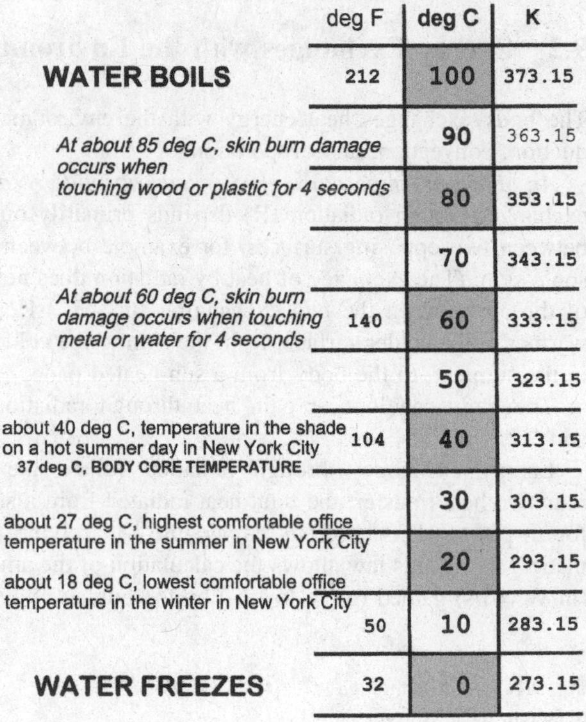

Fig. 9.3 Temperature scales in common use

	deg F	deg C	K
WATER BOILS	212	**100**	373.15
		90	363.15
At about 85 deg C, skin burn damage occurs when touching wood or plastic for 4 seconds		80	353.15
		70	343.15
At about 60 deg C, skin burn damage occurs when touching metal or water for 4 seconds	140	**60**	333.15
		50	323.15
about 40 deg C, temperature in the shade on a hot summer day in New York City 37 deg C, BODY CORE TEMPERATURE	104	**40**	313.15
		30	303.15
about 27 deg C, highest comfortable office temperature in the summer in New York City		**20**	293.15
about 18 deg C, lowest comfortable office temperature in the winter in New York City			
	50	**10**	283.15
WATER FREEZES	32	**0**	273.15

$$Q_R = a\,S\left(dT_O^4 - eT^4\right).$$

a   Stefan-Boltzmann Radiation Constant in $J/(sm^2\,K^4)$
S   body surface area participating in the energy exchange, in m^2.
d   absorption coefficient of the receiving surface—see below.
T_O temperature of the receiving surface, in K.
e   emission coefficient of the emitting surface—see below.
T   body surface temperature of the emitting surface, in K.

The absorption coefficient d depends on skin color; for solar rays (with wave-lengths $0.3 < \lambda < 4$ µm) it ranges from 0.6 for light skinned people to 0.8 for dark skinned persons. The wave lengths of radiation emitted from the human body are in the infrared range, $3 < \lambda < 60$ µm. Hence, it radiates like a black body, with an emission coefficient e close to 1, independent of the actual color of the radiating human skin. An overall estimate for energy transfer through radiation is about 4.5 W/m^2 of surface and degree of temperature difference*.

9.2.1 Conduction Heat Exchange

Conduction is heat transfer by molecular contact. Energy exchanges through both conduction and convection (see below) follow Newton's law of cooling: the amount of heat transferred is proportional to the area of human skin participating in the process, and to the temperature difference between skin and the adjacent layer of the external medium.

Heat exchange through conductance (K) exists when skin contacts a solid body. Energy flows from the warmer body to the colder one; as the temperatures of the contact surfaces become equal, the energy exchange ceases.

The amount of energy exchanged by conduction is

$$Q_K = h\,S(t_m - t).$$

h thermal conductivity coefficient- see below.
S body surface area participating in the heat exchange, in m^2.
t_m temperature of the medium with which S is in contact, in °C.
t temperature of the body surface S, in °C.

The rate and amount of heat exchanged depend on the ability of the contacting material to transfer heat, as indicated by the thermal conductivity coefficient. At the same low temperature, wood or cork "feel warmer" than metal because their heat conductivity is low whereas the metal accepts body heat easily and conducts it away. The quantity of exchanged energy also depends on the tightness with which the bodies touch. Insulating material between skin and object affects the amount of energy transferred by conduction; more on this below in the discussion of clothing insulation.

9.2.2 Convection Heat Exchange

Convection is the transfer of heat by circulation of particles of a gas or liquid. Exchange of heat through convection (C) takes place when the human skin is in contact with air or with a fluid, such as water. Heat energy migrates from the skin to a layer of colder gas or fluid next to the skin; heat passes to the skin if the surrounding medium is warmer.

As long as the temperatures of the skin and the surround differ, there is some natural movement in the air or fluid: this is called free convection. As the medium moves along the skin surface, it removes the boundary layer whose temperature has become close to skin temperature; so, relative motion helps to maintain a temperature differential that facilitates convective heat exchange. Forced action (by wind or an air fan, for example, or while swimming in water rather than floating motionless) can increase the movement: this induced convection enlarges energy transfer.

The process of exchanging heat via convection Q_C (gain indicated by +, loss by –) is similar to conduction:

$$Q_C = c\ S(t_m - t).$$

In air, the convective heat transfer coefficient c_a is between 2 and 25 W/(m^2 °C), depending on the relative movement of the medium*. Forced convection in air can be as high as 250 W/(m^2 °C); an example is "wind chill", discussed below. For the nude body in still water, the convective transfer coefficient c_w is about 230 W/(m^2 °C) for $t_m < t$ when the body is at rest; the value can more than double when the human swims at any speed because of the turbulence of the water layer near the body produced by the swimming motions*.

9.2.3 Evaporation Heat Exchange

Evaporation is the change of a liquid (here, water) to vapor; this phase change requires heat. As evaporation (E) occurs on the skin or in the airways, the energy is extracted from the body, which makes it cooler. For the human, heat exchange by *evaporation* E is only in one direction: the body loses heat. (There is no condensation of water on the body, which would add heat.) Evaporation of water (sweat) on warm skin requires an energy of about 2427 J/cm^3 (580 cal/cm^3), which reduces the heat content of the body by that amount.

The heat lost by evaporation Q_E from the human body is a function of participating wet body surface, humidity and vapor pressures*.

$$Q_E \sim h_r S(p_a - p)$$

h_r relative humidity of the surrounding air
S   body surface area participating in the heat dispersion
p_a vapor pressure in the surrounding air
p   vapor pressure at the skin.

Of course, Q_E is zero for $p \leq p_a$ since heat loss by evaporation can only exist if the surrounding air is less humid than the air directly at the skin. Therefore, movement of the air layer at the skin (see convection) increases the actual heat loss through evaporation if this replaces humid air by dryer air.

Some evaporative heat loss occurs even in a cold environment because there is always evaporation of water in the warm airways. The evaporation rate increases with enlarged ventilation at heavier work. Also, secretion of sweat onto the skin surface occurs in physical work even in cold environs. The nude body loses, at rest, in the cold 3–6 W/m^2 from the respiratory tract and, at work, up to 10 W/m^2 from the skin*.

9.3 Heat Balance

Heat balance between the body and its surrounds exists when the heat H developed in the body, change in heat storage S in the body, and heat exchanged with the environment by radiation R, conduction K, convection C, and evaporation E are in equilibrium:

$$H + S + R + K + C + E = 0 \, (\text{zero}).$$

The quantities S, R, K, and C are counted as positive if the body gains energy, negative if the body loses energy. The quantity H is always positive, quantity E can only be negative.

Figure 9.4 indicates schematically how the different kinds of heat transfer affect the body. In general, heat loss by radiation diminishes as the environment gets warmer while heat loss by evaporation increases.

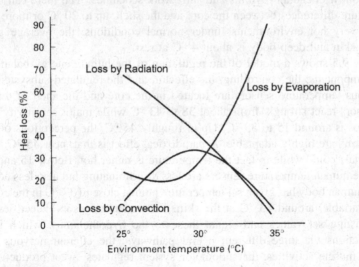

Fig. 9.4 Cooling the body in a warm environment Adapted from Kroemer et al. (2003), Kroemer Elbert et al. (2018)

9.4   Regulation and Sensation of Body Temperature

The body produces heat in its "metabolically active" tissues: primarily at skeletal muscles, but also in internal organs, fat, bone, connective and nerve tissues.

Heat energy is circulated throughout the body by the blood. The heart generates the flow of blood and controls its pressure and volume per time. The vasomotor actions of constriction, dilation, and shunting modulate the local flow of blood—see Chap. 6.

Heat exchange with the environment takes place mostly through the skin and at the body's respiratory surfaces.

In a cold environment, the body must conserve heat. This is primarily done by cooling the skin via (unconscious) reduction of blood flow to the skin and by increasing insulation via clothing.

In a hot environment, the body must dissipate heat and also prevent heat gain from the environment. This is mainly done by warming the skin via (unconsciously) increasing blood flow to the skin, by sweat production and evaporation, and by choosing suitable clothing.

The body attempts to prevent undercooling (*hypothermia*) or overheating (*hyperthermia*). The control system strives to keep the temperature of the body core constant, close to 37 °C, with only slight variations throughout the day which mostly follow circadian rhythms and shift work schedules. Yet, there can be large temperature differences between the core and the shell, up to 20 °C or more in very cold or very hot environments; under normal conditions, the average gradient between skin and deep body is about 4 °C at rest.

Figure 9.5 shows a model of the regulation of the human energy balance with three components: the controlling, the effector, and the regulated subsystems.

Various temperature sensors are located in the core and the shell of the body. Hot sensors react strongly from about 38 to 43 °C while major sensitivity to cold conditions is around 15 to 35 °C. Up to roughly 45 °C, the perceptions of "cool" and "warm" are highly adaptable. A paradoxical effect is that, near 45 °C, sensors may signal "cold" while in fact the temperature is rather hot. Below 15 and above 45 °C the human temperature sensors are less discriminating but also less adapting.

The human body has given set temperature points: close to 37 °C in the core and, highly variable, around 33 °C at the skin. Temperature sensors detect deviations from existing set values and signal these to the hypothalamus, which initiates counteractions via three different neural pathways: the efferent nervous system changes muscle activities, the sudomotor system regulates sweat production, and the vasomotor system controls blood flow.

Muscular activities can generate only more or less heat but cannot cool the body. So, if internal heat production must be diminished, muscular activities will be reduced, possibly to the extent that no work is being performed anymore.

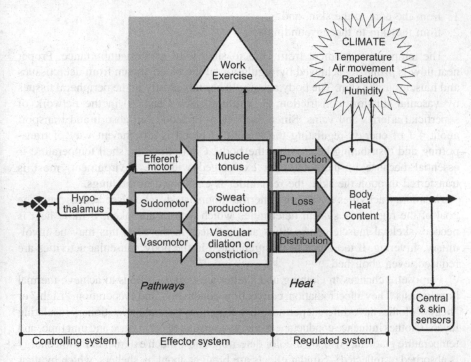

Fig. 9.5 Model of the regulation of body heat content

Conversely, if more heat must be generated, the work or exercise level will be augmented by increased muscular activities. (Given the low efficiency of muscular work, it generates much heat—see Chap. 7.)

In contrast, sweat production only influences the amount of energy lost but cannot bring about a heat gain. Vascular activities can affect the heat distribution through the body and control heat loss or gain, but they do not generate energy.

Muscular, vascular, and sweat production functions regulate the body heat content in direct interaction with the external climate. The climate itself is defined by humidity, radiation, temperature, and air (or fluid) movement, as discussed later.

9.4.1 Achieving Thermal Homeostasis

The body seeks to avoid overheating or undercooling of the core organs in the brain and the trunk, even at the cost of overheating or undercooling the shell. To assure proper temperature at the core, the human regulatory system must generate two suitable temperature gradients:

1. from the core to the skin, and
2. from the skin to the surroundings.

The temperature gradient from core to shell is of greatest importance. Proper heat flow is *primarily* achieved by regulation of the blood stream from deep tissues and muscles to the skin. The body accomplishes this mostly in the peripheral tissues by vascular dilation, constriction, or shunting (see Chap. 6) in the network of superficial arteries and veins. Since each gram of blood can absorb and transport about 4 J of energy, regulating the stream of blood is an efficient way of transporting and distributing heat within the body. Controlling the shell temperature is essential because of the total heat exchanged with the environment, most is transferred through the skin; the remainder is exchanged in the lungs.

Secondary activities to establish thermal homeostasis take place at muscles. The goal of the regulatory system determines which actions are taken: if more heat is needed, skeletal muscle contractions are initiated; in the cold, this may be involuntary shivering. If too much heat is generated in the body, muscular activities are reduced, even abolished.

Purposeful changes in clothing and shelter are *tertiary* actions to achieve thermal homeostasis. They affect radiation, convection, conduction, and evaporation. "Light" or "heavy" clothes have different permeability and ability to establish stationary insulating layers. Clothes influence conductance, energy transmitted per surface and unit time, and temperature gradient. Also, their color determines how much external radiation energy is absorbed or reflected*. Similar effects are brought about by shelters, which by their material, distance from the body, form, and color determine whether heat is gained or lost by the body through radiation, convection, and evaporation.

9.5 Measurements of Body Temperatures

One way of assessing the exchange of heat energy with the environment is to perform direct calorimetry, where a person is placed into an energy-tight compartment which allows the measurement of all heat energies exchanged—see Chap. 8. However, this is a tedious procedure which severely limits the ability of the person to work "normally".

Most methods to assess the heat balance are based on temperature measurements within or at the body. The so-called safe temperature range of the core is between 35 to 40 °C. Below 33 and above 42 °C, serious impairment of functions occur, especially in the brain and heart. Changes in core temperatures, particularly severe deviations from the "safe" temperatures, indicate work and environmental overloads or dysfunctions of the energy regulatory mechanisms in the body.

A variety of techniques exist to measure the temperature in body parts. A traditional site is the rectum, where temperature probes are usually inserted 5 to 10 cm behind the sphincter. Compared to the temperature in the hypothalamus, the rectal temperature is usually about 0.5 °C lower, provided that a steady state has been maintained for about 30 min. The rectal temperature of a resting individual is slightly higher than the temperature of arterial blood and about the same as in the liver. Brain temperature rises more quickly in response to heat influx than does rectal temperature.

Temperatures measured at the ear drum follow actual brain temperatures rather closely. Inserting a temperature probe into the esophagus or the stomach allows measurement of deep body temperatures.

Temperature measurements in the mouth, in the armpit or on the skin are less accurate but rather easily done and socially more acceptable.

Skin temperatures may be very different from the core temperature; they can vary substantially by location and over time, such as during exposure to heat or cold. Under cold conditions, with the core temperature maintained at approximately 37 °C, the trunk may have skin temperatures of about 36 °C; the thighs may be at 34 °C, the upper arms and knees at 32 °C, the lower arms at 28 °C, while the toes and fingers may be at 25 °C; such a combination can be perceived as quite comfortable.

Proper clothing can avoid chilling of body segments below acceptable temperatures. Obviously, fingers and feet need special protection in cold conditions.

The temperatures of the neck and head do not vary much. Here, the "core" is close to the shell and very little volume below the skin is available for vasomotor and sudomotor functions. To keep the core warm, skin temperature must be maintained at a rather constant and high level.

Measurement of temperatures at skin sites other than on neck and head provides information that is only loosely related to core temperatures. Hence, the concept of an "average skin temperature" is a difficult one; nevertheless, it has been employed by assigning weighting factors to the measurements taken at various body surfaces, depending on the locations and proportions of these surface areas compared to the total body. For example, the well-established Hardy and Dubois procedure is to multiply measurements taken at the head by 0.07, the arms by 0.14, the hands by 0.05, the trunk by 0.35, the thighs by 0.19, the lower legs by 0.13, and at the feet by 0.07. The results are summed for an "average" skin temperature*.

Infrared thermography or scanning radiometers are among the advanced techniques to measure skin temperatures at specific points. However, in order to establish general concepts of (average, suitable, dangerous ...) temperature for the skin (or of body, core, organ...), the underlying problems of definition, sampling and statistical treatment need to be addressed. For example, one approach to estimating the body's heat content is to presume an "average" skin temperature t_s, to measure the rectal temperature t_r and the body mass m in kg, then the Burton equation yields

$$\text{Heat content} = 3.47 \, m(0.65 t_r + 0.35 t_s).$$

In this equation the specific heat of the body is assumed to be 3.47 kJ/(kg °C), but that value may vary considerably, depending on individual body composition. Also, the ratios of t_r and t_s are assumed to be 0.65 and 0.35, respectively, but they can range from 0.9 and 0.1 in warm environments and during exercise, and vary between 0.6 and 0.4 at rest in a cool environment*.

9.6   Assessing the Thermal Environment

The thermal environment is a combination of four physical factors:

1. ambient air (or water) temperature,
2. ambient humidity,
3. air (or water) movement, and
4. temperatures of surfaces that exchange heat by radiation.

9.6.1   Ambient Temperature

Measurement of ambient temperature is traditionally performed with thermometers, usually filled with alcohol or mercury; thermistors, thermocouples and other techniques can be used as well. In any case, it must be ensured that other climate factors (humidity, air movement, and heat radiation) do not affect the measure of the ambient air temperature. The usual way is to keep the sensor dry and shield it in a reflecting bulb from radiated energy. Hence, air temperature is commonly measured with a so-called dry bulb thermometer and often termed dry bulb temperature or, simply, dry temperature.

9.6.2   Air Humidity

Air humidity, the content of water vapor in air, can be measured with a hygrometer: originally a human or horse hair that changes its length with wetness, now an instrument whose electrical conductivity (resistance or capacitance) changes with the existing humidity. Another type of hygrometer is the psychrometer, which consists of one wet-bulb and one dry-bulb thermometer; evaporation cools the wet thermometer more than the dry one. The difference is proportional to the humidity of the air, as higher vapor pressure reduces evaporative cooling. A hygrometer is called natural if there is no artificial air movement around it, forced if there is.

Air humidity may be expressed either in absolute or in relative terms. Saturation, the highest absolute content of vapor in the air, is reached when any further increase would lead to the development of water droplets, falling out of the gas. That so-called dew point depends on air temperature and barometric pressure: higher temperature and pressure allow more water vapor to be retained than lower conditions. However, one mostly speaks of relative humidity, which indicates the actual vapor content (usually in percent) in relation to the possible maximal content at the given air temperature and air pressure.

9.6.3 Air Movement

Air movement is measured with various types of anemometers, usually based on electrical or mechanical principles, such as a windmill. One may also measure air movement with two thermometers, one wet and one dry (similar to what can be done to assess humidity), relying on the fact that the wet-bulb thermometer shows more increased evaporative cooling with higher air movement than the dry-bulb thermometer. Air movement helps particularly in convective heat exchange because it moves "fresh" (less moist) air to skin surfaces. Here, turbulent air movement is as effective as laminar movement in heat transfer. (See below for the discussion of "wind chill".)

9.6.4 Radiant Heat

Radiant heat exchange depends primarily on the difference in temperatures between the individual and the surroundings; it also depends on the emission properties of the radiating surface and on the absorption characteristics of the receiving surface. While there is no problem in measuring surface temperatures (which are the major factors in human-environment energy transmission by radiation), one easy and direct way to assess the amount of energy transferred is to place a thermometer inside a black globe, which absorbs practically all radiated heat energy.

9.6.5 The Combined Effects of Climate Factors

Several approaches to the concept of an "effective temperature" scale* have been devised to express the combined effects of the four environmental factors in one model, chart, or index. The most often used index is the Wet Bulb Globe Temperature (WBGT) which weighs the effects of all climate parameters with sufficient validity, reliability, and usability for most warm and hot environments. (There is some concern, however, about the adequacy of the WBGT for extreme conditions, such as in a hot desert, and for combinations of high humidity with little air movement*.) The WBGT index exists in two versions: one applied to outdoors conditions, the other indoors. The WBGT requires measurement of

- WB: natural *Wet Bulb* temperature of a sensor in a wet wick exposed to air;
- GT: *Globe Temperature* at the center of a black sphere of 15 cm diameter; and
- DB: *Dry Bulb* temperature measured with a sensor shielded from radiation.

Outdoors, the WBGT considers the effect of heat radiation, usually of solar origin (this formula also applies to a sunlit person indoors):

$$WBGT(\text{outdoors}) = 0.7\,WB + 0.2\,GT + 0.1\,DB$$

Indoors, without intensive radiated heat, solar or other, a simpler version applies which does not require a DB measurement:

$$WBGT(\text{indoors}) = 0.7\,WB + 0.3\,GT$$

Professional and regulatory agencies provide WBGT recommendations for "safe" work in hot environments*. Table 9.1 lists recommended upper WBGT temperatures for US workers of 70 kg body weight, acclimatized and wearing suitable clothing.

Table 9.1 "Safe" WBGT values for US workers. The WBGTs of the workplace and of the restarea are assumed to be similar (adapted from OSHA 1999, 2016)

Hourly work / rest ratio	Intensity of work (metabolic rate in Watt)		
	Light, <230 W	moderate, 230 – 400 W	heavy, >400 W
Continuous work	30.0	26.7	25.0
75% work, 25% rest	30.6	28.0	25.9
50% work, 50% rest	31.4	29.4	27.9
25% work, 75% rest	32.2	31.1	30.0

9.7 Reactions of the Body to Hot Environments

In hot environments, the body produces heat which must be dissipated by convection, conduction, radiation and evaporation. The body tries to achieve the difficult task of dispelling heat into a warm environment primarily by regulating its own blood distribution and metabolic rate.

9.7.1 Redistribution of Blood

The easiest way to accomplish outward heat energy flow is to have the skin temperature above the temperature of the immediate environment. To achieve this, the body directs blood flow to the skin: the skin vessels are dilated and the superficial veins fully opened. This can enlarge the blood flow fourfold above the resting level. The increased conductance of surface tissues facilitates energy loss through radiation, conduction, and convection because all are proportional to the temperature differential between skin and environment.

However, in a hot environment it is difficult to increase the skin temperature above the ambient temperature. If not enough heat can be transferred via a temperature differential, the body's sudomotor system activates sweat glands so that evaporation of the produced sweat may cool the skin. Recruitment of sweat glands from different areas of the body varies among individuals. Large differences in the ability to sweat exist among individuals: most have at least 2 million sweat glands in the skin, but others have fewer. The overall amount of sweat developed and evaporated depends very much on clothing, environment, work requirements, and on the individual's acclimatization.

9.7.2 Reduction of Muscle Activities

If heat transfer to a hot environment by blood distribution and sweat evaporation proves insufficient to keep the body sufficiently cool, the body must reduce its muscular activities in order to diminish the amount of energy generated through metabolic processes. This is the final and necessary action of the body if otherwise the core temperature would exceed a tolerable limit. When the body has to choose between unacceptable overheating and continuing to perform physical work, the choice will be in favor of core temperature maintenance, which means reduction or cessation of work or exercise.

9.7.3 Indications of Heat Strain

There are several signs of excessive heat strain on the body. The first one is the sweat rate. The normal so-called insensible perspiration is about 50 mL/h. Sweat production increases depending on the heat that must be dissipated. In strenuous exercises and hot climates, several liters of sweat may be produced in one hour; sweat losses of up to 12 L in 24 h have been reported under extreme conditions.

Sweat begins to drip off the skin when the sweat generation has reached about one-third of the maximal evaporative capacity. Of course, sweat running down the skin, instead of being evaporated, contributes very little to heat transfer.

Other signals of heat strain include increased circulatory activities. To boost blood flow, cardiac output must be enlarged, mostly brought about by higher heart rate. This may be associated with a reduction in systolic blood pressure.

The water balance within the body provides another sign of heat strain. Dehydration indicated by the loss of only one or two percent of body weight can critically affect the ability of the body to control its functions. Therefore, the fluid level must be maintained, best by frequently drinking small amounts of water.

Sweating, which extracts water from the plasma, augments the relative salt content of the blood, but sweating also diminishes the overall salt content of the body because sweat contains salts, particularly NaCl. Normally, with western diets it is not necessary to add salt to drinking water since the salt in the food is sufficient to resupply salt lost with the sweat.

Among the first reactions to heavy exercise in excessive heat are sensations of discomfort and perhaps skin eruptions ("prickly heat" rash) associated with sweating. As a result of sweating, so-called heat cramps may develop, which are muscle spasms related to local lack of salt. They may also occur after quickly drinking large amounts of fluid.

Heat exhaustion usually results from dehydration and overloading the circulatory system. Associated effects are fatigue, headache, nausea, dizziness, often accompanied by giddy behavior. Heat syncope signals a failure of the circulatory system, demonstrated by fainting. Heat stroke indicates an overloading of both the circulatory and sweating systems and is associated with hot dry skin, increased core temperature, and mental confusion. Table 9.2 lists symptoms, causes, and treatment of heat stress disorders*.

Table 9.2 Heat stress disorders (adapted from OSHA 1999, 2016)

	Symptoms	Causes	Treatments
Transient heat fatigue	Decreases in productivity, alertness, coordination and vigilance.	Not acclimatized to hot environment.	Gradual adjustment to hot environment.
Heat rash, "prickly heat"	Skin rash in areas of heavy perspiration; discomfort or temporary disability.	Perspiration not removed from skin, possibly inflamed sweat glands.	Periodic rests in cooler area; shower; dry skin.
Heat collapse, "fainting"	Blackout; collapse.	Shortage of oxygen in the brain	Lay down, cool down.
Heat cramps	Painful spasms of skeletal muscles.	Loss or excess of salt; large quantity of water drunk quickly.	Adequate salt with meals; salted liquids if recommended by a physician.
Heat exhaustion	Extreme weakness or fatigue; thirst; giddiness; nausea; headache; pale or flushed complexion; body temperature normal or slightly higher; moist skin; in extreme cases vomiting and/or loss of consciousness.	Loss of water and/or salt; loss of blood plasma; strain on the circulatory system.	Rest in cooler area; salted liquids if recommended by a physician.
Heat stroke	Skin hot, usually dry and often red or spotted; core temperature 41 °C or higher and rising; irrational behavior; mental confusion; deliriousness; convulsions; unconsciousness possible. Death or permanent brain damage may result unless treated immediately.	Breakdown of the thermo-regulatory system; stoppage of sweating. The body's ability to remove excess heat is almost eliminated.	Remove to cool area; soak clothing with cold water; fan body; *get medical treatment immediately*.

9.7.4 Acclimatization to Heat

Continuous or repeated exposure to hot conditions brings about a gradual adjust-ment of body functions resulting in a better tolerance of the climatic stress and in improvement of physical work capabilities.

The process of acclimatization to heat shows increased sweat production, low-ered skin and core temperatures as well as reduced heart rate, compared with the reactions of the person at first exposure to the hot climate. Within about a week, acclimation is pronounced and, after about two weeks, full acclimatization is achieved. Interruption of heat exposure of just a few days reduces the lingering effects of acclimatization, which is entirely lost within about two weeks after return to a moderate climate.

Specifically, heat acclimatization is brought about by improved control of the vascular flow, by an augmented stroke volume accompanied by a reduced heart rate, and by higher sweat production. The improvement in sudomotor action is most prominent. Perspiration on the face and the feeling of "sweating" diminish with heat acclimation even though total sweat production may have doubled after several days of exposure to the hot environment. More volume and better regulation of sweat distribution are the primary means of the human body to bring about dissi-pation of metabolic heat.

Vasomotor improvements intertwine with sudomotor advancement. The reduced skin temperature (lowered through sweating) allows a redistribution of the blood flow away from the skin surfaces, which need more blood during initial exposure to heat. Acclimation re-establishes normal blood distribution within a week or two. Cardiac output must remain rather constant even during initial heat exposure, when an increase in heart rate and a reduction in stroke volume occur. Both rate and volume are reciprocally adjusted during acclimation since arterial blood pressure remains essentially unaltered. There may also be a (relatively small) change in total blood volume during acclimation, particularly an increase of plasma volume during the first phase of adjustment to heat.

A healthy and well-trained person acclimates more easily then somebody in poor condition, but training cannot replace acclimatization. If strenuous physical work must be performed in a hot climate, then such work should be part of the accli-mation period.

On average, the female body has smaller mass than the male, meaning a smaller heat "sink". Women usually have relatively more body fat and accordingly less lean body mass than men; their surface area is smaller, and their blood volume is smaller as well. Nevertheless, there appear to be no appreciable differences between females and males with respect to their ability to adapt to a hot climate.

Adjustment to heat will take place whether the climate is hot and dry, or hot and humid. Acclimatization seems to be unaffected by the type of work performed, either heavy and short or moderate but continuous. It is important to replace fluid (and possibly salt) losses during acclimation and throughout heat exposure.

9.8   Reactions of the Body to Cold Environments

In a cold environment, the body must conserve its produced heat. The most effective actions are behavioral: putting on suitable clothing to cover the skin, seeking shelter, or using external sources of warmth. The human body has only limited natural means to regulate its temperature in response to a cold surround: these are mainly re-distribution of the blood flow and increasing metabolic rates.

9.8.1   Redistribution of Blood

To conserve heat, the body lowers the temperature of its skin: this reduces the temperature difference against the cold outside and, consequently, decreases the outflow of heat. Lowering the skin temperature is done by displacing the circulating blood from the periphery towards the core, away from the skin. This can be rather dramatic; for example, the blood flow in the fingers may be reduced to a very small percentage of what existed in a moderate climate.

Of the total circulating blood volume in the body, around 5 L, normally about 5% flows through the blood vessels in the skin. The body employs several procedures to regulate the distribution of blood, apparently under the control of the sympathetic nervous system (see Chap. 3), in addition to local reflex reactions to direct cold stimuli. One way to control blood flow is cutaneous vasoconstriction, which cuts off many of the blood vessel pathways; so, less blood flows towards the superficial skin surfaces. At the same time, the body re-routes blood from the superficial veins of the extremities near the skin to deep veins. Many deep veins are anatomically close to arteries, which carry warm blood from the heart. Heat exchange between the cooler blood in the veins and the warmer blood in the arteries has two effects: cooling of the body core is reduced because of the venous return with warmed blood, and cooled arterial blood supplying the extremities and skin keeps them cooler, which reduces the conductance of the surface tissues and diminishes the energy flow towards the environment.

An interesting phenomenon associated with exposure to cold is the "hunting reaction": after initial cutaneous vasoconstriction has taken place, suddenly a dilation of blood vessels can occur which allows warm blood to return to the skin, often the hands, which re-warms that body section. Then vasoconstriction returns again, and this sequence may be repeated several times.

The body's automatic reactions to a cold environment demonstrate the over-riding need to keep the core temperature sufficiently high. Displacing blood volume from the skin to more central circulation is very efficient in keeping the core warm and the surfaces cold. Peripheral vasoconstriction can bring about a six-fold increase in the insulating capacity of the subcutaneous tissues. The associated danger is that the temperature in the peripheral tissues may approach that of the environment. Cold fingers and toes are mostly just unpleasant but serious damage is possible if tissue temperature gets close to freezing.

The blood vessels of neck and head do not undergo much vasoconstriction so they stay warm even in cold environments, with less danger to the tissues. However, the resulting large temperature difference to the environment brings about a large heat loss, which can be prevented by wearing a helmet, cap, scarf and the like to create an insulating layer.

Incidentally, the development of "goose bumps" of the skin helps to retain a layer of stationary air close to the skin, which is relatively warm and has the effect of an insulating envelope, reducing convective energy loss at the skin.

Heat loss by convection increases when the air moves more swiftly along exposed skin surfaces. In the 1940s, the US Army performed experiments to determine the effects of cold air movement on the cooling of water and how subjects perceived such "wind chill" on their exposed skin: this led to tables of so-called wind chill temperatures. Around the turn of the century, Canadian and American scientists performed more realistic investigations on convective energy loss through exposed skin depending on air temperature and velocity.

Table 9.3 lists the metric wind chill temperature equivalents which reflect the effects of wind flow at various temperatures. These wind chill temperatures are based on the cooling of naked skin, not on the cooling of a clothed person. Also, these numbers do not take into account air humidity and solar radiation.

Table 9.3 shows, for example, that at zero-degree air temperature, a wind of 5 km/h cools exposed skin in the same way as calm air of -2 °C would do; if the wind increases to 50 km/h, it cools exposed skin as calm air at a temperature of -8 °C would do. If the actual air temperature is -15°, a 50 km/h wind would cool skin like calm air at a temperature of -29°; if the temperature is actually -35°, the same 50 km/h wind cools skin as a calm air temperature of -56° would do and frostbite (on uncovered skin) can occur in as little as three minutes.

9.8.2   Increased Metabolic Heat Production

The other major body reaction to a cold environment is to increase the generation of metabolic heat. Often, first the overall muscle tone increases in response to body cooling. The enlarged firing rates of motor units (see Chap. 2) without generating

Table 9.3 "Equivalent windchill temperatures" for naked skin depending on air temperature and air velocity

Wind Calm km/h	ACTUAL AIR TEMPERATURE °C												
	10	5	0	-5	-10	-15	-20	-25	-30	-35	-40	-45	-50
	EQUIVALENT WINDCHILL TEMPERATURE °C												
5	9	4	-2	-7	-13	-19	-24	-30	-36	-41	-47	-53	-58
10	9	3	-3	-10	-15	-21	-27	-33	-39	-45	-51	-57	-63
15	8	2	-4	-11	-17	-23	-29	-35	-41	-48	-51	-60	-66
20	7	1	-5	-12	-18	-24	-31	-37	-43	-49	-56	-62	-68
25	7	1	-6	-12	-19	-25	-32	-38	-45	-51	-57	-64	-70
30	7	0	-7	-13	-20	-26	-33	-39	-46	-52	-59	-65	-72
35	6	0	-7	-14	-20	-27	-33	-40	-47	-53	-60	-66	-73
40	6	-1	-7	-14	-21	-27	-34	-41	-48	-54	-61	-68	-74
45	6	-1	-8	-15	-21	-28	-35	-42	-48	-55	-62	-69	-75
50	6	-1	-8	-15	-22	-29	-35	-42	-49	-56	-63	-70	-76
55	5	-2	-9	-15	-22	-29	-36	-43	-50	-57	-63	-70	-77
60	5	-2	-9	-16	-23	-30	-37	-43	-50	-57	-64	-71	-78

Tissue may freeze after exposure longer than 30 min	<30 min	Freezing risk in 10 minutes	Freezing risk in 3 minutes

actual movements cause a feeling of stiffness. Then, suddenly shivering (thermogenesis) begins, caused by muscle units firing at different frequencies of repetition (rate coding) and out of phase with each other (recruitment coding). Since no mechanical work is done to the outside, the entire effort is transformed into heat production, allowing an increase in the metabolic rate to up to 4 times the resting rate. If the body does not become warm, shivering may become violent when motor unit innervations synchronize so that whole muscles are involved. While such shivering may generate heat that is five or more times the resting metabolic rate, it can be maintained only for a short period at a time.

Of course, muscular activities also can be done voluntarily, either by increasing the dynamic muscular work already in progress or by contracting additional muscles, moving body other segments such flexing the fingers. Since the energy efficiency of the body is very low (see Chap. 7), dynamic muscular work may easily increase the generation of metabolic heat to ten times or more over that at rest.

9.8.3   How Cold Does It Feel?

In a cold environment, an individual's decision whether or not to stay in the cold depends on the person's assessment of how cold body surfaces or the body core actually are. It creates a dangerous situation if a person either fails to perceive or fails to react to the body's signals that it is becoming too cold or, worse, if the body temperature becomes so low that further cooling is below the threshold of perception.

The perception of the body getting cold depends upon signals received from surface thermal receptors, from sensors in the body core, and from a combination of these signals. As skin temperatures decrease below about 36 °C, the intensity of the cold sensation increases; cold sensation is strongest near 20 °C. However, at lower temperatures, the perception of cold diminishes. It is often difficult to separate feelings of cold from pain and discomfort.

The conditions of cold exposure may greatly influence the perceived coldness. It can make quite a difference whether one is exposed to cold air (with or without movement) or to cold water, whether or not one wears protective clothing, and what one is actually doing. When the temperature plunges, each downward step can generate an "overshoot" sensation of cold sensor receptors which react very quickly not only to the difference in temperature, but also to the rate of change. Yet, if the temperature stabilizes, cold sensations become smaller as one adapts to the condition.

Exposure to very cold water accentuates the overshoot phenomenon observed in cold air. This may be due to the fact that, at the same low temperature, such as 0 °C, the thermal conductivity of water and hence the convective heat loss are about twenty-five times greater than in air. In experiments, subjects (wearing a flotation suit) were immersed into cold water of 10 °C; their temperatures at groin, back and rectum were continuously recorded, and the subjects rated how cold they perceived these areas to be. The results of the experiment showed that the subjects were unable to reliably assess how cold they actually were; neither their core nor surface temperatures correlated with their cold sensations*.

Altogether, the results of many experiments and experiences indicate that the subjective sensation of cold is a poor, even dangerously unreliable indicator of core and surface temperature of the body. Measuring ambient temperature, humidity, and air movement and exposure time and reacting to these physical measures is a better strategy than relying on subjective sensations.

9.8.4   Indications of Cold Strain

If vasoconstriction and metabolic rate regulation cannot prevent serious heat energy loss through the body surfaces, the body will suffer some effects of cold stress. The skin is, as mentioned, first subjected to cold damage while the body core is protected as long as possible.

As skin temperature diminishes to about 15–20 °C, manual dexterity begins to fail. Tactile sensitivity is severely diminished as the skin temperature falls below 8 °C. At local temperatures of 8–10 °C, peripheral motor nerve velocity is decreased to near zero; this generates a nervous block, so local cooling is accompanied by rapid onset of physical impairment. If tissue temperature approaches freezing, ice crystals develop in the cells and destroy them, a result known as frostbite.

Strong reductions in skin temperatures cause a fall in core temperature. Severe decrease of core temperature is dangerous. Vigilance begins to drop at core temperatures below 36 °C. At core temperatures of 35 °C, one may not be able to perform even simple activities. When the core temperature falls even more, the mind becomes confused, with loss of consciousness occurring around 32 °C. At core temperatures of about 26 °C, heart failure may occur. At very low core temperatures, such as 20 °C, vital signs disappear, but the oxygen supply to the brain may still be sufficient to allow revival of the body from this stage of hypothermia.

Severe cooling of skin and central body makes it difficult or impossible to perform activities, even if they could save the person ("cannot light a fire") leading to apathy ("let me sleep") and finally profound hypothermia and death.

In cold water, hypothermia can occur quickly. A person can endure up to 2 h in water at 15 °C, but becomes helpless in water of 5 °C in about half an hour; then death is imminent. Obese persons with more insulating adipose tissue are at advantage over slender ones. The survival time in cold water can be increased by clothing which provides some insulation. Floating motionless results in less metabolic energy generated and spent than when swimming vigorously; furthermore, the swimming motions destroy the stationary layer of warmed water close to the body which, while intact, reduces heat loss by conduction.

Table 9.4 lists the primary reactions of the body to cold and warm environments.

9.8.5   Acclimatization to Cold

Acclimatization to a cold environment is not very pronounced. In fact, it is debatable that true physiological adjustments to moderate cold take place when appropriate clothing is worn, because the actual cold exposure of the body is slight and does not require appreciable increases in metabolism or other major adjustments in vasomotor or sudomotor systems.

Table 9.4 Summary of the body's main thermoregulatory actions

Environment	Purpose	Means
Cold	Prevent heat loss	Reduce skin temperature
Hot	Avoid heat gain, Facilitate heat loss	Increase skin temperature, Increase sweat production
Cold	Increase body heat content	Increase muscle activities
Hot	Avoid increase in body heat content	Reduce muscle activities

When first exposed to cold temperature, some changes in local blood flow appear, particularly increased blood flow through the hands or in the face. In laboratory exposure to extremely cold conditions, with little shelter offered by clothing, some hormonal and other changes have been observed.

On average, the female body has smaller mass (heat "sink") than the male; women usually have relatively more body fat and accordingly less lean body mass; their surface area and their blood volume are smaller as well. Under cold stress, females have slightly colder temperatures at their (thinner) extremities but show no difference in core temperature. While women possibly have a slightly higher risk of cold injuries to the extremities, altogether there are no great differences between females and males with respect to their ability to adapt to cold climates. Any slight statistical tendencies can be counteracted easily by ergonomic means.

9.9   Working in Heat or Cold

Hot and cold climatic conditions (as well as air pollution and high altitude)* affect persons' abilities to perform short or long, light or heavy work in various ways.

9.9.1   Effects of Heat

When exposed to whole-body heating, the human body must maintain its core temperature near 37 °C. It does so by raising its skin temperature, increasing blood flow to the skin, accelerating heart rate and enlarging cardiac output. The change in blood routing reduces the blood that can be supplied to muscles and internal organs. Yet, if muscles must perform work, their raised metabolism poses increased demands on the cardiovascular system.

9.9.1.1   Cardiovascular Effects

The pumping capacity of the heart is between about 25 ("average" adults) and 40 (elite athletes) L/min. The blood vessels in skin and internal organs can accept up to 10 L/min, and all muscles together up to 70 L/min. Since the available cardiac output is half or less of these 80 L/min, the ability of the heart to pump blood is the limiting factor for muscular work in a hot climate.

9.9.1.2   Effects on Muscles

An increase of muscle temperature above normal does not affect the maximal isometric contraction capability of muscle tissue, but it reduces the dynamic output of muscles. Muscle overheating accelerates the metabolic rate; this can make the muscle ineffective if it must work over some longer period of time. The loss of power and endurance owing to excessive muscle temperature can be counteracted by cooling muscles before exercise. Such cooling reduces the cardiovascular strain and blood lactic acid concentration, and depletes muscle glycogen at a lower rate.

9.9.1.3   Dehydration

When working in a hot environment, the body loses water. The body can adapt to heat, but not to dehydration. Acute water loss incurred in a few hours or less, called hypohydration, does not reduce isometric muscle strength (or reaction times) if the water loss is less than 5% body weight. However, fast and large water loss (such as introduced by diuretics) generates the risk of heat exhaustion, which is primarily the result of fluid volume depletion. Dehydration reduces the body's capacity to perform aerobic or endurance work.

To counteract water loss, one must drink fluid. Plain water is best. If strenuous activities last longer than one or two hours, diluted sugar additives may help to postpone the development of fatigue by reducing muscle glycogen utilization and improving fluid-electrolyte absorption in the small intestine. Regular, liberally salted food during meals, as customary in North America, is normally sufficient to counteract salt loss. In fact, salt tablets have been shown to generate stomach upset, nausea, or vomiting in up to 20% of all athletes who took them.

9.9.1.4   Effects on Mental Performance

It is difficult to evaluate the effects of heat (or cold) on mental or intellectual performance because of large subjective variations and a lack of practical objective testing methods. However, as a rule, mental performance of an unacclimatized person deteriorates with room temperatures rising above 25 °C; that threshold increases to 30 or even 35 °C if the individual is acclimatized to heat. Brain functions are particularly vulnerable to heat; keeping the head cool improves the tolerance to elevated deep body temperature. A high level of motivation may also counteract some of the detrimental effects of heat. Thus, in laboratory tests, onset of performance decrement in perceptual motor tasks may occur in the low 30 °C WBGT range while very simple mental performance is often not significantly affected by heat as high as 40 °C WBGT.

9.9.1.5 Working in the Heat: Summary

Short-term maximal muscle strength exertion is not affected by heat. However, the ability to perform high-intensity endurance-type physical work is severely reduced before acclimation to heat, which takes normally up to two weeks. Even after acclimatization, the demands on the cardiovascular system for heat dissipation and for blood supply to the muscles continue to compete. The body prefers heat dissipation, with a proportional reduction of performance capability. Dehydration further reduces the ability of the body to work; hypohydration poses acute health risks. Mental performance is usually not affected by heat as high as 40 °C WBGT.

In consideration of these effects on the ability of the human to perform work in a hot environment, a number of technical and administrative measures can be taken*. The most important ones are listed in Table 9.5. They concern both the demands of physical effort and the control of environmental factors. (Specific design features of the thermal environment are discussed later in this chapter.)

9.9.2 Effects of Cold

In a cold environment, as in a hot climate, the body must maintain its core temperature near 37 °C. When exposed to cold, the human body first responds by peripheral vasoconstriction, which lowers skin temperature, to decrease heat loss through the skin. Such reduction in blood flow occurs in all exposed areas of the body although least strongly at the head. If control of blood flow away from the periphery is insufficient to prevent heat loss, shivering sets in. Shivering is a regular muscular contraction mechanism but without generating external work since all energy is converted to heat. Muscular activities of shivering and of physical work require increased oxygen uptake which is associated with increased cardiac output.

9.9.2.1 Cardiovascular Effects

The necessary increase in cardiac output, while heart rate remains at regular level, is brought about mostly by increasing blood pressure to augment the stroke volume. Yet, keeping the heart rate low as a reaction to cold exposure opposes the natural response associated with physical exercise, which is to increase the heart rate to help enlarge cardiac output.

Table 9.5 Control measures applied to work in hot environments (adapted from CCOPHS 2016)

Engineering Controls	
Reduce energy demands on workers	Mechanize tasks.
Reduce exposure to radiated heat from hot objects	Insulate hot surfaces. Use reflective shields, aprons, remote controls.
Reduce convective heat gain	Lower air temperature. Increase air speed. Increase ventilation. Provide cool recuperation booths.
Increase sweat evaporation	Reduce humidity. Use a fan to increase air movement.
Supply suitable clothing	Provide loose clothing that permits sweat evaporation but stops radiant heat. In extreme conditions use cooled protective clothing.
Administrative Controls	
Acclimatization	Provide sufficient acclimatization before full workload.
Duration of work	Shorten exposure time and use frequent rest breaks.
Rest area	Provide cool (air-conditioned) rest areas.
Water	Provide cool drinking water.
Pace of Work	Allow workers to set their own pace of work.
First aid and medical care	Define emergency procedures. Assign person(s) trained in first aid to each work shift. Train workers in recognition of symptoms of heat exposure.

9.9.2.2 Effects on Body Temperature

The two opposing cardiac responses to cold and exercise affect body temperature. At light work in the cold, core temperature tends to fall after about one hour of activity. Cold sensations in the skin regularly initiate reactions leading to lowered skin temperature, yet areas over active muscles can remain warmer due to the heat generated by muscle metabolism. Thus, in the cold, relatively much heat is lost through convection (and evaporation). Which of the opposing physiological cold responses predominates depends on the special conditions regarding ambient temperature, type of body activity, and the clothing insulation.

While one can feel the coldness of air in the upper respiratory tract, normally the respiratory passages can warm inhaled air enough to preclude cold injuries to lung tissues. Discomfort and constriction of airways may be felt when inspiring very cold air through the mouth; yet, air temperature is hardly ever too cold for exercise and physical work.

9.9.2.3 Effects on Work Capacity

For sub-maximal work in the cold, oxygen consumption is increased as compared to working at normal temperatures. (Shivering can cause some increase of oxygen cost at low work levels.) At higher exercise intensities, oxygen cost in the cold is about the same as at normal temperatures. However, often an extra effort is required to "work against heavy clothing" when worn to insulate against heat loss.

A cold climate does not affect the ability for maximal exercise for up to about five hours as long as the physiological stimuli provoked by exercise override those of cold. However, if core temperature gets lowered, maximal work capacity is reduced; apparently mostly by suppressing heart rate and thus reducing the transport, in the blood stream, of oxygen to the working muscles.

Cold exposure can decrease muscle temperature, which reduces muscle contraction capability and endurance, causing an early onset of fatigue. If muscles are cold, their isometric strength is impaired.

9.9.2.4 Dehydration

Dehydration occurs surprisingly easily in the cold, partly because sweating is increased in response to the increased energy demands of working in the cold, and partly owing to a suppressed thirst sensation. Further, urine production is increased in the cold, which can trigger water loss through more frequent urination. While dryness of cold air may cause respiratory irritation and discomfort, severe dehydration through the lungs does not occur since exhaled air is cooled on its way out to nearly the temperature of the inhaled air. Cooling water vapor causes condensation onto the surface of the airways; this explains the common experience of a "runny nose" in the cold.

9.9.2.5 Effects on Mental Performance and Dexterity

If the core temperature of the body drops below about 36 °C, vigilance is reduced. Central nervous system coordination suffers at about 35 °C; apathy sets in, and loss of consciousness occurs near 32 °C. While muscle spindles are initially more active as·muscle temperature drops, at about 27 °C their activity is reduced to about 50 percent and is completely abolished at about 15 °C. In the hands, joint temperatures below 24 °C and nerve temperatures below 20 °C severely reduce the ability for fine motor tasks. Manual dexterity is reduced as finger skin temperatures fall below 15 °C. Tactile sensitivity is reduced below 10 °C. A nervous block occurs if nerve temperature falls below 10 °C; movement becomes impossible and motor skills are completely lost. At about 5 °C, skin receptors for pressure and touch cease to function, the skin feels numb*. Frostbite is the result of ice crystals destroying tissue cells.

9.9.2.6 Working in the Cold: Summary

Strong isometric muscle exertions are impaired only if the muscles are cold. The ability for light work is reduced in the cold. Endurance activities are lessened only if core or muscle temperatures are lowered and if dehydration occurs. Clothing worn for insulation may hinder work. Dexterity and mental performance suffer in extreme cold.

For continuous work in temperatures below the freezing point, appropriate clothing is necessary and warming shelters such as heated tents, cabins, or rest rooms should be provided. The work should be paced to avoid excessive sweating, and drinking water should be provided.

Proper equipment design, safe work practices and appropriate clothing minimize the risk of cold injury and facilitate suitable work outcome.

9.10 Designing the Thermal Environment

Various combinations of climate components (ambient temperature, humidity, air movement and heat radiation) can subjectively appear as similar. The WBGT index, discussed earlier, is currently the most used indicator of climate factors that, combined, have equivalent effects on the human.

For work outdoors, clothing is the most effective and generally used solution to keep the worker safe and comfortable—yet, in some areas with hot climates, minimal or no clothes are worn when climate and customs permit. Outdoors, few technical means are available beyond temporary shelters, such as tents, which can help to reduce the effects of wind and solar radiation.

Clothes provide a layer of thermal resistance between the environment and the human body. Clothing determines the surface area of uncovered skin. More exposed surface allows better dissipation of heat in a hot environment but can lead to excessive cooling in the cold. Fingers and toes need special protection in cold conditions because they are so far away from the warm body core. Colors of the clothes are important in a heat radiating environment, such as in sunshine, because dark colors absorb heat radiation whereas light colors reflect it.

The insulating value of clothing is defined (such as in ISO 9920) in clo units: 1 clo = 0.115 m^2 °C/W. This is the insulating value (the reciprocal of thermal conductivity) of "normal" clothing worn by a sitting subject at rest in a room at about 21 °C and 50% relative humidity.

> With appropriate clothing and light work, many people in North America feel that comfortable WBGT ranges are from about 21 to 27 °C in a warm climate or during the summer, and from 18 to 24 °C in a cool climate or during the winter. Skin temperatures in the range of 32–36 °C are generally comfortable, associated with core temperatures between 36.7 and 37.1 °C. Preferred ranges of relative humidity are between 30% and 70%.

Indoors, available technology allows far-reaching (and often costly) control of the climate components. The most important factor is air temperature, but humidity also plays a major role; both interact with air movement and air replacement. Table 9.6 lists recommended ranges for air temperature and humidity for offices in North America. Air temperatures at floor level and at head level should not differ by more than about 6 °C. Differences in temperatures between body surfaces and side walls should not exceed approximately 10 °C. Air velocity should not exceed 0.5 m/s, and preferably remain below 0.1 m/s*.

Table 9.6 Temperature and relative humidity in offices in North America (adapted from ANSI/ASHRAE 2017; CAN/CSA 2017)

Conditions	Acceptable		Preferred	
	Relative Humidity	Office Room Temperature	Relative Humidity	Office Room Temperature
Summer, clothing ~0.5 clo	~30 %	24.5 to 28 °C	~50 %	21 to 23 °C or slightly higher
	~60 %	23 to 25.5 °C		
Winter, clothing ~1.0 clo	~30 %	20.5 to 25.5 °C	~50 %	21 to 23 °C
	~60 %	20 to 24 °C		

Preferred temperature ranges suit least 80% of individuals

What is of importance to the individual is not the climate in general (the so-called macroclimate), but the climatic conditions with which one interacts directly. Every person prefers a "personal microclimate" that feels comfortable under certain conditions of adaptation, clothing, and work; hence, a suitable microclimate is strongly individual and variable*. The comfortable personal microclimate strongly depends on clothing worn and on the type and intensity of work performed. It depends on gender. It depends on the surface-to-volume ratio of the body, which for example in children is much larger than in adults, and on the fat-to-lean body mass ratio. It also depends somewhat on age: with increasing years, persons tend to be less active and to have weaker muscles, to have a reduced caloric intake, and to start sweating at higher skin temperatures.

The individual choice of clothing is influenced by

- efficacy in terms of interaction with the physical environment to generate a preferred "personal microclimate";
- practicality in terms of providing ease of wear and tasks performance, in some jobs combined with mechanical protection; and
- appearance, often strongly affected by societal customs and fashions.

Individual thermocomfort varies by acclimatization: the status of the body (and mind) of having adjusted to changed environmental conditions. A macroclimate which may feel rather uncomfortable and seem to restrict one's ability to perform work during the first day of exposure may become quite agreeable during the following few weeks. Relatedly, seasonal changes in climate and work, of clothing and of attitude play major roles: in the summer, most people are willing to accept warmer, more humid and draftier conditions, both indoors and outdoors, than in the winter.

With so many variables of climate factors, kinds of work, clothing habits and willingness to adjust to given conditions, it is not surprising that, as sketched in Fig. 9.6, some persons deem a certain WBGT temperature too cold, and others consider it too warm, under the same conditions that most people find comfortable.

Fig. 9.6 Opinions about climates

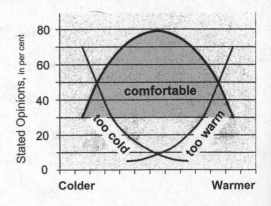

Notes

The text contains markers, *, to indicate specific references and comments, which follow.

Body surface: To estimate the area A (in m^2) of body surface from the weight of the body (in kg) and its height (stature, in m), the simplified "1916 Dubois and Dubois" equation is often used: $A = 0.202 \times weight^{0.425} \times height^{0.725}$ (Parsons 2003).

The body exchanges heat energy with the environment: Parsons (2003, 2014) provides a thorough review and detailed equations.

Kelvin: The unit kelvin originally was called a "degree" like other temperature scales. In 1968, the unit name was changed to simply "kelvin" (symbol: K) to indicate that it is not relative to an arbitrary reference point like the Celsius and Fahrenheit scales but rather an absolute unit of measure.

Energy transfer through radiation: de Dear et al. (1997), Silva and Coelho (2002).

Convective heat coefficient: de Dear et al. (1997), Parsons (2003, 2014).

Convective heat loss when swimming: Nadel (1984).

Vapor pressure: The units for pressure are $1000\ N\ m^{-2} = 1$ kPa $= 7.52$ mmHg $= 7.52$ Torr.

Heat loss from the respiratory tract and the skin: Nadel and Horvath (1975).

Heat exchange via clothing: Parsons (2003, 2014) and Bensel and Santee (2006) provide detailed information on the properties of clothing.

Calculating an "average" skin temperature: Lenhardt and Sessler (2006), Parsons (2003) provides a listing of weighting factors.

Specific heat of the body: Livingstone (1968), Parsons (2003, 2014).

Effective temperature scales: Eissing (1995), Mairiaux and Malchaire (1995), Malchaire et al. (2001), Parsons (2003, 2006).

Adequacy of the WBGT for extreme conditions: Parsons (2006, 2014).

"Safe" work in hot environments: OSHA (1999, 2016).

Heat stress disorders: OSHA (1999, 2016).

Cold sensations: Hoffman and Pozos (1989).

Working in hot and cold climatic conditions (as well as air pollution and high altitude): Bernard (2002), Astrand et al. (2003), Kroemer et al. (2003); Kroemer Elbert et al. (2018).

Technical and administrative control measures: CCOHS (2016).

Effects of cold body temperatures: Heus et al. (1995).

Climate in a built environment, such as in offices: ANSI-ASHRAE Standard 55, latest edition; Parsons (2008, 2014).

Individual microclimate: Fanger (1970), Wang et al. (2019a, b).

Summary

The body must maintain its core temperature near 37°C with little variation despite major changes in internally developed energy (heat), in external work performed, in heat energy received from a hot environment or in heat energy lost to a cold environment.

Heat energy may be either gained from or lost to the environment by radiation, conduction, and convection, but can only be lost by evaporation.

The major avenues of the body to control heat transfer between its core and skin are the vasomotor pathways (control of blood flow) and the sudomotor pathways (sweat production).

Muscular activities are the major means to control heat generation in the body.

Sweating is the ultimate means to cool the body surface.

In a hot environment, the body tries to keep the skin hot to prevent heat gain and to achieve heat loss.

In a cold environment, the body tries to keep the skin cold to avoid heat loss.

The thermal environment is determined by combinations of

- *air temperature (affecting convection and evaporation)*
- *air humidity (mostly affecting evaporation)*
- *air movement (affecting convection and evaporation)*
- *temperature of solids in contact with the body (affecting conduction)*
- *temperature of surfaces distant from the body (affecting radiation).*

The combined effects of the physical climate factors can be expressed in form of a climate index such as the WBGT. Certain ranges of humidity, temperatures, and air velocity are perceived as comfortable for given tasks and clothing. Every person has individual and variable preferences for a "personal microclimate" under given conditions of adaptation, clothing, and work.

Glossary

Absolute humidity The content of water vapor in air.

Acclimatization Continuous or repeated exposure to hot (or cold) climate which brings about a gradual adjustment of body functions resulting in a better tolerance of the climatic stress and in improvement of physical and psychological work capabilities.

Air humidity See absolute humidity and relative humidity.

Air temperature Measured with a sensor kept dry and shielded by a reflecting bulb from radiated energy; hence, often called "dry bulb temperature" or simply "dry temperature".

Body core The concept of central body components in the head and trunk, which are surrounded by the body shell (see there).

Body shell The concept of body components which surround the body core (see there), are close to the skin and are exposed to the environment.

Climate The ambient conditions of air temperature, air humidity, air movement, and heat radiation.

Clothing See thermal properties of clothes.

Conduction heat transfer Heat transfer by molecular contact between two solid bodies.

Convection heat transfer Transfer of heat by circulation of particles of a gas or liquid.

Dehydration The process or result of the body losing water.

Dew point The air temperature at which the air has the highest absolute content of vapor (saturation), depending on the barometric pressure.

Dry bulb temperature See air temperature.

Dry temperature See air temperature.

Energy flow Movement of energy (heat) from a warmer body to a colder one; as the temperatures of the contact surfaces become equal, the energy exchange diminishes.

Energy transfer See energy flow.

Ergonomics The application of scientific principles, methods and data drawn from a variety of disciplines to the design of engineered systems in which people play significant roles.

Evaporation The change of a liquid (such as water) to vapor; this phase change requires energy.

Evaporation heat loss As evaporation (see there) occurs on the skin or in the lungs of the human, energy is extracted from the body; so, it becomes cooler.

Heat transfer See energy flow.

Humidity The content of water vapor in air.

Hygrometer An instrument to measure humidity (see there).

Hyperthermia Overheating of the body.

Hypodehydration Fast dehydration (see there).

Hypothermia Abnormally low body temperature.

Macroclimate The ambient conditions of air temperature, air humidity, air movement, and heat radiation in the general environment.

Microclimate The conditions of the macroclimate (see there) with which an individual interacts directly.

Psychrometer An instrument to measure humidity (see there).

Radiation heat transfer Heat exchange through radiation; depends primarily on the temperature difference between two opposing surfaces.

Relative humidity The ratio (usually in percent) between the actual vapor content and the possible maximal content at given air temperature and air pressure.

Saturated air Air with the highest possible water vapor content; at the dew point (see there).

Thermal properties of clothes The net properties of items worn on the body that affect thermal insulation, transfer of moisture and vapor, heat exchange (by conduction, convection, radiation, evaporation), and air penetration.

Thermometer An instrument to measure temperature.

WBGT The Wet Bulb Globe Temperature index; this is the apparent temperature considering the effects of temperature, humidity, wind speed, and (usually solar) radiation on humans.

Wind chill The cooling effects of cold air movement on exposed skin.

References

ANSI/ASHRAE (2017) Standard 55–2013: Thermal environmental conditions for human occupancy. American National Standards Institute/American Society of Heating, Refrigerating, and Air-Conditioning Engineers, New York/Atlanta

ASHRAE (2017) ASHRAE Handbook Fundamentals - IP Units. American Society of Heating, Refrigerating, and Air-Conditioning Engineers, Atlanta

Astrand PO, Rodahl K, Dahl HA, Stromme SB (2003) Textbook of work physiology. Physiological bases of exercise. 4th ed. Human Kinetics, Champaign

Bensel CK, Santee WR (2006) Use of personal protective equipment in the workplace. Chapter 33. In: Salvendy G (ed) Handbook of human factors and ergonomics. 3rd ed. Wiley, New York

Bernard TE (2002) Thermal Stress. Chapter 12. In: Plog BA, Quinlan PJ (eds) Fundamentals of Industrial Hygiene. 5th ed. National Safety Council, Itasca

CAN/CSA (2017) Standard Z412–17. Office ergonomics. Canadian Standards Association, Toronto

CCOHS (2016) Working in hot environments. Canadian Centre for Occupational Health and Safety, Hamilton

de Dear RJ, Arens E, Hui Z, Oguro M (1997) Convective and radiative heat transfer coefficients for individual human body segments. Int J Biometereology 40(3):141–156

Eissing G (1995) Climate assessment indices. Ergonomics 38(1):47–57

Fanger PO (1970) Thermal comfort: analysis and applications in environmental engineering. McGraw-Hill, New York

Heus R, Daanen HAM, Havenith G (1995) physiological criteria for functioning of hands in the cold. Applied Ergonomics 26(1):5–13

Hoffman RG, Pozos RS (1989) Experimental hypothermia and cold perception. Aviat Space Environ Med 66:964–969

Kroemer KHE, Kroemer AD (2005) Office ergonomics. Korean ed. Kukje, Seoul

Kroemer KHE, Kroemer HB, Kroemer-Elbert KE (2003) Ergonomics. How to design for ease and efficiency. 2nd amended ed. Prentice Hall, Englewood Cliffs

Kroemer Elbert KE, Kroemer HB, Kroemer Hoffman AD (2018) Ergonomics: How to design for ease and efficiency, 3rd edn. Academic Press, London

Lenhardt R, Sessler DI (2006) Estimation of mean body temperature from mean skin and core temperature. Anesthesiology 105(6):1117–1121. https://doi.org/10.1097/00000542-200612000-00011.PMID:17122574;PMCID:PMC1752199

Livingstone SD (1968) Calculation of mean body temperature. Can J Physiol Pharmacol 46:15–17

Mairiaux P, Malchaire J (1995) Comparison and validation of heat stress indices in experimental studies. Ergonomics 32(1):58–72

Malchaire J, Piette A, Kampmann B, Mehnert P, Gebhardt H, Havenith G, de Hartog E, Holmer I, Parsons K, Alfano G, Griefahn B (2001) Development and validation of the predicted heat strain model. Ann Occup Hyg 45(2):123–135

Nadel ER (1984) Energy exchanges in water. Undersea Biomedical Research 11(4):149–158

Nadel ER, Horvath SM (1975) Optimal evaluation of cold tolerance in man. Chapter 6A. In: Horvath SM, Kondo S, Matsui H., Yoshimura H. (eds) Comparative studies on human adaptility of Japanese, Caucasians, and Japanese Americans, Volume 1. Japanese Committee of International Biological Program, Tokyo

OSHA (1999) TED 01-00-015. Occupational exposure to heat, Occupational Safety and Health Agency, Washington

OSHA (2016) OSHA Technical Manual. Section III: Chapter 4. Heat stress. Occupational Safety and Health Agency, Washington

Parsons KC (2003, 2014) Human Thermal Environments. 2nd and 3rd eds. Taylor & Francis, London

Parsons KC (2006) Heat stress standard ISO 7243 and its global application. Ind Health 44:369–379

Parsons KC (2008) Overview of design regulations for thermal comfort –reviews and highlights of latest thermal comfort standards. http://www.cibse.org/content/Groups/Natural_Ventilation/Thermal_Comfort_18_Nov_2008/Ken%20Parsons.pdf

Silva MCG, Coelho JA (2002) Convection coefficients for the human body parts determined with a thermal mannequin. 277–280. In: Proceedings of Room Vent 2002, 8-11 September 2002. Copenhagen

Wang R, Zhao C, Li W, Qi Y (2019) Research on thermal comfort equation of comfort temperature range based on Chinese thermal sensation characteristics. In: Karwowski W, Trzcielinski S, Mrugalska B (eds) Advances in manufacturing, production management and process control. AHFE 2019. Advances in intelligent systems and computing, vol 971. Springer, Cham

Wang R, Li W, Zhao C, Qi Y (2019) Study on the perception characteristics of different populations to thermal comfort. In: Karwowski W, Trzcielinski S, Mrugalska B (eds) Advances in manufacturing, production management and process control. AHFE 2019. Advances in intelligent systems and computing, vol 971. Springer, Cham

Chapter 10
Body Rhythms
and Work Schedules

Overview

The human body follows an endogenous regular rhythm of physiological functions throughout the 24 h day. During waking hours, the body is prepared for physical work whereas at night, sleep is normal. Attitudes and behavior also change regularly during the day. The circadian rhythms can be modified by an imposed new set of time signals and of activity–rest regimen, such as associated with shift work schedules. Shift work should be arranged to least perturb the natural physiological, psychological, and behavioral rhythms. Disturbing the natural circadian rhythms can have negative health and social effects and cause reductions in work performance.

The Model

Daily rhythms are systems of temporal programs encrypted within the human organism. They strongly affect well-being and task performance. They should be kept intact for continued normal functioning, both physically and psychologically, by selection of suitable work schedules.

Introduction

Human body functions and behavior follow internal rhythms. Under regular living conditions, these natural internal temporal programs are well established and persistent. One is the female menstrual cycle of about 28 days; another is the set of daily fluctuations, called circadian (or diurnal) rhythm.

© Springer Nature Switzerland AG 2020
K. H. E. Kroemer et al., *Engineering Physiology*,
https://doi.org/10.1007/978-3-030-40627-1_10

The circadian rhythms (from the Latin *circa*, about, and *dies*, the day) are endogenous regular physiological occurrences, which appear, for example, in body temperature, heart rate, blood pressure, and hormone excretion, as sketched in Fig. 10.1. They appear also in common psychological and behavioral patterns.

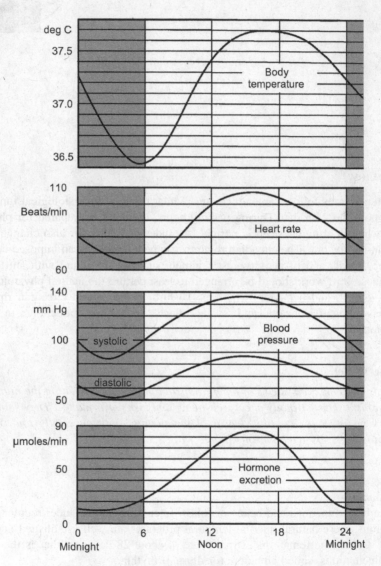

Fig. 10.1 Typical variations in body functions during the 24 h day

Daily rhythms are patterns of temporal programs within the human organism. They are evident under rigorous experimental conditions, they persist under varying external settings, they are valid by actual observation and experience, and they are important for well-being and performance.

Within the body, self-sustained pacemakers, "internal clocks" running on a cycle of about 24 h, control the circadian rhythms. Several rhythmic patterns are coupled with each other, such as core temperature, blood pressure, and sleepiness. The inclinations to do certain activities, such as to rest and sleep and many other aspects of social behavior, follow customs and sequences during the day. These activities relate to the "chrono-biological" rhythms during the 24 h day. Another well documented self-sustained rhythm is the (typically) 28 day menstrual cycle.

Strong external events can put an existing circadian rhythm out of order. Such imposing events contain a "time setter", called zeitgeber (German, "time giver"). Among the zeitgebers are daily light/darkness, true clocks, some pharmaceutical drugs, and temporally established activities, such as office hours, meal times and other regular events and habits. The strengths of these markers vary.

10.1   Menstrual Cycle

Synchronized activities of the hypothalamus, pituitary, and the ovary regulate the female menstrual cycle. The typical 28 day time period is usually divided into four phases: (1) menstruation, (2) pre-ovulatory or follicular, (3) ovulatory, and (4) post-ovulatory or luteal. The dominant hormonal changes occur in the release of estrogen (associated with spikes in levels of luteinizing hormone and follicle-stimulating hormone) peaking just before ovulation, and progesterone (peaking roughly 10 days after ovulation, if fertilization does not occur). Hormonal release is low during the menstruation phase.

The hormonal changes during the menstrual cycle influence a woman's physiological state (such as body temperature) and possibly psychological state; however, observable events in attitude or performance are usually insignificant. The bulk of existing research relies on self-reported changes in mood and physical complaints in the course of the menstrual cycle. Fairly little information is available on objective measures of performance. There is neurophysiological evidence that estrogen and progesterone affect brain function, yet the two hormones have antagonistic effects on the central nervous system, with estrogen stimulating and progesterone inhibiting. Varying hormone production during the menstrual cycle may weakly affect the capacity to perform certain tasks*, but the extent to which hormones actually determine performance depends on how any potential decrease in work capacity may be offset by increased effort.

The occurrence of physical complaints and sometimes negative moods in many women before menstruation is fairly well established, but the precise nature of the so-called pre-menstrual syndrome is not yet determined. There is some evidence that menstruation can bring about negative social behaviors such as irritability and nervousness; however, these can be mediated by social and psychological factors.

In summary, out-dated hypotheses of reduced performance* during the pre-menstrual and the menstrual phases are not supported by objective data.

10.2   Circadian Rhythms

Maintenance of physiological functions in spite of external disturbances is one prerequisite for human health. This state of balanced control is called homeostasis. However, a closer look at this supposedly "steady state" of the body reveals that many of its functions are in fact not constant but show rhythmic variations; for example, ups and downs in body temperature or hormone secretion follow each other regularly in similar patterns, day after day. The specific periods of different rhythms are quite diverse, such as the heart beating about once every second, body temperature having its peak value every 24 h, or a woman's menstrual cycle reoccurring every 28 days. Rhythms with a cycle length of 24 h are called *circadian*, those that take less than 24 h to repeat are called *ultradian*; those which repeat in more than 24 h, *infradian*.

Among the circadian rhythms, well-known physiological variables are body temperature, heart rate, and blood pressure as depicted in Fig. 10.1. Most variables show a high value during the day and lower values during the night, although hormones in the blood tend to be more concentrated during the night, particularly in the early morning hours. Many variations in amplitude during the circadian circle are fairly small, approximately ±1 °C for oral temperature; others are in the range of approximately ±15% about the average, such as heart rate and diastolic blood pressure. However, some body functions oscillate considerably: triglycerides vary by nearly ±80% in the blood serum, while the sodium content of the urine vacillates even more. The amount by which the functions change during the course of the day, and the temporal locations of rhythm extremes during the day, can be quite different among individuals and may change even temporally within a person*.

One way to observe the daily rhythms and to assess their effects on performance is simply to watch an individual's freely chosen behavior. During daytime, a person is normally awake, active, and eating but, at night, sleeping and fasting. However, physiological events do not exactly follow that activity pattern. For example, body core temperature decreases for several hours after the person has fallen asleep; the temperature is usually lowest between 3 and 5 o'clock in the morning. Then the temperature rises even before one awakens and gets up; it continues to increase, with some variations, until late in the afternoon. Thus, body temperature is self-governed, and not a passive response to regular daily behavior, such as waking, eating meals, performing work, and doing other social activities.

Interactions among external activities and internal rhythms and their zeitgebers do exist. Variations due to exogenous influences may occur in observed rhythmic events; for example, skin temperature (particularly at the extremities) increases with the onset of sleep, whenever this happens. Turning the lights on boosts the activity level of birds, regardless of when this occurs. Thus, skin temperature or activity levels do not necessarily indicate the internal rhythm but may just mask the underlying physiological patterns of the body, which are robust, self-regulated, and predisposed to remain in existence even if daily activities change.

Under regular circumstances, there is a well-established phase coincidence between the external activity signs and the internal events. For example, during the night, the low values of physiological functions in humans, such as core temperature and heart rate, are primarily due to the circadian rhythm of the body; however, they are further helped by nighttime inactivity and fasting. During the day, peak activity usually coincides with high values of internal functions. Thus, normally, the observed circadian rhythm is the result of internal and external events which concur. If that balance of concurrent events is disturbed, negative consequences in well-being, health or performance may result.

When a person is completely isolated from exogenous factors and there are no regular activities or other external zeitgebers, the internal body rhythms are "running free", meaning they are independent of external time cues and are internally controlled. Many experiments have consistently shown that human circadian rhythms persist when free-running, but their time periods usually are slightly different from the regular 24 h duration: most rhythms run freely at about 25 h, some take even longer. Since the earth continues to rotate at 24 h relative to the sun, this observation suggests that body rhythms are independent from external stimuli and follow their own built-in clocks. Yet, if a person is subjected again to regular daily time markers and activities, the internal rhythms resynchronize on their 24 h cycles.

10.2.1 Models of Oscillatory Control

The commonly used oscillator model of the human circadian system assumes that various overt rhythms are jointly controlled by a few basic internal oscillators which, however, may have different controlling power. The basic oscillators, probably located in the hypothalamus, are in turn controlled by external stimuli (usually related to the earth's rotation), and also influence each other. If their intrinsic periods are close together, they synchronize. Yet, such internal coordination may fail; for example, when artificial zeitgebers occur within the entrainment range of one oscillator but outside of another. In this case, internal desynchronization takes place: rhythms controlled by one oscillator remain entrained, but those controlled by another oscillator begin to run free. For example, the sleep/wake cycle stays at 24 h, while the temperature rhythm may free-run at a period of 25 h*.

Two types of experiments can serve to investigate the constancy of rhythms or their temporal isolation: removal of external indicators (natural or artificial) or manipulation of artificial timer markers. Experiments without time markers serve to evaluate purely internal control. Under constant experimental conditions, such as in a dark isolated cave without external pacemakers, most human circadian rhythms free-run at about 25 h periods, desynchronized from the 24 h zeitgeber.

Other experiments rely on manipulation of strong artificial time markers of various types to evaluate the effects of internal and external factors. Manipulation of the zeitgeber allows laboratory simulation of jet lag or shift work. Experiments with strong artificial time setters have shown that these play a major role in entraining or synchronizing the internal rhythms so that they follow the new periodic time cues. Synchronization of the internal rhythms to time events is possible with cycle durations between 23 and 27 h. (At shorter or longer periods of time cues, the circadian rhythms are free-running, though often not completely independent of the time cues.) It appears to be easier to set internal clocks "forward", as it occurs in the spring when daylight-savings time is introduced in North America and in Europe, or when subjected to jet-lag when traveling east-ward, than to retard the internal clock.

10.2.2   Individual Diurnal Performance Rhythms

As evident from their responses in self-assessment questionnaires, some people are "morning types", early risers, who claim to be especially alert and productive early in the day. These persons seem to have consistently shorter free-running periods than most, entrained to have early peaks in body temperature, heart rate and melatonin excretion. "Evening types", in contrast, are people with long internal rhythms and late peaks. While it is not certain whether females have, on average, a free-running period that is a bit shorter than that of males, young children appear to be more morning-oriented than adults; however, many adults become early risers again as they age*.

Given the robust oscillations in physiological functions during the day, one expects corresponding changes in mood and performance—but the regular organization of the day (getting up, working, eating, relaxing, sleeping) also affects, and usually strongly so, a person's attitudes and work habits. Experimentally, one can separate the effects of internal circadian rhythms and of external daily organization, but the combined effect of internal and external factors, for example in regard to task performance, is of great practical interest.

Early in the 1900s, it was thought that the morning hours were best for mental activities, and the afternoon more suitable for physical work. However, individual traits such as motivation, skill and habits, or the specifics of tasks and work environment, may overshadow the effects of circadian variations.

Performance in everyday work, especially when it is monotonous, often shows a pronounced reduction just after noon. However, this "post-lunch dip" in performance is not accompanied by a similar change in physiological functions; for example, body temperature does not change appreciably at that period of the day. The interruption of work for lunch might bring about an increased lassitude, a status of deactivation, associated with increased blood glucose resulting from food digestion—so, the post-lunch dip may be caused by the exogenous effects of work break and food intake, "masking" the endogenous circadian effects. In activities with medium to heavy physical work, usually no such dip occurs after lunch (except when food and beverage ingestion was very heavy and if true physiological fatigue had been built up during the pre-lunch activities).

In summary, some of the many different activities performed during the day decidedly follow a circadian rhythm, to varying degrees. For some people, exogenous masking effects may be more pronounced than on others. For example, information processing in the brain (including immediate or short-term memory demands), mental arithmetic activities, or visual searches, may be strongly affected by personality or by the length of the activity and by motivation. Thus, it appears that one cannot make "normative" statements about diurnal performance variations or abilities during regular working hours—with the additional understanding that physical fatigue (tiredness) resulting from work already performed is likely to reduce performance, as discussed in Chap. 8.

10.3 Sleep

Two millennia ago, Aristotle thought that during wakefulness a particular substance ("warm vapors") in the brain built up which needed to be dissipated during sleep. In the 19th century, there were two opposing schools of thought: one that sleep was caused by some "congestion of the brain by blood," the other that blood was "drawn away from the brain." Behavioral theories were common in the 19th century, for example positing that sleep was the result of an absence of external stimulation, or that sleep was not a passive response but an activity to avoid fatigue from occurring. Early in the 20th century, a common assumption was that various sleep-inducing substances accumulated in the brain, an idea taken up again in the 1960s. In the 1930s and 1940s, various "neural inhibition" theories were discussed, including "sleep-inducing centers" in the reticular formation of the brain. Other theories about the function of sleep assume some kind of recovery from the wear and tear of wakefulness. Alternative theories propose that sleep is simply a form of instinct or non-behavior to occupy the unproductive hours of darkness and a means to conserve energy. Restoration, energy conservation and occupation of time appear to explicate certain characteristics of sleep but they do not explain sleep completely or sufficiently. Thus, there is no simple explanation for what sleep does and why we must sleep—but it is evident that humans need sleep*.

It is expedient to presume that two central clocks of the body regulate alertness, sleepiness, and many physiological functions: one clock controls sleep and wakefulness, the other clock regulates physiological functions such as body temperature. Normally, the internal clocks synchronize to increase physiological activities during wakefulness and decrease physiological activities during sleep. However, this congruence of the two rhythms may be disturbed, for instance by night shift work, where one must be active during nighttime and sleep during the day. If such patterns continue, the physiological clocks can adjust to the external requirements of the new sleep/wake regimen: the formerly well-established physiological rhythm may flatten out and, after some time, re-establish itself according to the new sleep/wake schedule.

10.3.1   Sleep Phases

The brain and muscles are the human organs that show the largest changes from sleep to wakefulness: their activities can be observed by electrical means.

Electrodes attached to the surface of the scalp serve to observe sleep cycles in the brain. They pick up electrical activities of the encephalon (cortex), which wraps around the inner brain (see Chap. 3): thus, the measuring technique is named electro-encephalography (EEG). The EEG signals provide information about the activities of the brain. Another common technique records the electrical activities associated with muscles (see Chap. 2) that move the eyes and of muscles located in the chin and neck regions. The recording of activities of muscle (Greek, *myo*) is called electro-myography (EMG); one of these, electro-oculography (EOG), records specifically the activities of eye muscles.

EEG signals can be described in terms of amplitude and frequency. The amplitude is measured in microvolts (mV): the amplitude increases as consciousness falls from alert wakefulness through drowsiness to deep sleep. EEG frequency is measured in cycles per second (Hz); the frequencies observed in human EEG range from 0.5 to 25 Hz. Frequencies above 15 Hz are called "fast waves," frequencies under 3.5 Hz "slow waves." Frequency falls as sleep deepens; "slow wave sleep" (SWS) is of particular interest to sleep researchers.

Certain EEG frequency bands have been given Greek letters. The main divisions are:

- Beta, above 14 Hz. Such fast waves of low amplitude (under 10 mV) occur when the cerebrum is alert or even anxious.
- Alpha, between 8 and 14 Hz. These frequencies occur during relaxed wakefulness when there is little information input to the eyes, particularly when they are closed.
- Theta, between 3.5 and 8 Hz. These frequencies are associated with drowsiness and light sleep.
- Delta, slow waves under 3.5 Hz. These are waves of large amplitude, often over 100 mV, and occur more often as sleep becomes deeper.

Certain specific occurrences in the EEG waves have been labeled, such as vertices, spindles, and complexes, which appear regularly in association with certain sleep characteristics.

The importance placed upon EMG and EEG events by sleep researchers has been changing over decades. Currently, EMG events of the eye muscles serve to primarily divide sleep into periods associated with rapid eye movements (REM) and those without (non-REM, or NREM). Non-REM conditions are further

Table 10.1 Sleep stages

Condition	Sleep Stage	Muscle EMG	Brain EEG	Percentage of sleep (averages)
Awake	None, awake	Muscles active	brain active, alpha and beta	na
Drowsy, transitional light sleep	1 (N1) non-REM	eyelids open and close, eyes roll	theta, loss of alpha, sharp vertex waves	5
True sleep	2 (N2) non-REM		theta, few delta, sleep spindles, k-complexes	45
Transitional true sleep	3 (N3) non-REM		much delta, slow wave sleep	7
Deep true sleep	4 (N3) non-REM		delta predominant, slow wave sleep	13
Sleeping	REM	Rapid Eye Movements, most other muscles relaxed	brain alert, much dreaming, alpha and delta	30

subdivided into four stages, as in Table 10.1, according to their associated EEG characteristics. A related classification maintains non-REM stages 1 and 2 as N1 and N2 but combines stages 3 and 4 as stage N3 "delta sleep", "slow wave" and "deep sleep".

After falling asleep, the brain usually goes quickly to sleep stages 3 and 4 (N3), where brain and skeletal muscles are not active. That sleep phase commonly lasts about 1.5 h. Thereafter, sleep becomes lighter. When it reaches stage 1, the EEG shows fast but low voltage brain activities while rapid eye movements often occur. In this REM sleep, in addition to the muscles of the eyes, often also muscles of the hands and feet are active (while most other skeletal muscles stay relaxed) and heartbeat and breathing speed up irregularly. The REM stage may last just a few minutes, after which the sleeper falls back into the deeper sleep stages. In non-REM sleep, regular and slow breathing and heart rates occur, and the EEG activity is slow but shows high voltage. Each stage appears to have distinct beneficial effects; this concept can explain why, even after sleeping one's usual duration, one still feels tired if the sleep was disturbed or interrupted.

Normally, several sleep cycles repeat during the night. They are probably co-organized between diverse brain regions, thus involving two or more oscillators. The REM/non-REM cycles occur in roughly 1.5 h timings; Fig. 10.2 shows a typical pattern of night sleep. While there is much within- and between-subjects

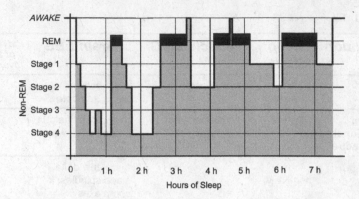

Fig. 10.2 Conceptual example of sleep stages during undisturbed night sleep (adapted from Kroemer 2009)

variability, commonly the portion of REM increases in the later sleep phases before natural awakening.

10.3.2 Sleep Loss and Tiredness

Surprisingly, it is still not entirely clear why humans (or animals) need sleep. A common opinion is that sleep has recuperative benefits, that it allows some sort of restitution or repair of tissue or brain following the "wear and tear" of wakefulness. However, what is meant by restitution or repair is usually not clearly expressed, nor fully understood. Certainly, sleep is accompanied by rest and, to a large extent, by energy conservation. But a human can attain similar relaxation when resting and awake. Regarding restitution of the body, it is true that certain hormones are released more during sleep than during wakefulness; prominent among these hormones is the human growth hormone. However, few other such beneficial sleep events have been noted. For example, protein in tissues is continuously broken down into its amino-acid building blocks or reconstituted from recent food intake. If such breakdown were excessive during wakefulness, then the rate of synthesis should be especially high during sleep. This is not the case: in fact, during sleep, protein synthesis is low and breakdown is increased, leading to a loss of protein through dissolution, not to an increase through restitution.

Many experiments have failed to show restorative physiological effects of sleep. In fact, even moderate sleep deprivation has little physiological effect (although there are clear signs of impairment of functions in the central nervous system). For example, moderate sleep deprivation does not impair muscle restitution or the physiological ability to perform physical work. Any reduction in performance of physical exercise either under sleep deprivation or during ebb times of circadian rhythms may be entirely due to reduced psychological motivation, because there is no significant decrease in physiological capabilities*.

In contrast, the restorative benefits of sleep to the brain are fairly well researched. Two or more nights of sleep deprivation bring about psychological performance detriments, particularly reduced motivation to perform (but apparently not a reduction of the inherent cognitive capacity), behavioral irritability, suspiciousness, speech slurring, and other performance reductions. These effects indicate some central nervous system (CNS, see Chap. 3) impairment owing to sleep deprivation and, consequently, a need for the brain to sleep. However, though one feels tired, even after up to two days of sleep deprivation mental performance is still close to normal on stimulating and motivating tasks whereas boring tasks show diminished performance. Performance an all tasks is reduced after more than two nights of sleep deprivation. And yet, even with sleep loss, the regular circadian rhythms persist: performance levels are lower during nighttime, when the body and brain usually rest, than during daytime.

One speculative explanation for the ability to perform mental tasks well even after two days of sleep deprivation is that, for regular tasks, there is an overcapacity of cerebral neural networks; so, many are not fully used*. According to this hypothesis, with sleep deprivation, some circuits become impaired, possibly as a result of missing "restitution"; consequently, the extra circuits are put into service. At first, this suffices to avoid overt effects on performance. With increasing deprivation, more circuits become impaired; finally, all available circuits become overloaded and performance drops.

The brain assumes a physiological state during sleep which is unique to sleep and cannot be attained while awake. During relaxed wakefulness, muscles can rest but the cerebrum remains in a condition of "quiet readiness," prepared to act on sensory input, without diminution in responsiveness. Only during sleep do cerebral functions show marked increases in thresholds of responsiveness to sensory input. In the deep sleep stages associated with slow wave non-REM sleep, the cerebrum is apparently functionally disconnected from subcortical mechanisms. Regardless of whether the cerebrum needs off-line recovery during sleep from the demands of waking activities or whether it simply disconnects and withdraws, the brain needs sleep to restitute, a process that cannot sufficiently take place during waking relaxation.

Apparently, not all of one's usual length of sleep-time is essential for brain restitution. For example, after sleep deprivation, not all lost sleep needs to be reclaimed; missing 4 h of sleep does not require that 4 additional hours be found for slumber later. Long-term studies with volunteer subjects have shown that a sleep period that is 1–2 h shorter than usual can be endured for many months without noticeable consequences. It seems that the first 5–6 h of regular sleep (which also contain most of the slow-wave non-REM sleep and at least half the REM sleep) are obligatory to retain psychological performance at normal level. Additional sleep (called facultative or optional) mostly serves to occupy unproductive hours of darkness, with dreams (mostly in REM but not totally confined to it) the "cinema of the mind"*.

10.3.3   Normal Sleep Requirements

While there are, as usual, variations among individuals, certain age groups show rather regular sleeping hours. For example, young North Europeans/American adults sleep, on average, 7.5 h with a standard deviation of about 1 h. Some people are well rested after 6.5 h of sleep, whereas others habitually take 8.5 h and more. Individuals naturally sleeping less than 3.5 h are very rare among middle-aged people; no true non-sleepers seem to exist among healthy persons*. If people can sleep for just a few hours per day, many are able to keep up their performance levels even if the attained total sleep time is shorter than normal. The lower limit for restorative sleep seems to lie around to 5 h of sleep per day, with even shorter periods still being somewhat useful; day-time napping can be helpful can supplement shorter night-time sleeps.

Figure 10.3 shows the effects of sleep loss on body temperature: the temperature keeps its phase but is elevated during the night and morning. This is indicative of changes in body functions associated with sleep loss. If a person does not get the necessary amount of sleep, the resulting problem is tiredness; the obvious cure is to get more sleep. However, people who cannot sleep in at will have a problem; night-shift workers, for example, generally find it difficult to make up for lost sleep in the busy (noisier and brighter) daytime.

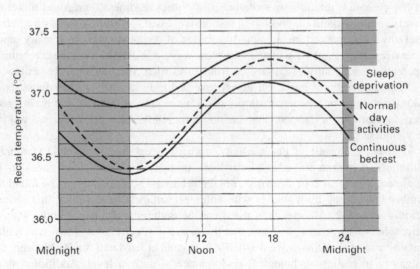

Fig. 10.3 Typical changes in body temperature associated with sleep deprivation, normal activities, and bed rest (adapted from Colligan and Tepas 1986)

10.4 Sleep Deprivation and Prolonged Periods of Work

If it is necessary to continue work for long periods of time, such as a full 24 h day or longer, this condition also results in sleep deprivation. The stress that is experienced as well as difficulties in task performance during such long working spells are likely to result both from the extended work itself and from the lack of sleep. That intermingling of causes makes a general discussion of this topic complicated. Furthermore, different types of work tasks may show varying effects; wakefulness or sleepiness appear in cycles during the 24 h of the day and hence have differing effects; and the motivation of the worker often plays an important role regarding performance.

10.4.1 Performing Tasks

On a short work task, lasting less than about half an hour, existing sleep loss affects performance less than if the work that must go on uninterruptedly for a longer period. Sleep deprivation strongly affects monotonous work; performance is likely to become worse with each repetition. In contrast, implementation of a task new to the operator is only minimally affected if the person has a sleep deficiency. Lack of sleep diminishes execution of a complex task more than performing a simple one. Tasks which are paced by the work itself deteriorate more with sleepiness than do operator-paced tasks.

After losing out on sleep, it usually takes longer to perform a job, but accuracy may not particularly suffer. A task that is interesting and appealing, even if it includes complex decision making, can be executed rather well even over lengthy periods of time; but if the task is unappealing, even disliked, making decisions takes longer. When people must stay awake for long periods of time, memory use degrades, both in long-stored memory information and short-term memory*.

10.4.2 Incurring and Recovering from Performance Decrement

In general, performance starts to degrade after one night without sleep; the deterioration becomes more pronounced after additional nights of sleep deprivation. After missing four nights of sleep, very few people are able to stay awake and to perform adequately even if their motivation is very high*. The following discussion assumes sleep deprivation of at least one night.

With increasing sleep-deprived time at work, so-called microsleeps occur more frequently. The person falls asleep for a few seconds, but these short periods (even if frequent) do not have much recuperative value because the subject still feels sleepy and performance continues to worsen. During long working times coupled with lack of sleep, periods of no performance (called lapses or gaps) commonly occur. These are short periods of reduced arousal or even of light sleep.

Naps lasting one to two hours can improve subsequent performance. However, if a person is awakened during a deep-sleep phase, sleep inertia (SI) can appear: feelings of grogginess, drowsiness, sluggishness, or disorientation together with impaired cognitive and sensory-motor performance. The SI experience* is situational and individual, strongest immediately after waking and then decreasing in intensity, but SI can linger up to half an hour. This period of impairment is of particular concern to persons "on call" who must make snap-decisions and take complex actions quickly after waking: physicians and nurses, pilots, rescue crews, fire fighters, soldiers, police personnel or motorists. Prior sleep loss can exacerbate SI, which is particularly important in extended work shifts. Countermeasures against SI can be pro-active or re-active: planning length and timing of sleep and naps (often difficult to do) and rousing actions taken after wake-up using caffeine, bright light, sound, temperature, face-washing, and food. A 2016 review of the SI literature* did not find any of these reactive countermeasures able to attenuate performance impairment immediately and strongly.

The timing of taking naps during sleep deprivation may impact effectiveness: for example, the common early afternoon nap has surprisingly little measurable effect on performance of subsequent work. On the other hand, naps of at least 2 h taken in the late evening or during nighttime, when the circadian rhythm is falling, have positive effects lasting several hours, provided that the amount of sleep deficit incurred until this moment is moderate, such as one night without sleep*.

Between 2002 and 2004, 34 resident physicians (ages 24–32 years; 23 males) were observed during their 3-week "on-call" rotation schedules. They worked 24–28 h extended-duration work shifts (EDWS), followed by "post-call" presence, in Intensive Care Units of a hospital in Boston, USA.

Daily sleep logs were collected and a 10 min psychomotor vigilance task (PVT) was administered about every 6 *h during each EDWS*. *Post*-call performance during EDWS was compared with sessions matched for time-of-day and weeks-into-schedule in the same resident physicians during a work schedule without EDWS.

The resident physicians obtained an average of 1.6 ± 1.5 h cumulative sleep overnight during EDWS on 92% of nights. PVT attentional failures were significantly reduced only after sleep longer than 4 h (versus no sleep). Despite this apparent improvement, the odds of incurring attentional failure were 2.7 times higher during post-call following longer-than-4 h sleep compared with matched sessions without EDWS.

The authors* concluded that even after more than 4 h of sleep overnight (8% of EDWSs), performance remained significantly impaired. This suggest that "strategic napping" is insufficient to mitigate severe performance impairment introduced by duty extended beyond 16 h.

Many people claim that naps, such as 15–30 min of rest taken after lunch, perhaps followed by some caffeine intake, are helpful; for some, such naps appear almost necessary to feel "ready and fit" for continued work. Yet, there is no clear scientific support for the usefulness of short naps (especially for people who missed a night's sleep) on task performance. Of course, task performance isn't the only measure of success for napping; simply feeling better may be sufficient justification. Perhaps the recuperative effects are too subtle, too individual, and too dependant on interactions between physiological, psychological, and habitual traits and specific task conditions, to be easily demonstrated in a scientific experiment.

If long periods of mainly mental activities are necessary, interrupting the work for some physical exercises may prevent performance deterioration. Also, white noise may be beneficial and stirring music may help. Caffeine use is widespread. Taking powerful drugs, such as amphetamines, can restore performance to a nearly normal level even after three nights without sleep*–but such drastic action should be reserved for truly critical tasks to ensure any benefits outweigh the risks.

A cross-sectional survey (July 2005–December 2007) of 4957 police officers in North America showed that wake-promoting agents were used by 28% of police. (20% of the officers reported using sleep-promoting drugs.) Use of wake-promoting drugs, high caffeine use (common), and smoking (rare) to stay awake were statistically associated with increased odds of a fatigue-related error, of stress and burnout.

Night-shift work was associated with independent increases in excessive sleepiness, near-crashes, and fatigue-related errors. Excessive sleepiness appeared as an interaction between night-shift work and wake-promoting drug use.

The authors* concluded that police who use sleep-promoting and wake-promoting drugs, especially when working night shifts, are vulnerable to adverse health, performance and safety outcomes.

Fortunately, recovery from sleep deprivation is quite fast. One good night's sleep usually restores performance to a normal level, even after extensive sleep deprivation.

10.5 Shift Work

Shift work is a work schedule in which two or more persons, or teams of persons, work in sequence at the same workplace. Often, each worker's shift repeats, in the same pattern, over a number of days. For the individual, shift work may mean either working regularly at the same time (*continual* shift work) or at varying times (*discontinuous* shift work, often as *rotating* shift work).

10.5.1 The Development of Shift Work

Shift work is not new. In ancient Rome, by decree, deliveries were to be done at night to relieve street congestion. Bakers traditionally worked through the late night and early morning hours. Soldiers and firefighters work both day and night shifts. Many alternate work systems have been used; for example, farmers used to start work at dawn and stop at dusk. "Peddlers, politicians, professors" and others have always tried to set their own work schedules, often highly variable, to suit specific circumstances and preferences.

With the advent of industrialization, long working days became common, with teams of workers relieving each other to maintain blast furnaces, rolling mills, glass works, and other workplaces where continuous operation was desired. Covering the 24 h period with either two 12 h work shifts, or with three 8 h work shifts, is common practice. Technological or economical concerns have made shift work generally accepted in many industries, trades, and services, in developed and in developing countries.

Since the industrial revolution of the late 18th century, when 12 h shifts were common, drastic changes in work systems have occurred. Early in the 20th century, 6 day workweeks with 10 h shifts were widespread. Today, many work systems rely on an arrangement of 8 h of work per day but with only 5 workdays per week, a set-up started in many countries in the 1960s.

The general trend has been to reduce the number of days worked per week, usually to allow two weekend days to be free; also, to decrease the number of hours worked per day. In 1988, the German Secretary of Labor proposed a system of a "compressed workweek" (see below) with 9 h work per day and only 4 days of work a week. In 1996, the average number of hours worked per week was just over 32 in Germany. In 2000, France formally adopted a 35 h work week. Many employees in northern Europe, particularly if working for the government, may leave their workplaces early on Friday afternoon to return after a long weekend on Monday morning. The topic of working hours per day/week has become a political and economic issue*.

As management sees it, many industrial and service systems require two or more work crews to utilize machinery and keep processes going through much or all of the 24 h day and over the 7 day week. Thus, new forms of discontinuous systems are developing, such as 3-shift systems that cover 24 h, weekly or irregular rotation, work on weekends and 6 or 7 consecutive similar shifts.

10.5.2 Shift Systems

Shift work is different from "normal" day work in such that work is performed regularly during times other than morning and afternoon; and/or at a given workplace, more than that one shift is worked during the 24 h day. A shift may be shorter or longer than an 8 h work period.

Many diverse shift systems exist*. For convenience, they are usually classified into several basic patterns, but any given shift system may comprise aspects of several patterns. Four particular features of shift systems are especially important*:

1. Does a shift extend into hours that would normally be spent asleep?
2. Is a shift worked throughout the entire 7 day week, or does it include days of rest, such as free weekend?
3. Is there one shift or are there two, three, or more shifts per day?
4. Do shift crews rotate or do they work the same shifts permanently?

These aspects, shown in Fig. 10.4, are of particular concern with respect to the welfare of the shift worker, the work performance, and the organizational scheduling.

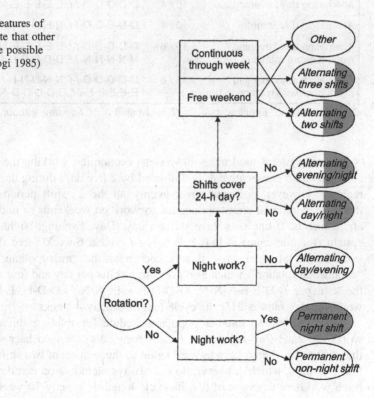

Fig. 10.4 Key Features of shift systems. Note that other shift attributes are possible (adapted from Kogi 1985)

Other critical identifiers of shifts and shift patterns are the starting and ending time of a shift; the number of workdays in each week; the hours of work in each week; the number of shift teams; the number of holidays per week or per rotation cycle; the number of consecutive days on the same shift, which may be a fixed or a variable number; and the schedule by which an individual worker either works or has a free day or free days.

Table 10.2 displays several permanent or a weekly rotating schedules, each with 5 workdays per week.

Table 10.2 Examples of shift systems with five workdays per week

System	Work days / free days	Shift sequence
Permanent day shift	5 / 2	D-D-D-D-D-*f-f*, D-D-D-D-D-*f-f*, …
Permanent evening shift	5 / 2	E-E-E-E-E-*f-f*, E-E-E-E-E-*f-f*, …
Permanent night shift	5 / 2	N-N-N-N-N-*f-f*, N-N-N-N-N-*f-f*, …
Rotating shifts:		
Alternating day / evening	10 / 4	D-D-D-D-D-*f-f*, E-E-E-E-E-*f-f*, …
Alternating day / night	10 / 4	D-D-D-D-D-*f-f*, N-N-N-N-N-*f-f*, …
Alternating day / evening / night; forward rotation	15 / 6	D-D-D-D-D-*f-f*, E-E-E-E-E-*f-f*, N-N-N-N-N-*f-f*, D-D-D-D-D-*f-f*, …
Alternating day / evening / night; backward rotation	15 / 6	D-D-D-D-D-*f-f*, N-N-N-N-N-*f-f*, E-E-E-E-E-*f-f*, D-D-D-D-D-*f-f*, …

D: Day shift   E: Evening shift   N: Night shift   *f: free day, without scheduled shift*

In most systems used today in western economies, working the same 8 h shift continues for 5 consecutive days, followed by 2 free days during the weekend. This regimen, however, does not cover evenly all the 21 shift periods of the week; therefore, additional crews are needed to work on weekends or under other "odd" arrangements. If one uses three shifts a day (Day, Evening, Night, *free*) the shift system (for one team) is D-D-E-E-N-N-*f-f* with a 6-work/2-free day ratio and a cycle length of eight days; this is known as the "metropolitan rotation". The "continental rotation", which also uses three shifts per day and four shift crews, has the sequence D-D-E-E-N-N-N, *f-f*-D-D-E-E-E, N-N-*f-f*-D-D-D, E-E-N-N-*f-f-f*. Its work/free day ratio is 21/7, its cycle length exactly 4 weeks.

Table 10.3 shows another popular schedule for rotating shifts. Each person works the same shift twice, then rotates forward to the next later shift and works that shift twice, then rotates forward again to the next set of two shifts. Then follow four free days, which, however, do not always include weekend days. The system has a work/free day ratio of 6/4, its cycle length is exactly 10 weeks; on average, each week has 33.6 work hours.

Table 10.3 "6/4" rotation shift schedule (adapted from Knauth 2007a, b)

Week	Mo	Tu	We	Th	Fr	Sa	Su	Mo	Tu	We	Th	Fr	Sa	Su
1, 2	D	D	E	E	N	N	free	free	free	free	D	D	E	E
3, 4	N	N	free	free	free	free	D	D	E	E	N	N	free	free
5, 6	free	free	D	D	E	E	N	N	free	free	free	free	D	D
7, 8	E	E	N	N	free	free	free	free	D	D	E	E	N	N
9, 10	free	free	free	free	D	D	E	E	N	N	free	free	free	free

D: Day shift   E: Evening shift   N: Night shift   *free:* day without scheduled shift

The ratio of work days versus free days in a complete cycle is an important characteristic of any shift system. Table 10.4 presents several other features that describe different shift systems.

Table 10.4 Characteristics of shift arrangements (adapted from Kogi 1985)

Cycle Length in days	$W + f$
Number of free days per year	$365\,f\ /\ (W + f)$
Number of days worked before the same set of shifts re-occurs on the same day of the week	$W + f$ if $(W + f)$ is a multiple of 7; or $7(W + f)$ if $(W + f)$ is not a multiple of 7

W: Sum of work days in one shift cycle.   *f*: sum of free days in one shift cycle

10.5.2.1   Flextime

The systems just described usually have traditional 8 h work shifts. A newer trend is toward "flextime," a flexible arrangement of work hours in a shift. Sliding time (akin to the German *Gleit-Zeit*) is a suitable description of this arrangement because it allows the employee to slide the prescribed number of working hours per shift (8 h, for example) within a longer block of time (such as 10 h) but such that the work time covers a "core" time (say, of 4 h) during which all workers must be present. Thus, by floating the day's working time across the core period, a worker can select the starting and closing times of work. Table 10.5 lists potential advantages and disadvantages of flextime.

Table 10.5 Advantages and disadvantages of flextime (adapted from Knauth 2007a, b, Kroemer et al. 2003, Kroemer Elbert et al. 2018)

POSITIVE
> Appeals generally to employees and employers
> Makes available work-free time at the employee's choosing
> Does not result in reduced employee pay
> Reduces commuting traffic problems
> Less fatiguing for workers
> Increases job satisfaction
> Recognizes and utilizes employees' individual differences
> Reduces tardiness
> Reduces absenteeism
> Reduces employee turnover
> Increases performance

NEGATIVE
> Makes it difficult to cover jobs at all required times
> Makes it difficult to schedule meetings or training sessions
> Reduces communication within the organization
> Increases energy and maintenance cost
> Requires more sophisticated planning, organization, and control
> Requires special recording of work time
> Requires additional supervisory personnel

10.5.2.2   Compressed Workweeks

Flextime is often, but not necessarily, combined with compressed workweeks, in which the regular total number of weekly work hours (such as 40 h/week) are condensed into fewer than the usual (say, 5) work days per week. This results in more work hours per day, but fewer work days. For example: the common 40 h of work per week may be performed in only 4 or even 3 days (instead of 5 days) per week. This allows the worker to have 3 or 4 free days each week. This option is attractive to many employees and employers: There are more work-free days, fewer trips to and from work, and fewer set-ups and close-downs at work. However, there are concerns about increased fatigue due to long workdays and reduced performance and safety—see Table 10.6.

The type of work to be performed determines, to a great extent, whether compressed workweeks can or should be used. Thus, long working days have been

Table 10.6 Advantages and disadvantages of compressed workweeks/extended workdays (adapted from Knauth 2007a, b, Kroemer et al. 2003, Kroemer Elbert et al. 2018)

POSITIVE
    Appeals generally to employees and employers
    Makes available more consecutive days away from the job at
       employee's choosing
    Reduces commuting problems and costs
    Affords more time per day for scheduling meetings or training
    Has fewer startup and warm-up periods
    Increases production rates
    Improves quantity and quality of services to the public

NEGATIVE
    Requires overtime pay
    Decreases job performance due to long work hours
    Tends to fatigue workers
    Furthers tardiness and early departure from work
    Increases absenteeism
    Increases on-the-job accidents
    Increases energy and maintenance costs
    Makes it difficult to schedule child care and family life during the
    workweek

mostly used in cases where one waits on "standby" during sections of the shift, such as firefighters do. Also, activities that require only few or small physical efforts, and which are diverse and interesting, have been done in long shifts. Examples include nursing, clerical and administrative work, technical maintenance, computer supply operations, and supervision of automated processes. Long shifts in manufacturing, assembly, machine operations, and other physically intensive jobs are less likely to be suited to a compressed workweek.

Existing information on the outcomes of compressed work weeks has been gathered mostly in psychological tests, from statements of employees, and by scrutinizing performance and safety records in industry. The results are contradictory, spotty, and apparently depend much on the given work conditions. In some cases, production and performance are high shortly after introduction of a compressed workweek but fall off after prolonged periods on such a schedule—however, other observations have not shown this trend. The people involved often indicate significantly increased satisfaction, probably more related to ease of arranging personal time than to improvements of the work. Efficacy of the organization may improve with the compressed schedule; however, usually these positive effects are not strong.

Intensive work, especially physical labor, during the whole length of a very long work shift (such as of 12 h) is likely to produce drowsiness and some reduction in cognitive abilities, motor skills, and in general performance during the course of the long work shift. It appears that there is an increased potential for a fatigued worker to take careless shortcuts to complete a job, and that work practices may become less competent in tedious and highly repetitive tasks during very long work shifts, as the workweek progresses*.

10.6   Suitable Shift Systems

By nature, the human is used to daylight activity, with the night reserved for rest. This appears to be an inherent feature, governed by endogenous circadian rhythms. Night work, then, seems to be "unnatural"; however, this does not necessarily mean that it is harmful. But it appears that working at night generates stress, severe in some cases, depending on the circumstances and the person.

Organizational criteria by which to judge the suitability of shift systems include the number of shifts per day, the length of every shift, or the times of the day during which work is needed or not needed; the coverage of the week by shifts; shift work on holidays; and so forth. Management usually makes the decisions on these "independent variables".

There are "dependent variables" as well which are important to managers: one is the performance of workers on shift schedules. Is the same work output to be expected regardless of the time of work during the 24 h day? Are specific activities better done at certain shifts? Do changes in shift work scheduling affect the worker's output? Is work on certain shifts more likely to be associated with accidents and injuries?

The well-being of the shift worker is another "dependent variable" of great importance. Do certain shift regimes disturb physiological or psychological health? For example, when a night shift needs to be worked, how does the inability to have normal night sleep affect the worker's health? What are the effects of shift work on social interactions with family, friends, and the society in general?

10.6.1   Health and Well-Being of Shift Workers

Evidently, circadian rhythm is remarkably resistant to sudden large changes. Since this internal system is so stable, theoretical considerations, common sense and personal experiences all suggest that the normal synchrony of behavior in terms of nightly rest and daily activity should be maintained as much as practical. Thus, work schedules should be arranged in accordance with the internal system or, if this is impossible (for example, if night work is necessary), to disturb the internal cycles as little as possible. One of the logical conclusions from this approach is that, if work activities contradict the internal rhythm, these should be kept as short as feasible so that the worker can return to the "normal" cycle quickly. For example, scheduled single night shifts should be interspersed between blocks of several normal workdays, instead of requiring a worker to work several night shifts in sequence, as in the metropolitan and continental rotations discussed earlier. Such a series of night shifts upsets the internal clock, while a single night shift would not disturb the entrained cycle as severely.

The other solution, both theoretically sound and supported by experience, is to entrain new circadian rhythms. It takes regular and strong zeitgebers to overpower the regular signals, especially light and darkness. For shift work, this means that the same setup (such as working the night shift) should be maintained for long periods of time (weeks, even months) and not be interrupted by different arrangements (in theory, not even on free days, such as on weekends). It appears that certain people are more willing and able to conform to such regular "non-day" shift regimens than others.

Health problems of shift workers* are often mentioned or suspected, but it is difficult to prove negative effects of a dangerous intensity. Workers on a steady evening shift have been found to sleep about half an hour longer than those on the day shift. The total sleep time of night shift workers is, on average, about half an hour shorter than the sleep of workers who are permanently on day shift. Further, night shift workers often complain about the reduced quality of sleep that they receive during the day, with noise particularly disturbing*.

Some researchers have found statistically significant health complaints as a function of shift work, particularly digestive disorders and gastro-intestinal complaints, yet other researchers have failed to prove significance. Overall, mortality of night shift workers seems to be about the same as of workers on other shifts. However, it is evident that persons who suffer from health disturbances are more negatively affected by night shifts than by other shift arrangements. It also appears that older workers who experience deteriorating health and difficulties in getting enough restful sleep (both phenomena seem to increase with age) may be more negatively affected by shift work than younger workers. In contrast, other older shift workers have relatively fewer health, sleep and social problems than their younger colleagues, probably due to less need for sleep, a shift to "morningness" and looser day-time social attachments.

10.6.2 Performance of Shift Workers

Due to reduced quantity and quality of sleep, many night workers suffer from a chronic state of partial sleep deprivation. Negative effects of sleep deprivation on behaviors have been well demonstrated, as discussed earlier. In some tasks, the interaction of circadian discrepancies between body state and work demand as well as of sleep deprivation may result in significant detriments to night work performance, including safety. During the first night shift or shifts, performance is most likely to be impaired in the very early morning hours, with the worst performance around 4 o' clock, coinciding with a low in body functions. Such impairment, which may be absent or minimal for cognitive tasks, varies in level similar to the effects of small amounts of alcohol. Accident injuries may be high on night shifts, but related statistics are usually confounded by variables other than the shift factor, such as the work task, worker age and skill, and schedules of shift work*.

Tolerance of shift work differs from person to person and varies over time. Three out of ten shift workers have been reported to leave shift work within the first three years*. Tolerance of those remaining on shift work depends on personal factors (age, personality, health troubles; the ability to be flexible in sleeping habits and to overcome drowsiness), on social-environmental conditions (family composition, housing conditions, social status), and of course on the work itself (workload, shift schedules, compensation). These factors interact, and their importance differs widely from person to person and changes over one's work life. Evening shift workers may be impacted particularly in their social and domestic relations, while night shift workers are more affected by insufficient sleep and suffer from being required to work while physiologically in a resting stage. However, physiological and health effects are not abundant in shift workers who have been on such assignments for years, possibly due to acclimation but also because persons who cannot tolerate these conditions abandon shift work soon after trying it.

10.6.3   Social Interactions of Shift Workers

A major problem associated with shift work is the difficulty of maintaining customary social interactions when the work schedule requires working or sleeping during times in which social relations usually occur. This complicates family relations as well as interaction with friends and participation in public events such as sports. Common daily activities are also affected, such as shopping or watching television.

Needs for social interactions are individually and culturally different. For example, parents of young children want to be home to be with their offspring and are unlikely to accept unusual work assignments that keep them away. In contrast, older persons who do not interact with their older children so intensely may be more inclined to work "non-normal" hours.

Caution is required, however, when transferring one's own living conditions and social expectations to different countries and cultures: certain events or conditions may or may not be present, may be regularly scheduled at different times, and may be considered of different value to individuals and the society at large. The siesta time of sunny and warmer climate zones is not commonly known in other regions. Some shops that are open continuously in the USA may be shuttered around mid-day, closed in the late afternoon, or locked on weekends in other regions. Family ties are more important in some cultures than in others and also vary among individuals. Television plays a large role in daily life of some groups of people and not in others. Thus, statements regarding the effects of shift work on social interactions that apply in one situation may not be pertinent in another.

10.7 How to Select a Suitable Work System

The foregoing discussions make it clear that, if at all possible, the working hours should be during daylight, in the morning and afternoon. However, in many cases this "normal" arrangement must be replaced by shift work which covers the late afternoon and evening hours, and/or the night.

As Fig. 10.5 illustrates, there is a seemingly inevitable fall in performance during night work, related to the circadian rhythm. This is of particular concern when night work period follows poor sleep and, especially, when the work shift is exceedingly long–more on this below. Poor sleep is likely if the assignment to night work is new.

A permanent shift assignment to evening or night shifts should allow the worker's internal rhythms to entrain on this imposed rest/work pattern. However, this approach is not as convincing as it might appear: if the worker returns to a daytime schedule during a free weekend (which most shift arrangements include) and interrupts the rhythm entrainment, the shift assignment is not truly constant and permanent. Furthermore, strong zeitgebers during the 24 h day (such as light and dark) remain intact even for a person on a regular late or night shift: these time markers hinder a complete re-entrainment of the internal functions. These conditions support the opposite conclusion, also well founded in theory: it is better to work only occasionally during darkness. If there looms only one evening or night

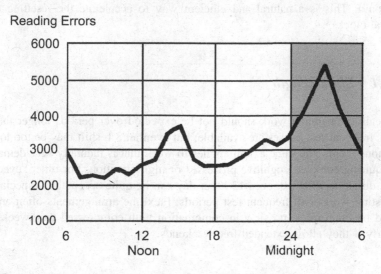

Fig. 10.5 Accumulated errors in gas meter readings, 1912–1931 (adapted from Lehmann 1962)

shift, most people are able to get through that unusual work session without much detriment, and they remain entrained on the usual 24 h cycle.

> Crews of airplanes who must cross time zones in their long distance flights and catch some sleep at their destination before returning experience several problems. The first is that the quality and length of sleep at the stopover location is often less satisfactory than at home. The resulting tiredness may be masked or counteracted by use of caffeine, even drugs. The second problem is the extended time of duty, which includes preflight preparations, the flight period itself, and the wrap-up after arrival at the stopover. Direction of flight may present another problem: negative effects are larger after an eastward flight than in westward direction. Further, crew members over 50 years of age are more affected than their younger colleagues.
>
> Recommendations for the shift arrangement for flight crews are fairly well established: In principle, flight crews should adhere to well-planned timing which duplicates, as far as possible, the sleep-wake activities at home. Accordingly, at the destination the crew goes to bed at their regular home time and also gets up at the regular home time. Thus, the crew members retain their regular circadian rhythm. Of course, their next flight duty should continue to be during their regular time of wakefulness*.

Entraining a new circadian rhythm needs time. Air travelers use a general rule that "it takes almost one day to adjust to each one hour of time change". Exposure to sunshine, or bright light of least about 1000 lx, at times when one feels sleepy but should be awake, counteracts the body's use of its own sleep-inducing hormone melatonin. This is a natural and efficient way to accelerate the resetting of the internal clocks*.

10.7.1   Shift Length

Physically demanding work should not be expected over periods longer than 8 h unless frequent rest pauses are available; but even an 8 h shift may be too long for strenuous labor. The same axiom applies to work that is mentally very demanding by requiring complex cognitive processes or high attention. For other "everyday" work, durations of 9, 10, even 12 h per day can be quite acceptable, especially for interesting work with frequent rest periods. Flextime arrangements often are welcomed by employees, possibly in combination with compressed workweeks, particularly if they allow extended free weekends.

In the USA, health care providers have habitually worked exceedingly long shifts. Nurses on shifts longer than 12 h can exhibit significantly decreased vigilance on the job, as well as significantly increased risks of making a medical error and of suffering an occupational injury. Physicians-in-training working traditional on-call shifts of more than 24 h are at greatly increased risk of making a serious or even fatal medical error, of experiencing an occupational injury, and of having a motor vehicle accident when driving after work.

It has been argued that large·inter-individual differences exist in the responses to acute sleep loss or chronic sleep deprivation, implying that physicians are particularly resistant to such effects. Another argument sometimes made is that patients need more continuity of care as provided by fewer shift changes, or even that resident physicians need these long shifts to train to function specifically when they are sleep-deprived*. Similar arguments can be brought forth regarding airplane crews on extremely long flights, for example, or concerning commercial truck drivers on long-lasting drives. There is no evidence that exceptionally many physicians (or nurses, or pilots, or drivers) are indeed resistant to the effects of sleep loss, and there are no means to monitor their resilience levels. Furthermore, an ethical question remains to be answered by both individuals and by their managers: how can shift schedules be optimized to assure the safety of patients (passengers, road users) from unintended misdeeds of tired operators?

As a part of the decision to select one of the many possible shift plans, establishing a set of criteria helps to arrive at justifiable and orderly judgments. Some general suggestions to consider include:

- Daily work duration normally should not be more than 8 h.
- The number of consecutive evening or night shifts should be as small as possible; it is best to intersperse only one single late shift between day shifts. (The alternative solution is to stay permanently on the same late shift.)
- A full day of free time should follow every single late shift.
- Each shift plan should contain consecutive work-free days, preferably including a weekend.

During evening or night shifts, workplaces should have high illumination, 2000 lx or more, to suppress production of the hormone melatonin, which causes drowsiness. Furthermore, environmental stimuli may help to keep the worker alert and awake, such as occasional "stirring" music and provision of (hot) snacks and of beverages (hot and cold, possibly caffeinated). The work task should be interesting and rewarding, because routine boring tasks are more difficult to perform efficiently and safely during the night hours.

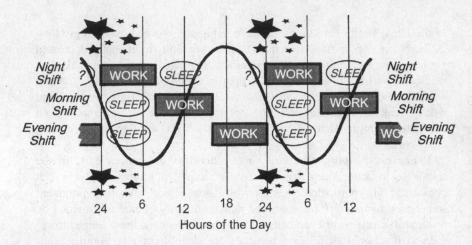

Fig. 10.6 Work-sleep sequences of persons working night, morning or evening shifts

The shift worker should use coping strategies for setting the biological clock, obtaining restful sleep, and maintaining satisfying social and domestic interactions. For example, sleep may be taken best in the morning directly after a night shift, not in the afternoon just before the workshift. Sleep time should be regular and kept free from interruptions. The shift worker should seek to gain the family's and friends' understanding of the need to rest. Certain times of the day should be set aside specifically and regularly to be spent with family and friends.

Figure 10.6 illustrates and summarizes the foregoing discussions. On the background of the daylight and darkness sequences of the day, it illustrates the circadian rhythm of wakefulness and rest. It shows that working during daylight hours complements the activity period of the internal rhythm while working during the late and night shifts disagrees with the natural inclination to sleep at these times. Sleep after an evening shift is still fairly easy to attain whereas restful sleep following a night shift is difficult to achieve during daylight.

Notes

The text contains markers, * , to indicate specific references and comments, which follow.

The menstrual cycle may affect the capacity to perform certain tasks: Patkai (1985) cites a study in which secretaries showed the highest typing speed before the onset of menstruation and during the first three menstrual days. The idea of a higher effort on these days was rejected by the secretaries since they considered themselves to be working at full capacity all the time.

Out-dated hypotheses of reduced performance: Gamberale et al. (1975), Smekal et al. (2007).

Circadian rhythm extremes during the day can be quite different among individuals and may change even within a person: Folkard and Monk (1985), Refinetti (2006) provides a history of research on circadian rhythms.

Desynchronization of oscillators; *for example, the sleep/wake cycle stays at 24 h, while the temperature rhythm may free-run at a period of 25 h:* Wever (1985), Czeisler and Gooley (2007).

"Morning types" and "evening types": Horne and Oestberg (1976), Di Milia et al. (2004), Refinetti (2016).

No explanation for what sleep does and we must sleep: Horne (1988a, 2006), Siegel (2003); The Royal Society of Medicine (2005).

Under sleep deprivation or during ebb times of circadian rhythms, there is no significant decrease in physiological capabilities: Refinetti (2006).

For regular tasks, there may be an overcapacity of cerebral neural networks; so, many are not fully used: Horne (1988b).

Dreams are the "cinema of the mind": Horne (1988b, p. 313).

No true non-sleepers seem to exist among healthy persons: Horne (1988, p. 186).

When people must stay awake for long periods of time, memory use degrades, both in long-stored memory information and short-term memory: Froeberg (1985).

After missing four nights of sleep, very few people are able to perform adequately even if their motivation is very high: Froeberg (1985).

Experience of sleep inertia: German *"schlaftrunken"*—"sleepdrunk".

Review of the sleep inertia literature: Hilditch et al. (2016).

Naps of at least 2 h taken in the late evening or during nighttime have positive effects lasting several hours: Gillberg (1985) Rogers et al. (1989).

"Strategic napping" is insufficient to mitigate severe performance impairment introduced by duty extended beyond 16 h: Hartzler (2014), St Hilaire et al. (2019).

Taking drugs, such as amphetamines, can restore performance to a nearly normal level even after three nights without sleep: Froeberg (1985).

Police Survey: Ogeil et al. (2018).

The topic of working hours per day/week has become a political and economic issue: Anonymous (14 November 2018) 100 years later, Germany calls the 8 h working day into question. DPA News/The local. de@thelocalgermany.

Many diverse shift systems exist: Described, for example, by Colquhoun (1985), Folkard and Monk (1985), Hurrell and Colligan (1985), Colligan and Tepas (1986), Tepas and Monk (1986), Monk et al. (1996), Smith et al. (1998), Hornberger et al. (2000), Folkard and Tucker (2003), Knauth (2006, 2007a, 2007b), Potterdell (2004), Monk (2006), Popkin et al. (2006), Kroemer (2017), Kroemer Elbert et al. (2018).

Four particular features of shift systems are especially important: Kogi (1985).

A fatigued worker may become less competent in tedious and highly repetitive tasks during very long work shifts: Folkard, Tucker (2003), Smith et al. (1998), Knauth (2006, 2007a).

Health problems of shift workers: Costa (2003, 2010, 2015), Bovin and Boudreau (2014), Pallesen et al. (2009), Booker et al. (2019).

Persons on night shift often complain about the reduced quality of sleep that they receive during the day: Carvalhais et al. (1988); Pallesen et al. (2009), Booker et al. (2019), Wong et al. (2019).

Accident injuries may be high on night shifts: Monk (1989, 2006).

Tolerance of shift work differs. Three out of ten shift workers have been reported to leave shift work within the first three years: Bohle and Tilley (1989).

Flight crew sleeping arrangements: Graeber (1988).

Exposure to sunshine or bright light is a natural and efficient way to accelerate the resetting of the internal clocks: Czeisler et al. (1990).

Sleep-deprived nurses and physicians: Van Dongen (2006), Czeisler (2006, 2009), Lockley et al. (2007).

Summary

Human body functions and human social behavior follow internal rhythms. Under regular living conditions, these temporal programs are well established and persistent. One such internal rhythm is the human female menstrual cycle of about 28 days; another is the set of daily fluctuations, called circadian rhythm. The menstrual cycle has little or no effects on work performance; however, circadian rhythm and execution of work are closely related.

The circadian rhythm is evident in body temperature, heart rate, blood pressure, and hormonal excretions and in the associated patterns of sleep (naturally during the night) and of daytime activities. This rhythm can be desynchronized and put out of order if time markers (zeitgebers) during the 24 h day are changed: activities required from the human at unusual times can have such effect. Sleep deprivation and tiredness influence human performance in various negative ways. Mental performance, attention, and alertness usually are reduced by moderate sleep deprivation, but execution of most physical activities is not. Furthermore, concerns exist that disturbing the internal rhythm, such as by certain types of shift work, might have negative health effects. Being excluded by shift work from participating in family and social activities is difficult for many persons.

Shift work is often desired or even necessary for organizational/economic reasons. Individual acceptance of working in shifts depends on a complicated balance of professional and personal concerns, including physiological, psychological, and social aspects. Of ten persons assigned to shift work, seven may be likely to stay on this schedule while the others drop out. Shift work may worsen pre-existing health conditions. Workers on permanent night shifts often complain about insufficient sleep and general fatigue.

Shift workers' task performance mainly depends on four factors: the type of work; the organization of work activities; the internal circadian rhythm of the body; and the individual's motivation and interest in the work. Each factor can govern, influence, or mask the effects of the other factors on task execution. Mental performance capabilities are particularly prone to deterioration during long-continued work and accompanying sleep loss. Long task duration, monotony, complexity, and repetitiveness all have decidedly negative effects. Performance is especially impacted during "low" periods of the circadian cycle such as in the early morning hours. Sleep-deprived persons do not perform well on either monotonous or complex tasks over long periods of time. Exceptionally high motivation may reduce or prevent performance decrements, at least temporarily.

A shift duration of eight hours of daily work is usually adequate, but shorter times for highly (mentally or physically) demanding jobs may be advantageous; longer times (such as 9, 10, even 12 h) may be acceptable for some types of work.

Suggestions for acceptable regimes of working hours and shift work include:

- *Job activities should follow entrained body rhythms.*
- *It is preferable to work during the daylight hours.*
- *Evening shifts are preferred to night shifts.*
- *If shifts are necessary, two opposing rules apply: either work only one evening or night shift per cycle, then return to day work, and keep weekends free; or stay permanently on the same shift, whatever that is.*

Glossary

Circadian rhythm Regular oscillations in body functions which repeat after about 24 h. This daily rhythm is endogenously (see there) generated and modulated by 24 h environmental repetitions, especially by light and darkness related to the Earth's rotation. (From the Latin *circa*, about, and *dies*, the day; see also diurnal rhythm).

Desynchronization Loss of synchronization (see there) within an organism between a rhythm and its zeitgeber (see there), or between two rhythms.

Diurnal rhythm Regular oscillations in body functions during the day (from the Latin *diurnus*, of the day; see also circadian rhythm).

Endogenous Originating inside an organism or system.

Entrain To synchronize a self-sustaining oscillation or oscillatorOscillator.

Ergonomics The application of scientific principles, methods and data drawn from a variety of disciplines to the design of engineered systems in which people play significant roles.

Exogenous Originating outside an organism or system.

Free-run The state of a self-sustaining oscillation in the absence of effective zeitgeber (see there) or other environmental events that may affect the period of the oscillation.

Infradian rhythm Regular oscillations in body functions with a frequency lower than circadian; that is, they repeat after more than 24 h.

Masking Disruption in the appearance of an overt rhythm caused by an external event without a direct effect on the period or phase of a pacemaker (see there).

Pacemaker An entity capable of generating an endogenous rhythm and of imposing this rhythm on one or more other entities.

Running free Being in the state of a self-sustaining oscillation in the absence of effective zeitgeber (see there) or other environmental events that may affect the period of the oscillation.

Self-sustaining oscillation An oscillation, caused by a pacemaker (see there), which continues without external support.

Synchronization The action of causing two or more processes to start at the same time and/or to proceed at the same rate.

Ultradian rhythm Regular oscillations in body functions with a frequency higher than circadian; that is, they repeat after fewer than 24 h.

Zeitgeber A time setter; agent or stimulus capable of resetting a pacemaker (see there) or synchronizing a self-sustaining oscillation (see there). (From the German *Zeit*, time; and *Geber*, giver).

References

Bohle P, Tilley AJ (1989) The impact of night work on psychological well-being. Ergonomics 34 (9):1089–1099

Booker LA, Barnes M, Alvaro P, Collins A, Chai-Coetzer CL, McMahon M, Lockley SW, Rajaratnam SM, Howard ME, Sletten, TL (2019) The role of sleep hygiene in the risk of shift work disorder in nurses. Sleep 1–8. https://doi.org/10.1093/sleep/zsz228

Bovin DB, Boudreau P (2014) Impacts of shift work on sleep and circadian rhythms. Pathol Biol (Paris) 62(5):292–301. https://doi.org/10.1016/j.patbio.2014.08.001 Epub 2014 Sep 22

Carvalhais AB, Tepas DI, Mahan RP (1988) Sleep duration in shift workers. Sleep Research 17:109

Colligan MJ, Tepas DI (1986) The stress of hours of work. Am Indus Hygiene Assoc J 47 (11):686–695

Colquhoun WP (1985) Hours of work at sea: watch-keeping schedules, circadian rhythms and efficiency. Ergonomics 28(4):637–653

Costa G (2003) Factors influencing health of workers and tolerance to shift work. Theor Issues Ergon Sci 4:263–288

Costa G (2010) Shift work and health: current problems and preventive actions. Safe Health Work 1(2):112–123

Costa G (2015) Sleep deprivation due to shift work. Handb Clin Neurol 131:437–446

Czeisler CA (2006) Work hours, sleep and patient safety in residency training. Trans Am Clin Climatol Assoc 117:159–189

Czeisler CA (2009) Medical and genetic differences in the adverse impact of sleep loss on performance: ethical considerations for the medical profession. Trans Am Clin Climatol Assoc 120:249–285

Czeisler CA, Gooley JJ (2007) Sleep and circadian rhythms in humans. Cold Spring Harb Symp Quant Biol. https://doi.org/10.1101/sqb.2007.72.064

Di Milia L, Smith PA, Folkard S (2004) Refining the psychometric properties of the circadian type inventory. Personality Individ Differ 36:953–1964

Folkard S, Monk TH (eds) (1985) Hours of work. Wiley, New York

Folkard S, Tucker P (2003) Shift work, safety and productivity. Occup Medicine 53:95–101

Froeberg JE (1985) Sleep deprivation and prolonged working hours (Chapter 6). In: Folkard S Monk TH (eds) Hours of work, Wiley, Chichester

Gamberale F, Strindberg L, Wahlberg I (1975) Female work capacity during the menstrual cycle: physiological and psychological reactions. Scand J Work Environ Health 1(2):120–127

Gillberg M (1985) Effects of naps on performance (Chapter 7). In: Folkard S, Monk TH (eds) Hours of Work, Wiley, Chichester

Graeber RC (1988). Aircrew fatigue and circadian rhythmicity (Chapter 10). In: Wiener EL, Nagel DC (eds) Human factors in aviation, Academic Press, San Diego

Hartzler BM (2014) Fatigue on the flight deck: the consequences of sleep loss and the benefits of napping. Accid Anal Prev 62:309–318. https://doi.org/10.1016/j.aap.2013.10.010. Epub 2013 Oct 19

Hilditch CJ, Dorrian J, Banks S (2016) Time to wake up: reactive countermeasures to sleep inertia. Ind Health 54(6):528–541. https://doi.org/10.2486/indhealth.2015-0236

Hornberger S, Knauth P, Costa G, Folkard S (eds) (2000) Shiftwork in the 21st century. Lang, Frankfurt

Horne J (1988) Why we sleep: the functions of sleep in humans and other mammals. Oxford University Press, Oxford

Horne JA (2006) Sleepfaring: a journey through the science of sleep. Oxford University Press, Oxford

Horne J, Oestberg O (1976) A self-assessment questionnaire to determine morningness-eveningness in human circadian rhythms. Internatl J Chronobiology 44:97–110

Hurrell JJ, Colligan MJ (1985) Alternative work schedules: flextime and the compressed workweek (Chapter 8). In: Cooper CL, Smith MJ (eds) Job stress and blue collar work. Wiley, New York

Johnson MP, Czeisler CA, Duffy JF, Brown EN, Ronda JM, Kronauer RE (1990) Exposure to bright light and darkness to treat physiologic maladaptation to night work. New York J Medicine 322(18):1253–1259

Knauth P (2006) Workday length and shiftwork issues (Chapter 29). In: Marras WS, Karwowski K (eds) The occupational ergonomics handbook, Interventions, controls, and applications in occupational ergonomics, 2nd ed. CRC Press, Boca Raton

Knauth P (2007a) Extended work periods. Indus Health 45:125–136

Knauth P (2007b) Schicht-und Nachtarbeit. In: Landau K (ed) Lexikon arbeitsgestaltung. Gentner, Stuttgart

Kogi K (1985) Introduction to the problems of shiftwork (Chapter 14). In: Folkard S, Monk TH (eds) Hours of work. Wiley, New York

Kroemer KHE (2009, 2017) Fitting the human, 7th edn. CRC Press, Boca Raton

Kroemer KHE, Kroemer HJ, Kroemer-Elbert KE (2003) Ergonomics: How to design for ease and efficiency. 2nd ed. Prentice Hall, New York

Kroemer Elbert KE, Kroemer HB, Kroemer Hoffman AD (2018) Ergonomics: how to design for ease and efficiency, 3rd edn. Academic Press, London

Lockley SW, Barger LK, Ayas NT, Rothchild JM, Czeisler CA, Landrigan CP (2007) Effects of health care provider work hours and sleep deprivation on safety and performance. Jt. Comm J Qual Patient Saf 33(Suppl 11):7–18

Monk TH (1989) Shiftworker safety: issues and solutions. In: Mital A (ed) Advances in industrial ergonomics and safety I. Taylor & Francis, Philadelphia, pp 887–893

Monk TH (2006) Shiftwork (Chapter 32). In: Marras WS, Karwowski K (eds) The occupational ergonomics handbook. Fundamentals and assessment tools for occupational ergonomics, 2nd ed. CRC Press, Boca Raton

Monk TH, Folkard S, Wedderburn AI (1996) Maintaining safety and high performance on shiftwork. Appl Ergonomics 27(1):17–23

Ogeil RP, Barger LK, Lockley SW, O'Brien CS, Sullivan JP, Qadri S, Lubman DI, Czeisler CA, Rajaratnam SMW (2018) Cross-sectional analysis of sleep-promoting and wake-promoting drug use on health, fatigue-related error, and near-crashes in police officers. BMJ Open 8(9): e022041. https://doi.org/10.1136/bmjopen-2018-022041. PMID: 30232109; PMCID: PMC6150149

Pallesen S, Bjorvatn B, Mageroy N, Saksvik-Lehouillier I, Waage S, Moen BE (2009) Measures to counteract the negative effects of night work. Scand J Work Environ Health 36(2):109–120

Patkai P (1985) The menstrual cycle (Chapter 8). In: Folkard S, Monk TH (eds) Hours of work, Wiley, Chichester

Popkin SM, Howard HD, Tepas DI (2006) Ergonomics of work systems (Chapter 28). In: Salvendy G (ed) Handbook of human factors and ergonomics 3rd edn. Wiley, New York

Potterdell P (2004) Work schedules (Chapter 3). In: Barling J, Kelloway EK, Frone MR (eds) Handbook of work stress. Sage, London

Refinetti R (2006, 2016) Circadian physiology, 2nd ed, 3rd ed. CRC Press, Boca Raton

Rogers AS, Spencer MB, Stone BM, Nicholson AN (1989) The Influence of a 1 h nap on performance overnight. Ergonomics 32(10):1193–1205

Siegel JM (2003) Why we sleep, Scientific American, November issue, pp 92–97

Smith L, Folkard S, Tucker P, Macdonald I (1998) Work shift duration: a review comparing eight hour and 12 hour shift systems. Occup Environ Med 55(4):217–229

Smekal G, von Duvillard SP, Frigo P, Tegelhofer T, Pokan R, Hofmann P, Tschan H, Baron R, Wonisch M, Renezeder K, Bachl N (2007) Menstrual cycle: no effect on exercise cardiorespiratory variables or blood lactate concentration. Med Sci Sports Exerc 39 (7):1098–1106

St Hilaire MA, Anderson C, Anwar J, Sullivan JP, Cade BE, Flynn-Evans EE, Czeisler CA (2019) Brief (<4 hr) sleep episodes are insufficient for restoring performance in first-year resident physicians working overnight extended-duration work shifts. Sleep 42(5):zsz041

Tepas DI, Monk TH (1986) Work schedules (Chapter 7.3). In: Salvendy G (ed) Handbook of human factors and ergonomics, 3rd ed. Wiley, New York

The Royal Society of Medicine (2005) History of sleep research. http://www.neuronic.com/rsm200sleep/presentations/sleepthenandnow.html

Van Dongen HP (2006) Shift work and inter-individual differences in sleep and sleepiness. Chronobiol Int 23(6):1139–1147

Wever RA (1985) Men in temporal isolation: basic principles of the circadian system (Chapter 2). In: Making shiftwork tolerable, Taylor & Francis, London

Wong IS, Popkin S, Folkard S (2019) Working time society consensus statements: a multi-level approach to managing occupational sleep-related fatigue. Ind Health 57(2). https://doi.org/10.2486/indhealth.sw-6

Chapter 11
Engineering Anthropometry

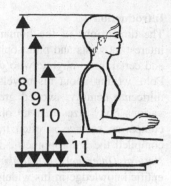

Overview

Anthropometric information describes the size of the human body. Traditional measurements mostly rely on bone landmarks to determine heights, breadths, depths, distances, circumferences, and curvatures. Newer assessment techniques rely on 3-dimensional topography of the body surface. Available current anthropometric information describes only a small number of civilian and military populations.

People come in a variety of sizes; their bodies are assembled in various proportions. Thus, designing to suit human bodies requires careful consideration; simply using some sort of statistical "average" will not do. Instead, for each body segment to be fitted, we must determine the specific dimensions that are critical for a specific design. For this, we usually select a minimal or a maximal value, or the range between them.

The Model
A full-body scanner captures the full geometry of an individual's body whereas most traditional measurements are isolated and not linked to each other. Statistical relations among recorded body dimensions can model and describe body contours, shapes, volumes, and masses, and motion envelopes. Specific measures explain size and mobility of body segments, especially of the hands. Descriptions of the overall size of the body, and of its motions, determine the sizes of equipment used for work and leisure, of workspaces, and of enclosures.

© Springer Nature Switzerland AG 2020
K. H. E. Kroemer et al., *Engineering Physiology*,
https://doi.org/10.1007/978-3-030-40627-1_11

Introduction

The dimensions of the human body and its segment proportions have been of interest to artists and philosophers, physicians and anatomists, rulers and generals, and certainly to anyone who designs and provides objects for human use. Marco Polo, writing about his travels from Italy eastward into China at the end of the thirteenth century, evoked great interest in Europe with his descriptions of the various body sizes that he observed. Physical anthropology as a recording and comparing science is often traced to his travel reports. Johann F. Blumenbach compiled the anthropometric information available in 1776 in his book, *On the Natural Differences in Mankind*. Alexander von Humboldt encompassed all scientific knowledge in his widely read five-volume *Kosmos* published from 1845 to 1862.

In the 19th century, anthropology split into special branches. Adolphe Quételet applied statistics to anthropological information in the mid-1800s. The science of biomechanics—see Chap. 4—emerged at the end of the 19th century. At that time, Paul Broca made extensive studies on the skull and drew far-reaching conclusions. The rapidly increasing diversity in anthropometric studies led to conventions of physical anthropologists, 1906 in Monaco and 1912 in Geneva, who agreed on standards for anthropometric techniques. In 1914, Rudolf Martin published the first edition of his text, *Lehrbuch der Anthropologie*, which became the authoritative guide and handbook for decades. From the 1960s on, new ergonomic tasks, developing engineering needs, advancing techniques in measuring and statistics brought about updates and re-directions in anthropometric methods and practices*. Automated full-body scanners, available since around 2000, can quickly capture individual body geometry and the recorded data allow 3D body modeling.

While compilations of measured data are available for some populations* on earth, information on the body sizes of many people is still missing. Global commercial interests und new efficient measurement techniques (see below) may lead to more complete anthropometric information.

11.1 Measurement Techniques

For measuring how tall a person is, traditional procedures employed four body postures: standing freely but stretched to maximum height; leaning backwards against a wall with the back flattened and stretched to maximum height; standing naturally upright, but not stretched; and lying on the back. The results of these measures of individual "stature" differ by a few millimeters when the subject either stands freely or leans; however "slumped" standing can reduce stature by several centimeters. Lying supine (routinely done with young children) results in the tallest measure. This example shows how standardization is necessary to assure uniformity in postures and results.

11.1.1 Terminology and Standardization

Traditional body measurements are defined by the two end-points of the distance measured, often bony landmarks such as in "elbow-to-fingertip length"; "stature" is formally defined as the vertical distance between the highest point of the skull and the vertical surface on which the upright person stands.

Figure 11.1 illustrates anatomical landmarks of the human body in the sagittal view; Fig. 11.2 shows landmarks in the frontal plane. Figure 11.3 contains reference planes, coordinate axes and descriptors; note that axis directions used in anthropometry may differ from directions used in other technical fields such as aeropace, for example.

For measurements, the subject is nude, or nearly so, and without shoes. To standardize the procedures, the subject's body is placed in defined poses: usually an upright posture, as already mentioned, with body segments at either 0, or 90 or 180° to each other. For example, the subject may be required to "stand erect; heels together; arms vertical, wrist and fingers straight and pointing downward." This is close to the so-called anatomical position—but differs from most of the postures naturally assumed when working. For measurements on a seated subject, the solid (flat and horizontal) surfaces of seat and foot support are arranged so that the thighs are horizontal, the lower legs vertical and the feet flat on their horizontal support; the upper body is upright like when standing.

For measurements that involve the head, it is usually positioned in the so-called Frankfurt plane: with both pupils at the same height, the right tragion (at the ear hole) and the lowest point of the eye socket in the skull (the rim of the right orbit) are placed on the same horizontal plane. This is simple to do with an isolated skull, but palpating and finding the orbit on a living subject is awkward. It is easier to position the head using the Ear-Eye Line (EE Line), which runs through the right ear hole and the junction of the right eyelids (the *external canthus*). This is illustrated in Fig. 11.4, which also shows the angle LOSEE between the EE Line and the line of sight (LOS). If one places the eye higher than the ear hole so that the EE Line tilts 11 degrees above the horizon, the head is approximately in the Frankfurt plane.

11.1.2 Classical Measuring Techniques

In conventional anthropometry, the measurer (usually a trained anthropologist) places the appropriate instrument at the point of interest on the subject's body. The measurement devices are quite simple. The most important device is the anthropometer, a graduated rod with a sliding branch perpendicular to the rod. The slider is placed on a body landmark and its distance from the fixed end of the rod appears on the scaled rod—see Fig. 11.5. The rod can be sectioned for short measurements and for transport and storage. A straight sliding caliper serves for short

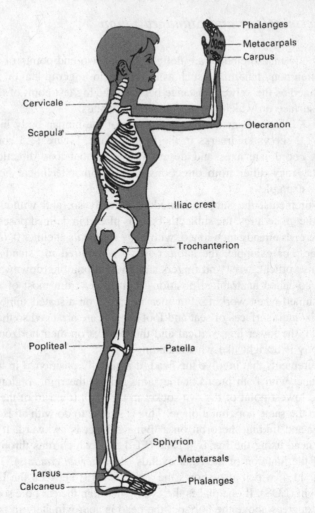

Fig. 11.1 Anatomical landmarks in the sagittal view

measurements, such as hand breadth or finger thickness; a spreading caliper has two curved branches, joined in a hinge. A cone provides measurements of the diameter around which fingers can close; circular holes of increasing sizes drilled in a thin plate indicate external finger diameters. Tapes, made of flat steel or of non-stretch woven material, serve for measurements of circumferences and curvatures. A graduated box is used for some foot measurements—also shown in Fig. 11.5. Scales, of course, measure weight.

Most instruments and procedures are elementary and the measured person easily understands their use. However, the processes can also be somewhat clumsy and certainly are time-consuming. Each measurement and tool must be selected in advance; what was not measured in the test session remains unknown unless the

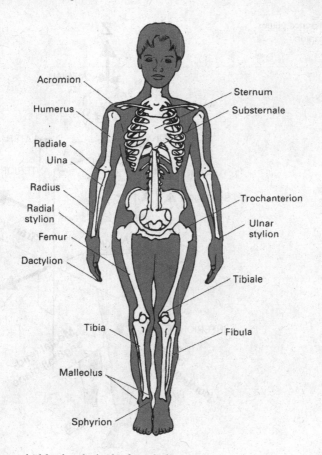

Fig. 11.2 Anatomical landmarks in the frontal view

subject is called back for more measuring. Another major disadvantage is that many of the conventional dimensions do not relate to each other in space. For example, as one looks at a subject from the side, stature, eye height, and shoulder height are located in different yet undefined frontal planes. Furthermore, certain parts of the body are sensitive to touch, such as the eyes.

Nevertheless, the classical "hands-on" anthropometric techniques remain practical for quick assessments on only a few subjects and can be used without modern resources or complex equipment even in remote locations.

11.1.3 New Measurement Methods

Three-dimensional full-body scans now have largely replaced conventional one-dimensional measures in large anthropometric surveys. They usually rely on a

Fig. 11.3 Reference planes, axes and descriptors

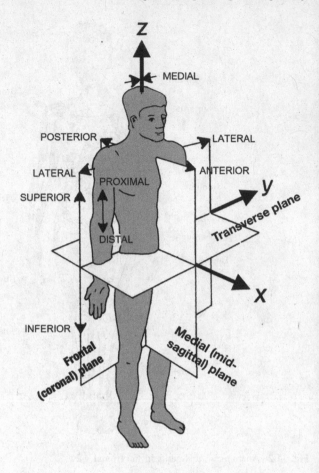

Fig. 11.4 Positioning the head in the Frankfurt Plane using the Ear-Eye Line

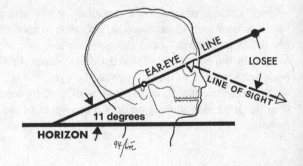

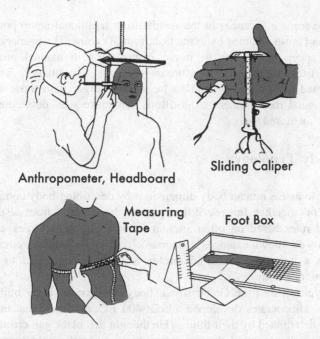

Anthropometer, Headboard

Sliding Caliper

Measuring Tape

Foot Box

Fig. 11.5 Traditional measuring instruments

low-power infrared laser as the distance-measuring device and apply topographic techniques to describe the body surface in minute detail, in three-dimensional space. The location of traditional landmarks under the visible surface, such as the top of the skull normally hidden under hair and skin, can be calculated from algorithms that derive the locations from points on the surface, or they can be indicated by markers placed on the skin.

Scanning systems to gather anthropometric data are expensive, but they provide advantages over the classic procedures. Instruments or hands of the measurer do not touch the subject. Scanners are fast yet collect vast amounts of data in three dimensions. Computerized data storage and processing allows retrieval, at any time, of 3D-descriptions of shapes of the body and its parts. Repeated scanning can record changes by motion, training or aging. Statistical body models can generate 3D virtual avatars* for many uses in ergonomic engineering, for design of computer graphics, for body models for tailoring, and for realistic simulations.

The subject's posture for whole-body scans is slightly different from that used in traditional surveys. Participants stand erect with the feet 30 cm apart, with arms straight and held slightly away from the body, fists clenched and pointing down. Subjects wear compression shorts and undergarments. Wig caps are placed on the participants' heads to model the head surface and compress the hair. Adhesive dot markers may be placed over selected landmarks previously identified on the subject's body.

There are some differences in the results from traditional anthropometric measurements and those obtained by extraction from 3D scans. The primary reasons for differences appear to be due to tissue compression in manual measurements, algorithms used to extract measurements from 3D images, and posture. Scan-generated measurements tend to be slightly larger than those obtained by classical, manual measurement. Algorithms can make scan data consistent with traditionally measured values*.

11.2   Body Typology

One can try to assess human body dimensions by describing body components and how these "fit" together. Images of the "perfect" body derive from aesthetic codes, canons, and rules based on often ancient (mostly Egyptian, Greek and Roman) concepts. A well-known example is Leonardo da Vinci's drawing, circa 1500, of a man within a frame of graduated circles and squares intended to signify the well-proportioned body.

Somatotyping (from the Greek *soma*, body) categorizes body builds into different types. Hippocrates developed, about 400 BC, a scheme that included four body types, determined by their fluids. (He thought that black gall created the moist type; yellow gall generated the dry type; slime made the cold type; and blood governed the warm type). In 1921, Ernst Kretschmer named three body types which supposedly related body build to personality traits: asthenic, pyknic and athletic body builds. (The term athletic was to indicate character traits, not sports performance.) In the 1940s, William H. Sheldon described a system of three body types of males, meant to predict their mental characteristics. Sheldon rated each person's appearance with respect to three morphic components: endomorphic, mesomorphic, and ectomorphic—see Table 11.1. Sheldon originally used intuitive judgment. His disciples brought actual body measurements into the typology system. Regrettably, these and following attempts at somatotyping (such as the Heath-Carter formula) do not provide reliable predictors of body sizes, strength, endurance, or other capabilities related to performance at work.

11.3   Anthropometric Data Sets

The military always had a particular interest in the body dimensions of soldiers—for a variety of reasons, among them the necessity to provide properly fitting uniforms, armor, and equipment. Furthermore, armies employ medical personnel willing and capable to perform body measurements on large samples "on command". Hence, anthropometric information about soldiers is extensive and reaches far into the past*. However, soldiers are a select sample of the general population: they are relatively healthy, fit, and young adults (and until recently, mostly males). Dimensions of the head, hand and foot measured on military populations are apparently similar to those of civilians* while other data are likely to differ.

Table 11.1 Body typologies

Description	Stocky, stout, soft, round	Strong, sturdy, muscular	Lean, slender, fragile
Kretschmer Typology	pyknic	athletic	asthenic (leptosomic)
Sheldon Typology	endomorphic	mesomorphic	ectomorphic

11.3.1 Normality

Anthropometric data usually appear in a reasonably normal (Gaussian) distribution, bell-shaped and symmetrical to the mean, as in Fig. 11.6. Hence, parametric statistics apply to most anthropometric information—but body weights (and muscle strength data) generally are not distributed normally.

Table 11.2 lists the most common statistical procedures*. The first and easiest check for normality of data uses the *mean, median*, and *mode*: if they coincide, normality is likely. Another simple approach is to calculate the mean m (average) by first using the complete range of data and then by leaving out the (say, ten) smallest and largest numbers: if both calculations end up with the same mean, normality is often the case. More formal calculations probe symmetry/skewness and peakedness. Regarding skewness: a result of 0 (zero) indicates a symmetrical distribution, a positive (negative) result points to skewness to the left (right). Regarding peakedness (*kurtosis*)*: a normal distribution has a kurtosis of 3; if the result is larger (or smaller) than 3 the distribution is more peaked (or flat).

11.3.2 Variability

"*A pioneer in the field of statistics, Sir Francis Galton [1822–1911] wrote that 'it is difficult to understand why statisticians commonly limit their interests to averages. Their souls seem as dull to the charm of variety as that of a native of one of our flat English counties whose retrospect of Switzerland was that, if its mountains would be thrown into its lakes, two nuisances could be got rid at once.' Basic to virtually all design problems is the fact that mankind is far more like Switzerland than a flat English county, and that, whatever the charms of variety may be, we need statistics to quantify this variety*": (Edmund Churchill, the eminent statistician in the field of engineering anthropometry during the second half of the 20th Century; as quoted —shortened—from page IX-5 of NASA/Webb 1978).

Fig. 11.6 The data
describing body height
(stature) of Americans follow
a normal distribution. About
95% of all males are between
162 and 188 cm tall; about
2.5% are shorter, another
2.5% are taller

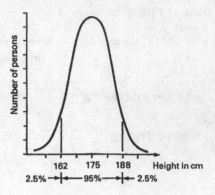

The standard deviation S (also called SD in some statistics references) describes the variability of anthropometric data: the larger S, the flatter (more spread out) is the distribution. That dispersion of a set of anthropometric data reflects the (true) variability of the underlying data, the accuracy of the measuring techniques, and the care taken in data handling, especially in their statistical treatment. For example, when comparing the descriptive statistics of two distinct though similar population samples, one may have a much larger coefficient of variation (CV in Table 11.2, see Eq. 11.15) of like dimensions than the other does. From this, the analyst may infer that the more dispersed sample reflects additional (and potentially suspicious) variability in the measuring technique and/or in data management.

The lowest and the highest values of a measured data set describe its full range; however, seldom is a design adaptable enough to fit the entire spread of measured dimensions and sizes. Instead, low and high cut-off values are selected, often as successive size sets such as in shoe or clothing size tariffs, at pre-determined percentiles (more on size tariffs below). Often (but not always correctly) the fifth percentile is taken as the lowest and the 95th percentile as the highest values to be considered in design; this should accommodate the central 90% of the population, excluding only the smallest 5% and the largest 5% extremes.

If the data distribution is normal (such as in Fig. 11.6), then known values for the mean m (which coincides with the 50th percentile, and with median and mode; see Eq. 11.1 in Table 11.2) and the standard deviation S (Eq. 11.5) can be used to calculate specific percentile points, as Eq. 11.12 and Table 11.3 describe.

11.3.3 Correlations

Some body dimensions relate closely with each other: for example, eye height follows stature closely. However, head measures do not show sizable relationships with stature, nor do circumferences, nor do breadths, and neither does weight. The

Table 11.2 Statistical formulas of particular use in anthropometry

Measures of Central Tendency		Equat. #
Mean, Average m	$m = \sum x / n$	**11.1**
Median	Middle value of values in numerical order	
Mode	Most often occurring value	
Measures of Variability		
Skewness (symmetry)	$\sum(x-m)^3 / nS^3$	**11.2**
Peakedness (kurtosis)	$\sum(x-m)^4 / nS^4$	**11.3**
Range	$x_{max} - x_{min}$	**11.4**
Standard Deviation S	$S = (variance)^{1/2} = [\sum(x-m)^2 / (n-1)]^{1/2}$	**11.5**
Coefficient of Variation CV	$CV = S / m;$ {in percent: $CV = 100S / m$}	**11.6**
Standard Error of the Mean SE	$SE = S / n^{1/2}$	**11.7**
Measures of Relations between variables x and y		
Correlation Coefficient r	$r = S_{xy} / (S_x \times S_y)^{1/2}$ $= \sum[(x-m_x)(y-m_y)] / [\sum(x-m_x)^2 \sum(y-m_y)^2]^{1/2}$	**11.8**
Coefficient of Determination R	$R = r^2$	**11.9**
Regression	$y = a + bx;$ with $b = rS_y / S_x$	**11.10** **11.11**
Percentile Value Location in a data distribution		
Percentile p	$p = m + kS$ (k from Table 11.3)	**11.12**
Sampling		
Sample size N	$N = q^2 S^2 / v^2$ (q from Table 11.10)	**11.13**
Ratio scaling		
Scaling factor E	$E = d_x / D_x = d_y / D_y$	**11.14**
Combining data sets		
Covariance COV	$COV(x,y) = r_{xy} S_x S_y$	**11.15**
Sum of two mean values	$m_z = m_x + m_y$	**11.16**
Standard deviation of the sum	$S_z = (S_x^2 + S_y^2 + 2r S_x S_y)^{1/2}$	**11.17**

(continued)

Table 11.2 (continued)

Difference of two mean values	$m_z = m_x - m_y$	**11.18**
Standard deviation of the difference	$S_z = (S_x^2 + S_y^2 - 2r\,S_x S_y)^{1/2}$	**11.19**
Factor k in a combined set z	$k_z = n_x k_x + n_y k_y$	**11.20**
Percentile p in a combined set z	$p_z = n_x p_x + n_y p_y$	**11.21**

statistical (Pearson) correlation coefficient r (Eq. 11.8 in Table 11.2) provides valuable information about the degree of relationship between sets of data.

Figure 11.7 contains scatter diagrams of the distributions of measures of two variables, x and y. If there is no relation between the variables, the value of the correlation coefficient r calculates to *zero* (top of Fig. 11.7). Values of 0.5 or less indicate weak relations whereas a correlation coefficient higher than 0.7 shows a strong dependency. If increasing values of x show exactly equivalent increases in y, r equals 1; if increases in x are followed by linear decreases in y, the value of r is -1 (bottom of Fig. 11.7).

Table 11.4 lists selected correlation coefficients r among body dimensions of US Air Force personnel, male and female, measured in the late 1980s: older tables on soldiers* showed similar relationships among body data. Unfortunately, correlation tables concerning civilian populations do not seem to be available.

Correlation coefficients provide valuable information about the relationship between sets of data. It is a useful practice in human factors engineering to require that the predictor variable explain at least 50% of the variation of the predicted value: accordingly, the coefficient of determination ($R = r^2$, see Eq. 11.9 in Table 11.2) must have a value 0.5 as a minimum. This means that the value of r must be at least 0.7.

This "0.7 convention" is important for basing decisions on a correlation and, relatedly, for the development and use of regression equations, which express the average of one variable as a function of another variable—see the discussions below under "Body Proportions" and "How to Get Missing Data". If the use of a single predictor variable is not sufficient to establish an overall correlation coefficient between predictor and predicted value of better than 0.7, additional predictor variables may be taken into the equation until the resulting "multiple regression equation" reaches a minimal cut-off point.

Table 11.3 Percentile values and associated k factors

BELOW MEAN				ABOVE MEAN			
per-centile	factor k	per-centile	factor k	per-centile	factor k	per-centile	factor k
0.001	−4.25	25	−0.67	50 (mean)	0	76	0.71
0.01	−3.72	26	−0.64	51	0.03	77	0.74
0.1	−3.09	27	−0.61	52	0.05	78	0.77
0.5	−2.58	28	−0.58	53	0.08	79	0.81
1	−2.33	29	−0.55	54	0.10	80	0.84
2	−2.05	30	−0.52	55	0.13	81	0.88
2.5	−1.96	31	−0.50	56	0.15	82	0.92
3	−1.88	32	−0.47	57	0.18	83	0.95
4	−1.75	33	−0.44	58	0.20	84	0.99
5	−1.64	34	−0.41	59	0.23	85	1.04
6	−1.55	35	−0.39	60	0.25	86	1.08
7	−1.48	36	−0.36	61	0.28	87	1.13
8	−1.41	37	−0.33	62	0.31	88	1.18
9	−1.34	38	−0.31	63	0.33	89	1.23
10	−1.28	39	−0.28	64	0.36	90	1.28
11	−1.23	40	−0.25	65	0.39	91	1.34
12	−1.18	41	−0.23	66	0.41	92	1.41
13	−1.13	42	−0.20	67	0.44	93	1.48
14	−1.08	43	−0.18	68	0.47	94	1.55
15	−1.04	44	−0.15	69	0.50	95	1.64
16	−0.99	45	−0.13	70	0.52	96	1.75
17	−0.95	46	−0.10	71	0.55	97	1.88
18	−0.92	47	−0.08	72	0.58	98	2.05
19	−0.88	48	−0.05	73	0.61	99	2.33

(continued)

Table 11.3 (continued)

20	**−0.84**	49	−0.03	74	0.64	99.5	2.58
21	−0.81	**50** (mean)	**0**	**75**	**0.67**	99.9	3.09
22	−0.77					99.99	3.72
23	−0.74					99.999	4.26
24	−0.71						

Any percentile value p (in a normal distribution of data) can be calculated from the mean m and the standard deviation S according to $p = m + kS$ (equation # 11-12)

Examples:

5th percentile is at $m - 1.64S$ because of $k = -1.64$
10th percentile is at $m - 1.28S$ because of $k = -1.28$
50th percentile is at m because of $k = 0$
60th percentile is at $m + 0.25S$ because of $k = 0.25$
95th percentile is at $m + 1.64S$ because of $k = 1.64$

11.3.4   Body Proportions

It may appear convenient to calculate proportional relationships (ratios or indices) between body dimensions. Of course, such procedure is justified and useful only when there is a sufficient correlation between the two variables, as just discussed. Of the body measures listed in Table 11.4, only overhead reach, wrist height, crotch height, sitting height, popliteal height, and span have correlation coefficients larger than 0.70 with stature. Not one of the listed circumferences, breadths, head or hand measures, or weight, associates highly with stature. Evidently, and unfortunately, the easily measured stature is not a good predictor for most other body dimensions*.

In general, long bone (link) dimensions relate well with each other. Also, many measures of height, of breadth and of depth dimensions, and of trunk circumferences, respectively, correlate well within their groups—but usually do not show a tight association with variables outside their grouping. Hence, if one wishes to predict a particular dimension from another, one must carefully check whether the correlation between the two data sets is sufficiently high.

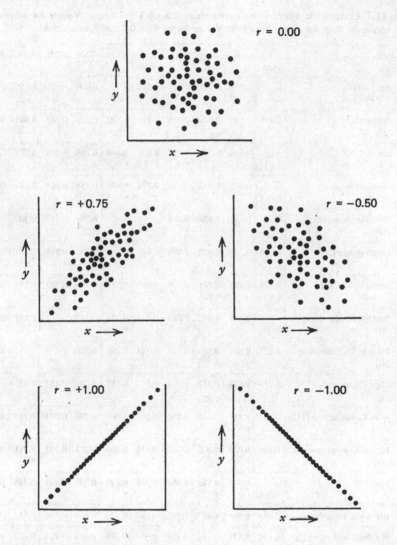

Fig. 11.7 Scatter diagrams of bivariate data distributions and correlation coefficients (adapted from Roebuck et al. 1975)

11.4 Variability of Anthropometric Data

Variability in anthropometric data primarily derives from five causes: data management, secular variations, intra-individual variations, inter-individual variations, and changing populations.

Table 11.4 Correlations between anthropometric data on US soldiers. Values for women are listed above the diagonal, for men below. Values larger than 0.7 carry an asterisk

		1 Age	2 W	3 Stat	4 OFR	5 WH	6 CH	7 SH	8 PH	9 SC	10 CC
1	Age [302]		0.219	0.041	0.017	0.044	−0.055	0.066	−0.074	0.155	0.193
2	Weight [125]	0.195		0.529	0.493	0.491	0.370	0.422	0.242	0.845*	0.806*
3	Stature [100]	−0.021	0.546		0.928*	0.848*	0.840*	0.755*	0.808*	0.377	0.222
4	Overhead Fingertip Reach[84]	−0.013	0.525	0.937*		0.704*	0.905*	0.554	0.868*	0.384	0.199
5	Wrist Height, standing [128]	0.028	0.527	0.856*	0.749*		0.625	0.754*	0.587	0.300	0.255
6	Crotch Height [39]	0.090	0.351	0.852*	0.890*	0.673		0.330	0.915*	0.267	0.093
7	Sitting Height [94]	0.026	0.447	0.741*	0.578	0.692	0.347		0.343	0.285	0.202
8	Popliteal Height, sitting [87]	−0.094	0.341	0.852*	0.883*	0.673	0.924*	0.383		0.188	0.023
9	Shoulder Circumference [91]	0.122	0.861*	0.399	0.413	0.334	0.250	0.326	0.256		0.808*
10	Chest Circumference [34]	0.279	0.873*	0.312	0.308	0.357	0.135	0.287	0.137	0.859*	
11	Waist Circumference [115]	0.364	0.849*	0.276	0.251	0.343	0.060	0.298	0.074	0.703*	0.839*
12	Buttock Circumference [24]	0.190	0.935*	0.401	0.380	0.412	0.204	0.373	0.191	0.781*	0.815*
13	Span [99]	−0.016	0.497	0.815*	0.908*	0.535	0.840*	0.398	0.844*	0.445	0.281
14	Biacromial Breadth [11]	0.034	0.496	0.487	0.506	0.295	0.370	0.407	0.394	0.633	0.419
15	Hip Breadth, standing [66]	0.209	0.831*	0.453	0.416	0.457	0.239	0.464	0.224	0.672	0.727*
16	Head Circumference [62]	0.125	0.508	0.342	0.312	0.302	0.224	0.303	0.240	0.433	0.421
17	Head Length [63]	−0.002	0.371	0.346	0.315	0.295	0.260	0.302	0.268	0.295	0.271
18	Head Breadth [61]	0.198	0.320	0.114	0.098	0.112	0.034	0.128	0.035	0.303	0.311
19	Hand Length [60]	0.032	0.453	0.650	0.724*	0.464	0.676	0.300	0.679	0.372	0.242
20	Foot Length [52]	−0.012	0.512	0.700	0.734*	0.537	0.687	0.383	0.697	0.409	0.299

(continued)

Table 11.4 (continued)

		*11*W C	*12*BC	*13*Sp	*14*BB	*15*Hi B	*16*HC	*17*He L	*18*He B	*19*Ha L	*20*FL
1	Age [302]	0.299	0.258	0.011	0.025	0.283	0.073	0.027	0.044	0.044	0.026
2	Weight [125]	0.767*	0.897*	0.438	0.440	0.778*	0.428	0.329	0.420	0.430	0.493
3	Stature [100]	0.167	0.361	0.787*	0.505	0.372	0.348	0.354	0.124	0.637	0.673
4	Overhead Fingertip Reach[84]	0.132	0.313	0.907*	0.535	0.294	0.337	0.345	0.095	0.737*	0.732*
5	Wrist Height, standing [128]	0.217	0.363	0.453	0.303	0.397	0.250	0.261	0.403	0.403	0.468
6	Crotch Height [39]	0.061	0.185	0.870*	0.418	0.146	0.287	0.302	0.043	0.706*	0.703*
7	Sitting Height [94]	0.142	0.351	0.336	0.384	0.438	0.246	0.255	0.159	0.256	0.330
8	Popliteal Height, sitting [87]	−0.031	0.063	0.840*	0.420	0.051	0.241	0.271	0.020	0.685	0.671
9	Shoulder Circumference [91]	0.697	0.726*	0.395	0.574	0.601	0.353	0.264	0.261	0.355	0.379
10	Chest Circumference [34]	0.781*	0.707*	0.167	0.304	0.603	0.393	0.191	0.246	0.186	0.288
11	Waist Circumference [115]		0.738*	0.109	0.214	0.673	0.223	0.117	0.229	0.127	0.170
12	Buttock Circumference [24]	0.859*		0.258	0.327	0.915*	0.313	0.226	0.220	0.258	0.323
13	Span [99]	0.201	0.352		0.565	0.203	0.345	0.338	0.083	0.827*	0.775*
14	Biacromial Breadth [11]	0.311	0.411	0.575		0.294	0.287	0.259	0.152	0.441	0.456
15	Hip Breadth, standing [66]	0.799*	0.902*	0.355	0.404		0.232	0.160	0.196	0.180	0.250
16	Head Circumference [62]	0.376	0.427	0.320	0.301	0.364		0.824*	0.497	0.342	0.360
17	Head Length [63]	0.222	0.301	0.304	0.235	0.259	0.820*		0.131	0.337	0.339
18	Head Breadth [61]	0.277	0.268	0.131	0.180	0.235	0.541	0.120		0.082	0.113
19	Hand Length [60]	0.166	0.320	0.810*	0.433	0.298	0.330	0.306	0.137		0.825*
20	Foot Length [52]	0.220	0.390	0.766*	0.445	0.377	0.333	0.304	0.161	0.806*	

Source: Cheverud, Gordon, Walker *et al.* (1990) with their numbering in brackets.

Note: Pairs of data that correlate above 0.700 are similar in the male and female groups; also, correlations below 0.300 are similar for both genders.

11.4.1   Data Management

Sloppiness in subject sampling, in performing measurements, in recording, sorting, analyzing, and reporting of anthropometric data are likely to result in data collections that are "unusual" in variability and central tendency (mean, median, and mode) as compared to other, more valid data. Thus, if one encounters a report of body sizes that is inconsistent with other information on a seemingly related or similar sample, it is advisable to check carefully. The "coefficient of variation" (Eq. 11.6 in Table 11.2) allows a quick first test: caution is in order if its value exceeds 5% for body heights, and 10% for breadths, depths, or reaches.

Another set of data-related problems stems from a false assumption of normality (that is, of a Gaussian distribution) in the basic data. Whereas many body size measurements are indeed normally distributed, some are not: body weights commonly form data sets that are skewed to one side, and so are measurements of muscle strength—see Chap. 2. Regrettably, statistics of body weight (and of strength) rely so regularly on the false assumption of normality that it is difficult to find the actual measured values in the literature. (Note that even this book follows the deplorable custom by describing compilations of body weights and strengths via *mean* (*average*) and *standard deviation* when, instead, the use of nonparametric indices would be correct, such as *median* or *mode*.) Figure 11.8 illustrates the case: Stature data display a normal distribution (similar to Fig. 11.6), but the associated body mass measures are severely skewed, showing a long tail of heavy weights. Obviously, it is numerically possible to calculate mean (here: 69 kg) and standard deviation values and then 5th and 95th percentiles, but doing so is basically false and practically misleading because the underlying distribution is non-normal*.

11.4.2   Secular Variations

The term "secular" refers to events that appear over long stretches of time. When looking at medieval armor displayed in a museum, we notice the apparently small sizes of soldiers centuries ago; evidently, they grow taller now. Today, we experience that many children grow, on average, to be taller than their parents are. However, "hard" anthropometric data on this development are only available for approximately the last hundred years when anthropometric surveys were done in reasonably consistent manners on sufficiently large samples. Figure 11.9 provides information on stature for a variety of young male civilian samples: the overall trend of increasing stature is apparent.

Similarly, Fig. 11.10 presents information on military data measured in the USA during about 120 years. Surprisingly, the military data show virtually no change from the 1860s until World War I; but there seems to be a pronounced increase thereafter. Why average stature was seemingly stable for nearly 60 years and then

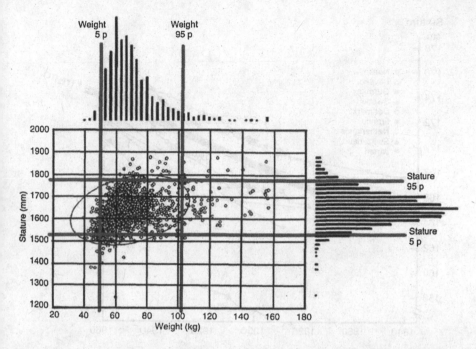

Fig. 11.8 Weight and stature distributions of American women. *Note* the skewed distribution of the weights associated with a normal distribution of heights (adapted from Robinette and Hudson 2006)

grew rapidly is open to speculation. Perhaps the recruitment of soldiers from the general population during the Civil War was different from the screening process during World War I; a massive influx of short immigrants may have changed the general population; it is even possible that the measurement techniques changed. (Even this brief discussion highlights some of the difficulties in comparing data that are disjointed both in time and in collection technique).

Nevertheless, the increase observed in stature during the last century is apparently "real". Data from major surveys in the USA, Europe and Asia indicated an increase in stature of about 1 cm per decade during the 20th century. Weight increases were even more dramatic, initially in the neighborhood of 2 kg for every ten years, followed by sudden substantial gains starting in the late 1900s leading to widespread obesity in the first decades of the 21st century.

Among the possible explanation of the increases in stature is the generally accepted idea that improvements in living conditions, both hygienic and nutritional, have allowed people to achieve their genetically possible stature more readily than in earlier times. If this is true, one would expect that an "average maximal height" should be approached asymptotically in the future, provided that the whole population benefits from improved living conditions. In fact, the 1988 survey of the US Army shows a slowing of the trend in stature increase: more recent data suggests that it may take about two decades now to gain another centimeter in height.

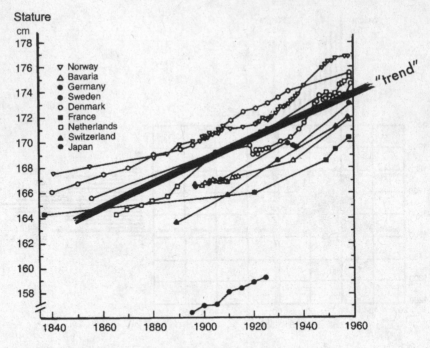

Fig. 11.9 Secular increase in stature of young European and Japanese males. Heavy line shows the apparent trend (adapted from NASA/Webb 1978)

Most secular developments of body dimensions are rather slow. Hence, the changes in body data should have little practical consequences for the design of tools, equipment, and work places, since most products have a relatively short "design life" for which secular changes in anthropometry of the users have no appreciable importance. However, the rather sudden increase in body weight, and, relatedly, in body breadths and circumferences does require some design adaptations, for example in seats and seating arrangements in trucks, airplanes and public transportation.

11.4.3   Intra-individual Variations

Some short-term variations in stature appear in just a few hours: immediately after rising from overnight sleep, one may be several centimeters taller than after a full day "on ones feet". This results mainly from flattening of the intervertebral disks, due to loss of fluid because of compression forces generated by gravity and body activities.

Other examples of intra-individual variations, which become apparent over periods of weeks or months, include variations in weight or circumferences, often associated with changes in nutrition and physical activities. During pregnancy,

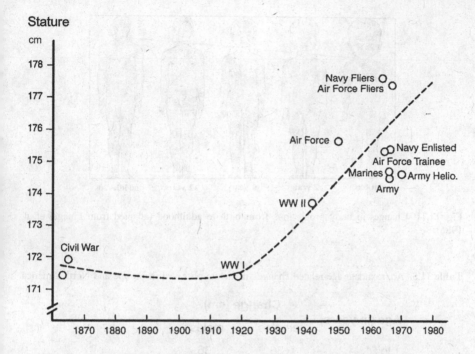

Fig. 11.10 Change in stature of US soldiers (adapted from NASA/Webb 1978)

women experience major changes in body size, proportions, and physical functions, which are generally transient.

Among the long-term intra-individual variables, the effects of aging on the body are rather obvious. During the growing years, stature, weight and all the other body dimensions increase, and the proportions change—see Fig. 11.11. In early adulthood, body measures and proportions become relatively stable for several decades. Then, with increasing age, certain dimensions usually reduce (such as body height) while circumferences and the external diameters of bones commonly increase. Table 11.5 lists, in approximate numbers, changes in stature with age.

11.4.4 Inter-individual Variations

Inter-individual variations are a result of DNA characteristics; roughly 10^9 possible chromosome combinations exist. An individual's genetic endorsement determines the cellular compositions (genotype) and biologically measurable characteristics (phenotype). Nutrition also has direct effects on body size; obviously, over-feeding leads toward obesity whereas lack of nourishment causes skinniness.

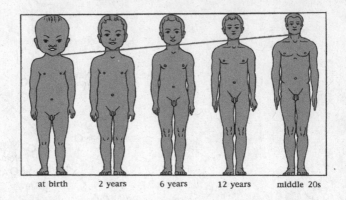

Fig. 11.11 Changes in body proportions from birth to adulthood (adapted from Fluegel et al. 1986)

Table 11.5 Approximate age-related changes in stature observed in Europe and North America

Age (in years)	Change (cm)	
	females	males
1 to 5*	+36	+36
5 to 10	+28	+27
10 to 15	+22	+30
15 to 20	+1	+6
20 to 35**	no change	no change
35 to 40	-1	no change
40 to 50	-1	-1
50 to 60	-1	-1
60 to 70	-1	-1
70 to 80	-1	-1
80 to 90	-1	-1

* Average stature at age 1 yr: females 74 cm, males 75 cm.
** Average maximal stature: females 161 cm, males 174 cm.

A variety of "recommended" body weight tables exists, usually subdivided by gender, age, and perhaps body build. The underlying assumption is that risks of morbidity and mortality from strokes and certain ailments of the heart, of diabetes, cancer and other health problems associate with body weights and weight/stature ratios.

Among the early height-weight tables, long used by physicians in North America, are those originally prepared in 1942 (and then continually updated) by a life insurance company. The underlying idea was that the weights of young adults were "ideal" and should be maintained throughout life. From the beginning, the validity of several assumptions regarding height-to-weight ratios was questioned*. A particular statistical concern is that body weight does not consistently correlate with stature: r is only about 0.5 among soldiers, as Table 11.4 shows, but even lower within the general population. Nevertheless, many people make judgments based on these presumably desirable weight-height ratios.

Since the middle 1990s, the so-called Body Mass Index (BMI) has been in wide use. It also relies on a postulated relation between measured body weight and height to be used as a screening tool as part of an overall health assessment. To calculate a BMI, one divides a person's weight by the squared stature—see Table 11.6. The result then is compared to a set of listed BMI ranges to assess whether the person* is underweight (BMI < 18.5), of normal weight (18.5–24.9), overweight (25.0–29.9) or obese (BMI of 30.0 and above). There are, however, some bothersome

Table 11.6 Body Mass Index (BMI) calculation

Metric:

To calculate BMI, divide weight (in kilograms, kg) by squared height (in meters, m):

$BMI = [weight (kg) / [height (m)]^2$

Example: weight = 68 kg, height = 1.65 m
Calculation: $68 / 1.65^2 = 24.98$

When height is measured in centimeters, multiply result by 10.000 (10^4):

$BMI = weight (kg) / [height (cm)]^2 \times 10^4$

Example: weight = 68 kg, height = 165 cm
Calculation: $10^4 \times 68 / 165^2 = 24.98$

Using the old US measures:

To calculate BMI, divide weight (in pounds, lbs) by squared height (in inches, in); then multiply with the conversion factor 703:

$BMI = weight (lb) / [height (in)]^2 \times 703$

Example: weight = 150 lbs, height = 65" (5'5")
Calculation: $703 \times 150 / 65^2 = 24,96$

issues, mostly related to the weak statistical linkage between weight and stature: the relation between the BMI number and body fatness varies strongly by gender, age, ethnic origin, and fitness. At the same BMI, women tend to have more body fat than men do; likewise, older people tend to have more body fat than younger adults. Highly trained athletes may have a high BMI caused by increased muscle mass rather than by increased body fatness. The criteria used to interpret the meaning of the BMI number* for children and teens are different from those used for adults for two reasons: the amount of body fat changes with age and differs between girls and boys.

Handedness is a well-known example of inter-individual variability: A general estimate is that about one individual among ten persons is left-handed. However, handedness needs a careful definition: for example, one may prefer the left hand for writing but the right one for swinging a hammer. Such information may be of some interest to, say, the manufacturers of special hand tools but otherwise does not imply that large differences in arm lengths, or arm circumferences, appear in nationwide anthropometric tables.

11.4.5   Changing Populations

Populations do not remain constant but change in age, health, gender balance and composition. For example, the work force in the industrialized countries today has many women in occupations that were dominated by men just a few decades ago. Occupations may change drastically as well: in many economies, "blue collar" workers in the traditional sense have become fewer as automation, desk work, and computer use have started to replace much physical labor.

Life expectancy has increased dramatically in many countries, mostly due to reduced infant mortality. In the USA, life expectancy increased by some 30 years since 1900; it now is near 80 years. As one consequence, the traditional assumption of a broad-based "population pyramid", where many young people support a few old ones, is no longer true. In 1971, the median age of the US population was about 28 years; in 1980, 30 years; in 1986, 32 years; it reached 36 years in the year 2000. Currently, half as many Americans are more than 65 years old as there are teenagers. It is estimated that in 2050, one in four Americans (and Europeans) will be over 65 years old*.

In many countries, the overall fertility rate declined: in the USA, from 3.7 births per woman in 1960 to 2.5 in 1970, then to about 1.7 in 2019. Such a low birth rate can lead to a reduction in the total population within just a few decades, but immigration keeps the number of US citizens growing.

Recent decades have shown much mobility of population groups. Central Europe, for example, saw a huge influx of "guest workers" and refugee families coming from African and Asian countries in the later part of the twentieth century. Such massive immigration can lead to distinct changes in the composition of the population: for instance, in Berlin (Germany) and in Paris (France), a large percentage of schoolchildren has immigrant parents. Americans are moving within their country as well: in 1980, for the first time ever, the majority of Americans lived in the southern and western states. The "sunbelt", prominently California, Florida, and Texas grew more in population than the other regions in the US; this migration reduced the population in the "snowbelt". Cities used to be magnets for the rural population; in the 1960s and 1970s the flow reversed, but since the 1980s some metropolitan areas have been growing again. However, such trends can change quickly, for example with changes in the economy.

As in many other countries, the population of the USA is a composite of many origins. For example, in 2005 most US citizens said they have Caucasian roots, predominantly German, English, or Irish. About 25 million citizens stated to be of African descent; about the same number claimed Hispanic origin. Recent immigration, especially from central and southern America and Asia, is again changing the composition of the US population quickly: estimates are that, by 2050, the Caucasian-derived population section will shrink from two thirds in 2005 (it was 85% in 1960) to less than half in 2050. Predictions are that Hispanics will make up about 24%, the percentage of African-Americans will stay at about 13% but the percentage of citizens with Asian origin will almost triple, to approximately 14% of the total population. Table 11.7 lists snapshots in time of the ancestral roots that Americans claimed in recent years.

Anthropometric data indicate statistically significant differences among several groups of ethnic origins*; there are also some specific differences in body sizes among various professions. For example, on average, US agricultural workers were shorter by about 2.5 cm in stature and had wider wrists than other workers. Female American agricultural and manufacturing workers showed larger waist circumferences than those in other occupations. Firefighters, police officers and guards were taller and heavier (males by 7 kg, females by over 10 kg) than Americans in all other occupations*. Truck drivers were heavier than the US general population: males by 13.5 kg and females by 15.4 kg*. Such differences may be of practical importance for the design and use of certain items in limited localities, though on a nationwide scale these differences among ethnic and professional groups are fairly small and of little practical significance. Variations in body size within groups are usually more striking than differences in averages between groups.

11.5 Anthropometry of Large Populations

Measured anthropometric data on large civilian populations are rather sparse whereas the body dimensions of military personnel are better known. Since the military is a sample of the overall population, it appears reasonable to infer dimensions of the general civilian population from military data. However, a major concern is that soldiers may be so highly selected (particularly, biased to be young and healthy, and historically male) that they constitute a special sample which does not accurately represent the overall population. Yet, if better information is not at hand, there may be no other choice than to use, with proper caution and insight,

Table 11.7 Ancestry claimed by US citizens (millions) (adapted from US Census Bureau)

Ancestry	2015	2000	1990	1980
German	46.4	42.8	57.9	49.2
African American	38.8	24.9	23.8	21.0
Mexican	34.6	18.4	11.6	7.7
Irish	33.5	30.5	38.7	40.2
English	24.8	24.5	32.7	49.6
American	22.8	20.2	12.4	7.9
Italian	17.3	15.6	14.7	12.2
Polish	9.4	9.0	9.4	8.2
French	8.3	8.3	10.3	12.9
Scottish	5.4	4.9	5.4	10.1
Puerto Rican	5.2	2.7	2.0	1.4
Norwegian	4.5	4.5	3.9	3.5
Dutch	4.3	4.5	6.2	6.3
Swedish	3.9	4.0	4.7	4.4
Chinese	3.9	2.3	1.5	0.9
Scotch-Irish	3.1	4.3	5.6	nda
Russian	2.8	2.7	3.0	2.8
Filipino	2.7	2.1	1.5	0.8
American-Indian	nda	7.9	8.7	6.7
French-Canadian	nda	2.3	2.2	0.8
Spanish-Hispanic	nda	2.2	2.0	2.7

(continued)

Table 11.7 (continued)

Welsh	nda	1.8	2.0	1.7
Danish	nda	1.4	1.6	1.5
Czech	nda	1.3	1.3	1.9
Portuguese	nda	1.2	1.2	1.0
Greek	nda	1.2	1.1	1.0
Japanese	nda	1.1	1.0	0.8
Vietnamese	nda	1.0	0.5	nda
Slovak	nda	0.9	1.9	0.8
Swiss	nda	0.9	1.0	1.0
All others	nda	below 1		

nda: no data available

military anthropometric data to approximate size data of the general population. Fortunately, at least the dimensions of the head, hand, and foot are virtually the same in military and civilian populations.

> To make sure that workspaces and implements fit their users, it is necessary to consider carefully their anthropometric descriptors. Two examples: a computer workstation that fits a college group in Bismarck, North Dakota (or in Amsterdam, NL), may not be of the correct size and adjustment ranges for, say, computer operators in southern Texas (or in Palermo, Sicily); school furniture appropriate for children in Palermo may not suit pupils in Amsterdam.

11.5.1 Available Body Size Data

Only a few populations on earth have been measured thoroughly and completely. Table 11.8 presents an overview of measured heights and weights (averages and standard deviations) of several ethnic, national, and geographic populations*. Unfortunately, in many cases the data a rather ancient and/or only few people were measured; hence, such statistics are not likely to describe the present underlying population, and use of such data requires considerable caution.

Table 11.8 Measured heights and weights of adults, published since 1989: averages (standard deviations) (adapted from Kroemer 2017)

	SAMPLE SIZE	STATURE (mm)	WEIGHT (kg)
ALGERIA: females (Mebarki & Davies, 1990)	666	1576 (56)	61 (13)
BRAZIL: males (Ferreira, 1988; cited by Al-Haboubi, 1991)	3076	1699 (67)	nda
CAMEROON: urban females urban males (35-44 years old) (Kamadjeu, Edwards & Atanga, 2006)	1156 558	1620 1721	64 75
CHINA: females (Taiwan) males (Taiwan) (Wang, Wang & Liu, 2002)	about 600 about 600	1572 (53) 1705 (59)	52 (7) 67 (9)
FRANCE: females males (IFTH & Goncalves, personal communication, 2006)	5510 3986	1625 (71) 1756 (77)	62 (12) 77 (13)
GERMANY: female army applicants male army applicants (Leyk, Kuechmeister & Juergens, 2006)	301 1036	1674 1795	64 75
GREAT BRITAIN: females males (Erens, Primatesta & Prior, 2001)	3870 3233	1611 1746	68 81
INDIA: females males (Chakarbarti, 1997) East-Ctr. India male farm workers (Victor, Nath & Verma, 2002) South India male workers (Fernandez & Uppugonduri, 1992) East India male farm workers (Yadav, Tewari & Prasad, 1997)	251 710 300 128 134	1523 (66) 1650 (70) 1638 (56) 1607 (60) 1621 (58)	50 (10) 57 (11) 57 (7) 57 (5) 54 (7)

(continued)

Table 11.8 (continued)

IRAN:			
female students	74	1597 (58)	56 (10)
male students	105	1725 (58)	66 (10)
(Mououdi, 1997)			
IRELAND:			
males	164	1731 (58)	74 (9)
(Gallwey & Fitzgibbon, 1991)			
Italy:			
females	753*	1610 (64)	58 (8)
females	386**	1611 (62)	58 (9)
males	913*	1733 (71)	75 (10)
males	410**	1736 (67)	73 (11)
*(Coniglio, Fubini, Masali et al., 1991) **(Robinette, Blackwell, Daanen et al., 2002)			
JAPAN:			
females	240	1584 (50)	54 (6)
males	248	1688 (55)	66 (8)
(Kagimoto, 1990)			
NETHERLANDS:			
females, 20-30 yrs old	68*	1686 (66)	67 (10)
females, 18-65 yrs old	679**	1672 (79)	74 (16)
males, 20-30 yrs old	55*	1848 (80)	81 (14)
males, 18-65 yrs old	593**	1808 (93)	86 (17)
*(Steenbekkers & Beijsterveldt, 1998) **(Robinette & Hudson, 2006)			
RUSSIA:			
female herders (ethnic Asians)	246	1588 (55)	nda
female students (ethn. Russians)	207	1637 (57)	61 (8)
female students (ethn. Usbeks)	164	1578 (49)	56 (7)
fem. factory workers (ethn. R.)	205	1606 (53)	61 (8)
fem. factory workers (ethn. U.)	301	1580 (54)	58 (9)
male students (ethn. Russians)	166	1757 (56)	71 (9)
male students (ethn. Usbeks)	150	1700 (52)	65 (7)
male factory workers (ethn. R.)	192	1736 (61)	72 (10)
male factory workers (ethn.mix)	150	1700 (59)	68 (8)
male farm mechanics (ethnic Asians)	520	1704 (58)	64 (8)
male coal miners (ethn. Russians)	150	1801 (61)	nda
male construction workers (ethnic Russians)	150	1707 (69)	nda
(Strokina & Pakhomova, 1999)			
SAUDI ARABIA:			
males	1440	1675 (61)	nda
(Dairi, 1986; cited by Al-Haboubi, 1991)			
SINGAPORE:			
males (pilot trainees)	832	1685 (53)	nda
(Singh, Peng, Lim & Ong, 1995)			

(continued)

Table 11.8 (continued)

SRI LANKA:			
females	287	1523 (59)	nda
males	435	1639 (63)	nda
(Abeysecera, 1985; cited by Intaranont, 1991)			
THAILAND:			
females	250*	1512 (48)	nda
Females	711**	1540 (50)	nda
males	250*	1607 (20)	nda
Males	1478**	1654 (59)	nda
*(Intaranont, 1991)			
(Abeysecera, 1985; cited by Intaranont, 1991)			
TURKEY:			
male soldiers	5108	1702 (60)	63 (7)
(Kayis & Oezok, 1991a)			
United States of America:			
females	about 3800	1625	75
males	about 3800	1762	87
(Ogden, Fryar, Carroll et al., 2004)			
U. S. Army soldiers:			
females*	2208	1629 (64)	62 (8)
males*	1774	1756 (67)	76 (11)
males**	1475	1760 (nda)	84 (nda)
*(Gordon, Churchill, Clauser et al., 1989)			
**(Gordon 2009)			
North American (Canada and USA)			
females, 18-26 yrs. old	1264	1640 (73)	69 (18)
males, 18-65 yrs. old	1127	1778 (79)	86 (18)
(Robinette, Blackwell, Daanen et al., 2002)			

nda: no data available.

Table 11.9 presents information on thirty-six measures of body sizes of Chinese, Russian, and North American adults, females and males, and of their weights. Figures 11.12, 11.13 and 11.14 illustrate the body dimensions.

Although the data listed in Table 11.9 appear rather comprehensive, their comparison and their use for design purposes needs careful consideration of specifics such as

- The Chinese persons (about 1200) were measured during the late 1990s, the location was Taiwan (not mainland China), and their ages ranged from 25 to 34 years.

Table 11.9 Body measurements and their engineering applications. All data in mm, except weight in kg (adapted from Kroemer KHE 2017)

DIMENSIONS, APPLICATIONS	POPULATIONS	FEMALES Mean (S)	MALES Mean (S)
1. Stature A main measure for comparing population samples. Reference for minimal overhead clearance. Add height for more clearance, hat, shoes, stride.	Chinese (Taiwan) Russians (Moscow) U.S. Army soldiers	1572 (53) 1637 (57) 1629 (64)	1705 (59) 1757 (56) 1756 (69)
2. Eye height, standing Origin of the visual field of a standing person. Reference for the location of visual obstructions and of visual targets such as displays; consider slump and motion.	Chinese (Taiwan) Russians (Moscow) U.S. Army soldiers	nda 1526 (57) 1520 (62)	nda 1637 (55) 1642 (67)
3. Shoulder height (acromion), standing Starting point for arm length measurements; near the center of rotation of the upper arm. Reference point for hand reaches; consider slump and motion.	Chinese (Taiwan) Russians (Moscow) U.S. Army soldiers	1285 (50) 1334 (54) 1335 (58)	1396 (53) 1440 (54) 1441 (63)
4. Elbow height, standing Reference for height and distance of the work area of the hand and the location of controls and fixtures; consider slump and motion.	Chinese (Taiwan) Russians (Moscow) U.S. Army soldiers	978 (38) 1010 (42) 1004 (45)	1059 (40) 1083 (48) 1083 (50)
5. Hip height (trochanter), standing Traditional anthropometric measure, indicator of leg length and the height of the hip joint. Used for comparing population samples.	Chinese (Taiwan) Russians (Moscow) U.S. Army soldiers	802 (41) nda 845 (45)	860 (48) nda 901 (49)
6. Knuckle height, standing Reference for low locations of controls, handles, and handrails; consider slump and motion of the standing person.	Chinese (Taiwan) Russians (Moscow) U.S. Army soldiers	708 (33) 731 (34) nda	757 (32) 773 (39) nda
7. Fingertip height (dactylion), standing Reference for the lowest locations of controls, handles, and handrails; consider slump and motion of the standing person.	Chinese (Taiwan) Russians (Moscow) U.S. Army soldiers	618 (32) 635 (32) 613 (35)	659 (30) 668 (37) 654 (37)
8. Sitting height Reference for the minimal height of overhead obstructions. Add height for more clearance, hat, trunk motion of the seated person.	Chinese (Taiwan) Russians (Moscow) U.S. Army soldiers	846 (32) 859 (32) 857 (33)	910 (30) 912 (32) 918 (36)
9. Sitting eye height Origin of the visual field of a seated person. Reference point for the location of visual targets such as displays; consider slump and motion.	Chinese (Taiwan) Russians (Moscow) U.S. Army soldiers	732 (31) 742 (29) 748 (30)	791 (29) 790 (33) 805 (33)
10. Sitting shoulder height (acromion) Starting point for arm length measurements; near the center of rotation of the upper arm. Reference for hand reaches; consider slump and motion.	Chinese (Taiwan) Russians (Moscow) U.S. Army soldiers	561 (27) nda 563 (29)	602 (26) nda 603 (31)
11. Sitting elbow height Reference for the height of an armrest, of the work area of the hand and of keyboard and controls; consider slump and motion of the seated person.	Chinese (Taiwan) Russians (Moscow) U.S. Army soldiers	252 (25) 236 (24) 232 (26)	264 (24) 243 (25) 245 (29)
12. Sitting thigh height (clearance) Reference for the minimal clearance needed between seat pan and the underside of a table or desk; add clearance for clothing and motions.	Chinese (Taiwan) Russians (Moscow) U.S. Army soldiers	nda 148 (14) 168 (14)	nda 151 (18) 181 (16)
13. Sitting knee height Traditional anthropometric measure for lower leg length. Reference for the minimal clearance below the underside of a table or desk; add height for shoe.	Chinese (Taiwan) Russians (Moscow) U.S. Army soldiers	471 (24 527 (24) 511 (27)	521 (29) 562 (25) 554 (28)
14. Sitting popliteal height Reference for the height of a seat; add height for shoe.	Chinese (Taiwan) Russians (Moscow) U.S. Army soldiers	379 (18) 423 (23) 388 (24)	411 (19) 468 (24) 430 (25)

(continued)

Table 11.9 (continued)

15. Shoulder-elbow length Traditional anthropometric measure for comparing population samples.	Chinese (Taiwan) Russians (Moscow) U.S. Army soldiers	*309 (18)* *nda* *334 (17)*	338 (19) nda 364 (18)
16. Elbow-fingertip length Traditional anthropometric measure. Reference for fingertip reach when moving the forearm in the elbow.	Chinese (Taiwan) Russians (Moscow) U.S. Army soldiers	*384 (27)* *nda* *440 (23)*	427 (27) nda 480 (23)
17. Overhead grip reach, sitting Reference for the height of overhead controls operated by a seated person. Consider ease of motion, reach, and finger/hand/arm strength.	Chinese (Taiwan) Russians (Moscow) U.S. Army soldiers	*1105 (44)* *1169 (46)* *1196 (62)*	1208 (49) 1276 (47) 1303 (68)
18. Overhead grip reach, standing Reference for the height of overhead controls operated by a standing person. Add shoe height. Consider ease of motion, reach, and strength.	Chinese (Taiwan) Russians (Moscow) U.S. Army soldiers	*1831 (67)* *nda* *1968 (98)*	2002 (79) nda 2141 (104)
19. Forward grip reach Reference for forward reach distance. Consider ease of motion, reach, and finger/hand/arm strength.	Chinese (Taiwan) Russians (Moscow) U.S. Army soldiers	*651 (33)* *702 (37)* *693 (43)*	710 (36) 759 (38) 757 (44)
20. Arm length, vertical A traditional measure for comparing population samples. Reference for the location of controls very low on the operator's side. Consider ease of motion, reach, strength.	Chinese (Taiwan) Russians (Moscow) U.S. Army soldiers	*669 (31)* *nda* *722 (37)*	738 (33) nda 787 (39)
21. Downward grip reach Reference for the location of controls low on the side of the operator. Consider ease of motion, reach, and finger/hand/arm strength.	Chinese (Taiwan) Russians (Moscow) U.S. Army soldiers	*nda* *nda* *607 (30)*	nda nda 663 (32)
22. Chest depth A traditional measure for comparing population samples. Reference for the clearance between seat backrest and the location of obstructions in front of the trunk.	Chinese (Taiwan) Russians (Moscow) U.S. Army soldiers	*213 (19)* *242 (21)* *247 (27)*	217 (19) 245 (20) 254 (26)
23. Abdominal depth, sitting A traditional measure for comparing population samples. Reference for the clearance between seat backrest and the location of obstructions in front of the trunk.	Chinese (Taiwan) Russians (Moscow) U.S. Army soldiers	*nda* *nda* *223 (32)*	nda nda 255 (37)
24. Buttock-knee depth, sitting Reference for the clearance between seat backrest and the location of obstructions in front of the knees.	Chinese (Taiwan) Russians (Moscow) U.S. Army soldiers	*530 (26)* *584 (29)* *591 (33)*	558 (31) 610 (30) 618 (31)
25. Buttock-popliteal depth, sitting Reference for the depth of a seat.	Chinese (Taiwan) Russians (Moscow) U.S. Army soldiers	*nda* *496 (29)* *485 (29)*	nda 517 (26) 503 (27)
26. Shoulder breadth (biacromial) A traditional measure for comparing population samples. Indicator of the distance between the centers of rotation of the two upper arms.	Chinese (Taiwan) Russians (Moscow) U.S. Army soldiers	*324 (25)* *360 (16)* *365 (18)*	369 (28) 397 (25) 416 (19)
27. Shoulder breadth (bideltoid) Reference for the lateral clearance required near shoulder level. Add space for ease of motion and tool use.	Chinese (Taiwan) Russians (Moscow) U.S. Army soldiers	*406 (24)* *412 (21)* *450 (29)*	460 (23) 458 (23) 510 (33)
28. Hip breadth, sitting Reference for seat width. Add space for clothing and ease of motion.	Chinese (Taiwan) Russians (Moscow) U.S. Army soldiers	*353 (23)* *372 (23)* *399 (33)*	360 (27) 362 (23) 379 (30)

(continued)

Table 11.9 (continued)

29. Span A traditional measure for comparing population samples. Reference for sideway reach.	Chinese (Taiwan) Russians (Moscow) U.S. Army soldiers	*1571 (62)* *1640 (75)* *1660 (83)*	1738 (69) 1782 (68) 1814 (85)
30. Elbow span (arms akimbo) Reference for the lateral space needed at upper body level for ease of motion and tool use.	Chinese (Taiwan) Russians (Moscow) U.S. Army soldiers	*801 (39)* *870 (38)* *nda*	894 (45) 935 (37) nda
31. Head length (depth) A traditional measure for comparing population samples. Reference for headgear size.	Chinese (Taiwan) Russians (Moscow) U.S. Army soldiers	*187 (6)* *nda* *190 (7)*	197 (7) nda 196 (7)
32. Head breadth A traditional measure for comparing population samples. Reference for headgear size.	Chinese (Taiwan) Russians (Moscow) U.S. Army soldiers	*161 (9)* *nda* *148 (5)*	167 (8) nda 154 (6)
33. Hand length A traditional measure for comparing population samples. Reference for hand tool and gear size. Consider manipulations, gloves, tool use.	Chinese (Taiwan) Russians (Moscow) U.S. Army soldiers	*167 (8)* *168 (8)* *181 (10)*	183 (10) 188 (9) 193 (10)
34. Hand breadth A traditional measure for comparing population samples. Reference for hand tool and gear size, and for an opening through which a hand must fit. Consider gloves, tool use.	Chinese (Taiwan) Russians (Moscow) U.S. Army soldiers	*75 (4)* *76 (3)* *78 (4)*	86 (5) 87 (5) 88 (4)
35. Foot length A traditional measure for comparing population samples. Reference for shoe and pedal size.	Chinese (Taiwan) Russians (Moscow) U.S. Army soldiers	*nda* *239 (11)* *246 (12)*	nda 266 (12) 271 (13)
36. Foot breadth A traditional measure for comparing population samples. Reference for shoe size, spacing of pedals.	Chinese (Taiwan) Russians (Moscow) U.S. Army soldiers	*nda* *88 (4)* *93 (5)*	nda 97 (6) 102 (5)
37. Weight (in kg) A traditional measure for comparing population samples. Reference for body size, clothing, strength, health, etc. Add weight for clothing and equipment worn on the body.	Chinese (Taiwan) Russians (Moscow) U.S. Army soldiers	*52 (7)* *60 (7)* *68 (11)*	67 (9) 71 (9) 86 (14)

compiled 20 March 2016

nda: no data available.

Sources of data:

Wang MJJ, Wang EMY, Lin YC (2002) Anthropometric data book of the Chinese people in Taiwan. Hsinchu, The Ergonomics Society of Taiwan.

Strokina AN, Pakhomova BA (1999) Anthropo-ergonomic atlas (in Russian, ISBN 5-211-04102-X). Moscow, Moscow State University Publishing House.

Gordon CC, Blackwell CL, Bradtmiller B et al. (2014) 2012 Anthropometric survey of U.S. Army personnel: Methods and summary statistics. Technical Report NATICK/ TR-15/007. Natick, U.S. Army Natick Soldier Research, Development and Engineering Center.

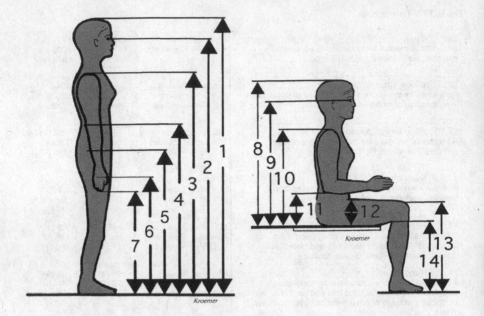

Fig. 11.12 Height measurements taken on subjects in erect postures, standing and sitting

- The Russian students (about 270) were measured during the late 1980s in Moscow, ranging in ages from 18 to 22 years.
- The US Army soldiers (more than 6000) were measured by 3D scanning between 2010 and 2012 in 12 US locations, their age spread was from 17 to 58 years.

11.6   How to Get Missing Data

Three avenues are open to obtain anthropometric information: first, searching the literature; second, conducting an anthropometric survey; third, using statistical procedures to deduce from existing data those that we need to know.

11.6.1   Finding Data in the Literature

The Internet is the most convenient source for retrieving new anthropometric information. Several categories of publications are most likely to provide reliable results. Foremost are those of anthropology research institutions, often associated with universities. The next group includes publications of government agencies (such as in the USA the *Centers for Disease Control and Prevention* CDC) or the

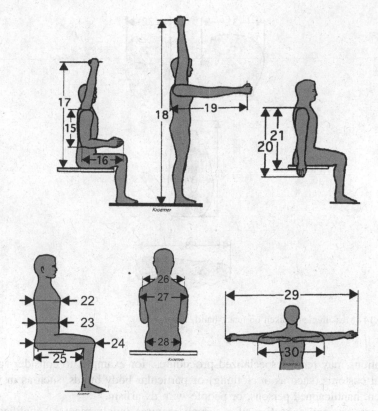

Fig. 11.13 Measurements of reaches, heights, depths, breadths, and spans

military, and of international institutions such as the United Nations and its sub-groups, for example the International Labour Organization (ILO), or the International Organization for Standardization (ISO). Another category includes journals and books that specialize on ergonomics/human factors topics. Trade groups (especially of clothing) are potential sources of data which, however, are usually collected to meet special interests. Finally, regional, national and international standards (such as ANSI and ISO) can contain anthropometric information*.

11.6.2 Conducting an Anthropometric Survey

Doing an anthropometric survey is a major enterprise and is best left to qualified anthropometrists. However, a few general remarks may be helpful for decision-making and planning.

The first task is to select the measuring technique: most likely either traditional measures "by hand" or three-dimensional computerized scanning. Special

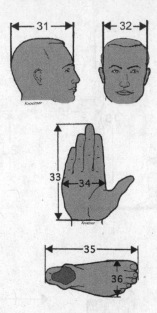

Fig. 11.14 Measurements taken on head, hand, and foot

populations may require specialized procedures, for example, to consider specific cultural customs (such as in clothing) or particular body builds, such as in young children, handicapped persons, or people with dwarfism.

Preparation and execution of a survey requires involvement of qualified staff over extended periods of time. Unbiased selection of a representative sample is a fundamentally important yet often difficult task. Special interests may determine subject selection and, hence, resulting data; for example, some survey data served military purposes whereas other measures were collected by clothing manufacturers and then deemed trade secrets. For various reasons, healthy young-to-middle-aged adults seem of higher interest than very young or old persons or those who are neither healthy nor wealthy.

The next step concerns the determination of the sample size. For expediency, one usually wishes to keep the sample as small as possible. Assuming a normal distribution of the variable to be measured, the required smallest sample size N derives from

$$N = q^2 S^2 / v^2$$

as per Eq. 11.13 in Table 11.2 with q taken from Table 11.10. S is the (known or estimated) standard deviation of the data and v is the desired accuracy of the measurement (v and S are expressed in the same units).

Table 11.10 Values of q for sample size determination

q	Statistic of interest
1.96	Mean
1.39	Standard deviation
2.46	50th percentile
2.46	45th and 55th percentile
2.19	40th and 60th percentile
2.52	35th and 65th percentile
2.58	30th and 70th percentile
2.67	25th and 75th percentile
2.80	20th and 80th percentile
3.00	15th and 85th percentile
3.35	10th and 90th percentile
4.14	5th and 95th percentile
4.46	4th and 96th percentile
4.92	3rd and 97th percentile
5.67	2nd and 98th percentile
7.33	1st and 99th percentile

An example: Table 11.11 shows that data are missing which describe the hand circumference at the knuckles. To get such information, we want to conduct measurements. We assume S to be 10 mm and want to measure with the accuracy of 1 mm, so v = 1 mm. With q = 1.96 (from Table 11.10), we calculate $N = 1.96^2 \ 10^2/1 = 384$. Accordingly, we must take measurements on at least 384 hands.

Simplified formulae* may be used to approximate minimal sample sizes: If 95% confidence and a precision of 1% of the mean is required, a sample of $N \geq (1.96 \ CV)^2$ is necessary to estimate mean values. Larger samples are required for estimating more widely spread data (with the same confidence and precision): larger spread means larger standard deviation S and thus larger CV.

Table 11.11 Hand and wrist measures, in mm (adapted from Kroemer 2017)

Hand Measures	Population	Men		Women	
		mean	SD	mean	SD
1. Length	British	180	10	175	9
	British, estimated 1986	190	10	175	9
	Chinese, Taiwan	183	10	167	8
	French	190	nda	173	nda
	Germans	189	9	174	9
	Germans, soldiers 2006	191	nda	176	nda
	Japanese	nda	nda	nda	nda
	Russians, Moscow	188	9	168	8
	US soldiers	193	10	180	10
	US Vietnamese	177	12	165	9
2. Breadth at knuckles	British	85	5	75	4
	Chinese, Taiwan	86	5	75	4
	French	86	nda	76	nda
	Germans	88	5	78	4
	Japanese	nda	nda	90	5
	Russians, Moscow	87	5	76	3
	US soldiers	88	4	78	4
	US Vietnamese	79	7	71	4
3. Maximal breadth	British	105	5	92	5
	Chinese	nda	nda	nda	nda
	French	nda	nda	nda	nda
	Germans	107	6	94	6
	Japanese	nda	nda	nda	nda
	Russians	nda	nda	nda	nda
	US soldiers	nda	nda	nda	nda
	US Vietnamese	100	6	87	6
4. Circumference at knuckles	British	nda	nda	nda	nda
	Chinese	nda	nda	nda	nda

(continued)

Table 11.11 (continued)

	French	nda	nda	nda	nda
	Germans	nda	nda	nda	nda
	Japanese	nda	nda	nda	nda
	Russians	nda	nda	nda	nda
	US soldiers	212	10	187	9
	US Vietnamese	nda	nda	nda	nda
5. Wrist circumference	British	nda	nda	nda	nda
	Chinese	nda	nda	nda	nda
	French	nda	nda	nda	nda
	Germans	nda	nda	nda	nda
	Japanese	nda	nda	nda	nda
	Russians	nda	nda	nda	nda
	US soldiers	176	9	154	8
	US Vietnamese	163	15	137	18

nda: no data available

11.6.3 Statistical Body Models

It is often desired to combine body dimensions, often called "stacking", such as determining total arm length from the lengths of lower and upper arm. For this purpose, two methods have been employed: one is fallacious whereas the other is suitable.

The percentile statistic is convenient for determining the location of any *one* given datum on its continuum range, such as 1504 mm for the 10th percentile value for female stature in Taiwan (STATp$_{10}$ = 1572 − 1.28 × 53; see Tables 11.2, 11.3, and 11.9). However, it is false to assume that all other body component measures of that imaginary 10th-percentile person must also be at their 10th percentile. For example, stacking p$_{10}$ leg length plus p$_{10}$ torso length plus p$_{10}$ head height does not add up to p$_{10}$ stature*: a person of 10th percentile stature may have relatively short legs but a long torso, or long legs and a short torso, or any other combination of torso and leg lengths.

A related fictitious model, beloved by politicians and journalists and even used by some misguided designers, is the so-called "average person", an apparition that consists of only "average" parts. Obviously, no so such ghost exists; statistically, stacking up 50th percentile body parts is just as erroneous as trying to do this with 10th or other solitary percentile sections.

However, regression equations are suitable to generate discrete body measures, provided that they employ as predictor variables other dimensions whose values are known for the population sample of interest. In this case, predicted values may be added or subtracted. For example, 10th percentile values for leg, trunk, and head heights each predicted from regressions do add up to 10th percentile stature.

11.6.4   Deducing Unknown Values from Existing Data

There are several statistical procedures to estimate data from existing related information: by ratio scaling, using regressions, and by combining data sets.

11.6.4.1   Estimation by Ratio Scaling

Ratio scaling* is one technique to estimate data from known dimensions, especially to deduce the mean of a dimension or its standard deviation. Ratio scaling relies on the assumption that, though people vary in size, their proportions are likely to be similar. If ever true at all, this conjecture may apply to body components that are interrelated. For example, many body lengths correlate highly with each other; also, many body breadths are related as a group, and so are many circumferences as a group. However, it is not true, even within their groups, that *all* body lengths (or breadths, or circumferences) correlate highly. Furthermore, many lengths are not correlated highly to breadths, and breadths are not highly related to many circumferences: see Table 11.4. Thus, one has to be very careful in attempting to derive one set of data from another.

For any ratio scaling, one should use only pairs of data that have a coefficient of correlation of at least 0.7 with each other. This "0.7 convention", already mentioned in this chapter, assures that the variability of the derived information is at least to 50% determined by the variability of the predictor. Certainly, one should not use ratio scaling if the base samples are likely to have body proportions different from those of the set to be predicted; for example, many Asian populations have proportionally shorter legs and longer trunks than populations in Europe and North America.

Provided that the data sets in population samples of x and y correlate highly, we can establish an estimated ratio scaling factor E of a desired dimension d_y in the y sample from the x sample data

- if the value d_x of that dimension in the x sample is known, and
- if values of the primary reference dimension in both samples, D_x and D_y, are known as well.

Equation 11.14 (in Table 11.2) defines the scaling factor E:

$$d_x/D_x = d_y/D_y = E$$

With $E = d_x/D_x$ known, we can calculate the desired dimension from

$$d_y = E \times D_y$$

in stepwise fashion, as follows:

Step 1   In the x sample, establish the scaling factor E between the desired dimension and a known reference dimension. The reference parameter must be common for both population samples.

Step 2   With E now known, the value of a dimension in population the y sample derives from multiplying the reference parameter in the y sample with E

Example On average, East German men (x sample) had an eye height (d_x) of 160.1 cm while their average stature (D_x) was 171.5 cm. *If West Germans (y sample) have a mean stature (D_y) of 179.5 cm, what is their approximate mean eye height d_y?*
Step 1:

$$E = d_x/D_x = 160.1/171.5 = 0.933.$$

Step 2:

$$d_y = E\,D_y = (0.933)\,179.5\,\text{cm} = 167.5\,\text{cm}.$$

This is estimated eye height for the West German sample.

For practical reasons, stature is often the reference value. However, as already mentioned, stature generally is related well with other body heights, but it is usually not correlated with body depths, breadths, circumferences, or body weight. Thus, ratio scaling must be done with great caution and careful consideration of the circumstances, especially accounting for statistical correlations.

11.6.4.2 Estimation by Regression Equation

Another way of estimating the relations among dimensions is through regression equations*. If two variables are involved, one often assumes that they relate linearly with each other. (However, this assumption should be verified). The general form of such a bivariate equation (Eq. 11.10 in Table 11.2) is

$$m_y = a + b\,m_x$$

To calculate m_y, the mean of the y sample, you must know the mean value m_x in the x sample and its coefficients a (the intercept) and b (the slope). If you want to use Eq. 11.10, keep in mind that you are assuming that the actual values of y are scattered about their mean in a normal (Gaussian) probability distribution.

11.6.4.3   Combining Anthropometric Data Sets

Occasionally one must add or subtract anthropometric values; for example, total arm length as the sum of upper and lower arm lengths.

If you add two measures (such as leg length and torso-head length), you generate a new combined distribution (here: stature). In doing so, you must take into account the existing covariation COV between the two measures. (For example, often but not always, a taller torso is associated with taller legs). The correlation coefficient r between the two data sets, x and y, and their standard deviations, Sx and Sy, describe the covariance (Eq. 11.15 in Table 11.2):

$$COV_{(x,y)} = r_{xy} S_x S_y$$

The sum of the two mean values of the two distributions Eq. (11.16) is:

$$m_z = m_x + m_y$$

The estimated standard deviation of m_z (Eq. 11.17) is:

$$S_z = \left(S_x^2 + S_y^2 + 2r\, S_x S_y \right)^{1/2}$$

The difference between two mean values (Eq. 11.18) is:

$$m_z = m_x - m_y$$

Its standard deviation (Eq. 11.19) is:

$$S_z = \left(S_x^2 + S_y^2 - 2r\, S_x S_y \right)^{1/2}.$$

Example 1   What is the average arm (acromion-to-wrist) length of an American Pilot? For a standing pilot, the 90th percentile of acromial shoulder height was 1532.0 mm and that of wrist height was 905.6 mm; at the 10th percentile, the values were 1379.5 mm and 808.6 mm, respectively. You estimate that the correlation between shoulder and wrist heights is 0.3.

First, calculate the average acromion (A) and wrist (W) heights to be able to estimate the standard deviations. The average of the 90th and 10th percentiles is at the 50th percentile.

$$m_A = (1532.0 + 1379.5)\,mm/2 = 1455.8\,mm$$

For a standing pilot, the 90th percentile is related to the standard deviation by (Eq. 11.12) p = m + kS, so

$$S_A = (p_{90} - m_A)/k$$

With k = 1.28 for the 90th percentile (from Table 11.3),

$$S_A = (1532.0 - 1455.8)\,mm/1.28 = 59.6\,mm$$

{or: S_A= (1379.5 − 1455.8) mm/(−1.28) = 59.6 mm using the 10th percentile values}
Likewise,

$$m_W = (905.6 + 808.6)\,mm/2 = 857.1\,mm$$
$$S_W = (905.6 - 857.1)\,mm/1.28 = 37.9\,mm$$

{or: S_W= (808.6 − 857.1) mm/(−1.28) = 37.9 mm using the 10th percentile values}
The average arm length (acromion-to-wrist, AW) is (per Eq. 11.18)

$$m_{AW} = m_A - m_W = 1455.8\,mm - 857.1\,mm = 598.7\,mm$$

The standard deviation of the average arm length m_{AW} is (per Eq. 11.19)

$$S_{AW} = [(59.6^2 + 37.9^2 - 2(0.3)(59.6)(37.9)]^{1/2}\,mm = 60.3\,mm$$

Example 2   What is the mass of the torso of a 75p female? You assume that the estimated mass of torso and head combined has a mean of 35.8 kg and a standard deviation of 5.2 kg. The estimated mass of the head, measured separately, has a mean of 5.8 kg with a standard deviation of 1.2 kg. The correlation between head and torso is assumed to be 0.1.
The mean torso mass is the difference of masses (Eq. 11.18):

$$mean_{torso} = 35.8\,kg - 5.8\,kg = 30.0\,kg.$$

The standard deviation is calculated from (Eq. 11.19)

$$S_{torso} = [(5.2^2 + 1.2^2 - 2(0.1)(5.2)(1.2)]^{1/2} \text{kg} = 5.2 \text{ kg}$$

The mass of a 75th percentile torso is (with $k = 0.67$ taken from Table 11.3)

$$mass_{torso75p} = 30.0 \text{ kg} + 0.67(5.2 \text{ kg}) = 33.5 \text{ kg}$$

11.6.4.4 Two-Sample Composite Population

On occasion it may be necessary to consider a population that consists of two distinct and known subsamples. For example: The task is to design for a user group "z" that consists of x % females and y % males; so, $x + y = z = 100\%$. A step-by-step procedure makes it easy to determine at what percentile of the composite population a specific value of z is:

Step 1: Determine k factors (as in Table 11.3) in the samples x and y.
For sample x: Eq. 11.12 (in Table 11.2) $p_x = m_x + k_x S_x$ yields $k_x = (p_x - m_x)/S_x$.
Similarly, for sample y: $k_y = (p_y - m_y)/S_y$.
Step 2: Obtain factor k in the combined population (per Eq. 11.20 in Table 11.2):

$$k_z = n_x k_x + n_y k_y.$$

Step 3: Determine percentile p_z associated with k_z from Table 11.3. If percentiles p are known in each group, one may simply add the proportioned percentiles (per Eq. 11.21 in Table 11.2):

$$p_z = n_x p_x + n_y p_y.$$

11.7 Using Anthropometric Data in Design

Anthropometric data describe the body sizes of people in standardized erect postures. Such information, as in Table 11.9, is basic for the design of workspaces and equipment, of tools and clothing, which must fit the human body. However, to utilize this information properly, anthropometric data need proper interpretation and, often, adjustment.

11.7.1 The "Normative" Adult

Many designers have acquired the questionable habit of designing for a so-called regular adult who possesses

- "normal" anthropometry, with all body dimensions (such as stature, hand reach or weight) close to their mean values;
- "normal" physiological functions, such as of the metabolic, circulatory and respiratory subsystems; whose nervous control functions, sensory capabilities and intelligence are all near "average".

In reality, however, hardly any person exists who is average in all or even many respects. Instead, "extra-ordinary" persons and population subgroups abound: very big or small individuals, temporarily or permanently impaired persons, women during their pregnancy, children and juveniles, or elderly people. So, instead of using the imaginary "normal adult" as the design prototype, ergonomists must consider the variability that exists naturally among people of different body sizes, genders, ages, and abilities*.

11.7.2 Body Positions and Motions at Work

It is difficult to maintain a given posture over long periods of time. Standing and sitting without moving, or even lying still, all quickly become uncomfortable and, with time, physically impossible; if enforced by injury or sickness, circulatory and metabolic functions become impaired, and bed sores appear. The human body is made for motion. Our articulations have various degrees of freedom to move—see, for example, Chap. 1 for data on maximal angular displacements in major body joints. Not moving has distinct and generally negative physical consequences.

In the past, the same erect body postures (sitting and standing) that subjects assume for anthropometric measurements often served as design models: probably because upright postures are easy to visualize and readily made into design templates. Furthermore, late in the 19th century, orthopedists had vigorously promoted the upright posture for standing and especially for sitting in a school or office. In spite of the counter-argument that the rigidly erect posture, such as commanded by military officers and school teachers, required tiring muscular efforts, most of the 20th century office furniture was constructed to enforce upright sitting. However, even in the 1890s, it was recognized that the human body needs to vary its posture; Fig. 11.15 shows one design of school furniture hat featured adjustable seats and footrests so that the pupil could stand or sit, even "perch" halfway between standing and sitting.

People move and stretch and reach to do their jobs, and on occasion they like to let their bodies unwind and "slump", especially when a supportive chair allows this, as illustrated in Figs. 11.16 and 11.17. Current work seats are designed to support upright as well as relaxed postures because they fit the body, especially the

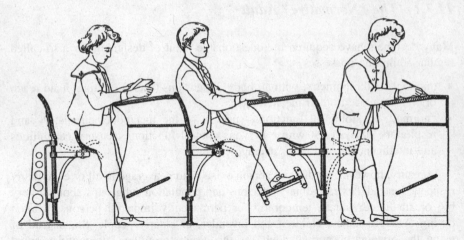

Fig. 11.15 Schindler's circa 1890 design of school furniture (adapted from Kroemer 2006)

Fig. 11.16 People sit and
move as as they wish (adapted
from Kroemer and Kroemer
2005)

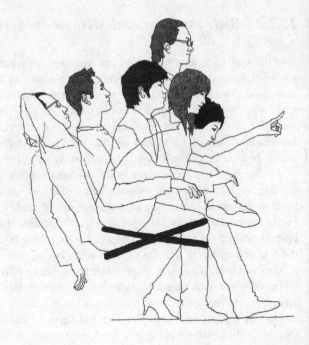

Fig. 11.17 Office chair that "fits" the body, especially the curvature of the back, and supports bones and tissues of the pelvic region and of the upper thighs. The seat is adjustable in height. The seat pan angle can decline and incline. The backrest can also assume different angular positions

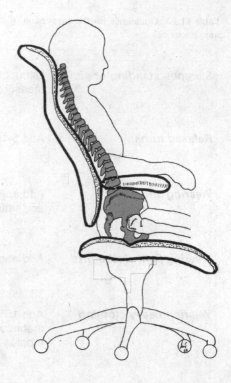

curvatures of the back, and they have easily adjusting heights and angles of seat pan and back rest.

The design of modern equipment reflects this everyday experience; in offices, in vehicles and airplanes, drivers and pilots and passengers want to sit comfortably, even if the available space is tight. Table 11.12 lists design suggestions for adjusting standard anthropometric information (as in Table 11.9) to accommodate functional motions and postures.

"Convenient" mobility is somewhere within the range of possible extreme postures. In everyday activities, most movements of the trunk, the head, the elbows, and the wrist occur in the midrange of mobility, as do motions of the hips and knees when sitting. In contrast, hip and knee movements of a person walking or standing are usually close to full stretch: Table 11.13 contains estimates of everyday motion ranges. Task requirements, work skills and habits affect these preferred ranges of motion—in fact, quite differing work postures are common for various tasks and in different regions, as a comparison of Figs. 11.17 and 11.18 readily demonstrates.

Table 11.12 Guidelines for the conversion of standard measuring postures to functional stances and dimensions

Slumped standing or sitting	Deduct 5-10% from relevant height measurements.
Relaxed trunk	Add 5-10% to trunk circumferences and depths.
Wearing shoes	Add approximately 25 mm to relevant standing and sitting heights; more for "high heels".
Wearing light clothing	Add about 5% to relevant dimensions.
Wearing heavy clothing	Add 15% or more to relevant dimensions. (Note that heavy clothing may strongly reduce mobility.)
Extended reaches	Add 10% or more to relevant reach measures for strong motions of the trunk.
Use of hand tools	Center of handle is at about 40% of hand length, measured from the wrist.
Forward bending of head, neck and trunk	Ear-Eye Line declines to near horizontal.
Comfortable seat height	Subtract up to 10% from popliteal height.

Table 11.13 Mobility ranges at work

Body joints/parts	Angles/positions	
Shoulder	Mostly mid-range, upper arm often hanging down	
Elbow	Mostly mid-range, at about 90 degrees	
Wrist	Mostly mid-range, about straight	
Neck/Head	Mostly mid-range, about straight	
Back	Near complete stretch, about erect	
	When walking or standing	**When sitting**
Hip (side view)	Near complete stretch, at about 180 degrees	Mostly mid-range, at about 90 degrees
Knee	Near extreme stretch, slightly less than or near 180 degrees	Mostly mid-range, at about 90 degrees

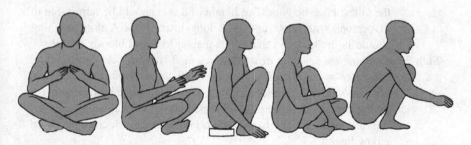

Fig. 11.18 On-the-ground work postures (adapted from Kroemer et al. 2003; Kroemer Elbert et al. 2018)

11.8   Designing to Fit the Body

Humans have all heads and trunks, arms and legs, but these body parts come in various sizes and in different proportions. As already discussed, compilations of anthropometric data (except weight) usually show normal (Gaussian) distributions. This fortunate fact allows the statistical information to be described by the simple statistical descriptors *mean* (same as *average*) and *standard deviation*.

Yet, misunderstanding and misuse have led to the false idea that one could "design for the average"; by its very definition, the mean value is larger than half the data, and smaller than the other half. Consequently, an average-sized product is inherently, by design, either too large or too small for its users. Furthermore, it is unlikely ever to encounter a person who displays mean values in several, many or even all dimensions*.

Useful steps in designing for fitting clothing, tools, workstations, and equipment to the body are as follows:

Step 1: *Select those anthropometric measures that directly relate to important design dimensions.* Some examples: hand length relates to handle size; knee height and hip breadth relate to the leg room in a console; shoulder and hip breadth relate to escape-hatch diameter; head length and breadth relate to helmet size; eye height relates to the heights of windows and displays; stature relates to the height of a door frame.

Step 2: *For each of these pairings, determine whether the design must fit only one given percentile (such as minimal or maximal) of the body dimension, or a range along that body dimension.* Some examples: the escape hatch must be big enough to clear the extreme largest values of shoulder breadth and hip breadth, enlarged by clothing and equipment worn. The handle size of pliers should probably best fit a smallish hand. The legroom of a console must accommodate the tallest knee heights. The height of a seat should be adjustable to fit persons with short and with long lower legs. A door opening should be higher than the tallest person to avoid bloody scalps.

Step 3: *Combine all selected design values in a carefully devised sample, computer model, mock-up, or drawing to ascertain that they are compatible.* An example: a tall legroom clearance height, needed for sitting persons with long lower legs, may be very close to, or even interfere with, the height of the working surface, which depends on elbow height.

Step 4: *Determine whether one design will fit all users—if not, several sizes or adjustment must be provided to accommodate the users.* Some examples: one extra-large bed size fits any sleeper; gloves and shoes must come in different sizes; seat heights should be adjustable.

Selecting appropriate percentile values of the critical dimensions is critical for successful design. There are two ways to determine percentile values. One approach is to use a graph of the distribution of data to measure, count or estimate critical percentile values. This works well whether the distribution is normal, skewed, binomial, or in any other form. Fortunately, normally distributed anthropometric data provide the second, even easier and more exact approach: to calculate percentile values from mean m and standard deviation S—see Table 11.2.

To determine a single (distinct) percentile point:
  (a) Select the desired percentile value to be calculated;
  (b) Determine the associated k value from Table 11.3;
  (c) Calculate the p value from $p = m + k\,S$. (Note that k and hence the product kS may be negative).

To determine a range:

1(a) Select upper percentile p_{max} that defines the upper end of the desired range;
1(b) Find related k_{max} value in Table 11.3;
1(c) Calculate upper percentile value $p_{max} = m + k_{max}\,S$;
2(a) Select lower percentile p_{min} that defines the lower end of the desired range; (Note that the two percentile values need not be at the same distance from the 50th percentile—in other words, the range does not have to be "symmetrical to the mean").
2(b) Find related k_{min} value in Table 11.3;
2(c) Calculate lower percentile value $p_{min} = m - kS_{min}$
  3. Determine range R = $p_{max} - p_{min}$.

11.8.1 Determining Clothing Size Tariffs

Often, it is useful to divide a distribution of body dimensions into certain groups, such as in establishing clothing sizes. An example is the use of neck circumference to establish selected collar sizes for men's shirts. The first step is to establish the ranges (see above) which shall be covered by the size groupings. The second step is to associate other body dimensions with the primary one, such as chest circumference, or sleeve length, with collar (neck) circumference. Establishing size tariffs can become a rather complex procedure* because the combination of body dimensions (and their derived equipment dimensions) depends on correlations among these dimensions, as already shown in Table 11.4.

Fig. 11.19 Convenient and
extended reaches (adapted
from Proctor and van Zandt
1994)

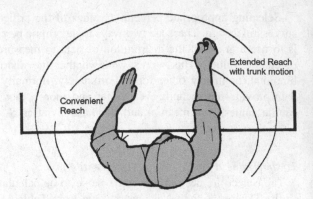

11.8.2   Determining the Workspace of the Hands

The human enjoys great hand dexterity, with shoulder, elbow and wrist joints
together providing extensive freedom of sweep. Classical anthropometric data
provide some information on reach capabilities with the body in standardized static
postures—see Chap. 1. Preferred work areas of the hands are in front of the trunk,
within curved range-of-motion envelopes that reflect the mobility of the forearm
from the elbow joint and of the total arm from the shoulder joint. Thus, these
envelopes often appear as partial spheres around body joints. Figure 11.19 depicts
the "convenient" reach area that requires mostly forearm movements while the
"extended" reach space requires motions of the completely stretched arm, and
possibly even trunk movement.

11.9   Human-Centered Engineering

In the field of ergonomics, the task of the human factors engineer concerns
designing, constructing, and operating the work processes, structures, machines,
devices and tools used in industry and in everyday life. The primary goals are to
facilitate achieving the task without unnecessary effort—see Fig. 11.20—while
assuring a safe work environment.

Anthropometric data provide the basic measurements needed to design equipment and tools to accommodate the human body. Together with knowledge about
mobility and physical capabilities (such as muscular, circulatory, metabolic—see
the related Chaps. 2, 6, and 7 in this book), the engineer can achieve usability and
usefulness of the design.

Using anthropometric information as design inputs can be as basic as making
door openings so high that nobody strikes their head on the frame—not even in the
Netherlands or in Finland with their tall populations. Other simple design tasks may
involve elevating a shorter operator to the proper working height, as in Fig. 11.21,
or adjusting the height of workbenches to fit different operators and work tasks,
shown in Fig. 11.22.

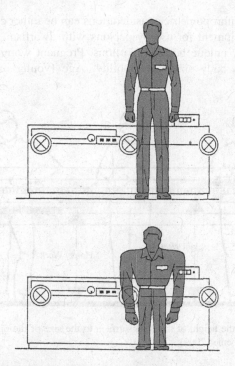

Fig. 11.20   Re-designing the operator or the machine? A lathe with its real and imagined operator (adapted from Eastman Kodak Company 1983)

Fig. 11.21   Providing a platform to stand on can be helpful to a short operator of a machine which was designed for a tall user—but note the danger of stumbling. Providing equipment that can be adjusted to fit the user is a better solution (adapted from ILO International Labour Office 1986)

In other cases, anthropometric considerations can be rather complicated, such as in constructing equipment for use by persons with dwarfism, because this population has distinctly unique body proportions. Pregnant women experience (temporary) changes in body size and mobility. Age (young or old) and medical

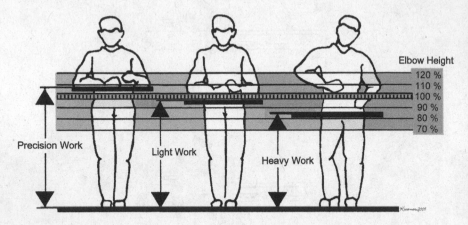

Fig. 11.22 Adjusting the height of the work surface to the size of the operator and to the work task (adapted from Kroemer 2009)

Fig. 11.23 Sitting comfortably—feeling good at work (adapted from Kroemer and Kroemer 2005; Kroemer 2017)

conditions may also affect anthropometric considerations. Environmental conditions can make the engineering design task complex, such as when building cockpits for high-performance aircraft or designing space suits. Often, the design intentions go beyond fit and usability and also include addressing user sentiments such as aesthetic appeal, satisfaction and comfort. Clearly, achieving any ergonomic design goal, from utility to pleasure, requires processes and products that "fit" the human (Fig. 11.23).

Notes

The text contains markers, *, to indicate comments and references, which follow:

Anthropometric methods and practices: Martin (1914), Hertzberg (1968), Roebuck et al. (1975), Gould (1981), Lohman et al. (1988), Gordon et al. (1989, 2014), Cheverud et al. (1990), Roebuck (1995), Li and Paquette (2019); ISO 7250-1:2008; ISO 15535:2012; ISO 20685:2010.

Compilations of measured data are available for some populations: Data in Garrett and Kennedy (1971), NASA/Webb (1978), Peebles and Norris (1998), Kroemer (1989, 2006, 2009), Kroemer et al. (2003, 2018), ISO/TR 7250-2:2010.

Avatars: Thaler et al. (2018), Pujades et al. (2019).

Make scan data more consistent with traditionally measured values: Han et al. (2010), Li and Paquette (2019).

Anthropometric information about soldiers is extensive and reaches far into the past: The NASA/Webb Anthropometric Sourcebook (1978) provides an overview of the data available up to the mid-1970s.

Dimensions of the head, hand, and foot ... are similar in military and civilian populations: McConville, Robinette and Churchill (1981).

The most common statistical procedures: For more information on the use of statistics in anthropometry see texts on statistics: for example, Hinkelmann and Kempthorne (1994, 2005), Williges (2007) and texts on anthropometry: such as Roebuck et al. (1975), NASA/Webb (1978), Gordon et al. (1989, 2014), Cheverud et al. (1990), Roebuck (1995), Marras and Karwowski (2006), Pheasant and Haslegrave (2006.

Regarding peakedness (kurtosis): Strictly, kurtosis measures outliers: https://www.spcforexcel.com/knowledge/basic-statistics/are-skewness-and-kurtosis-useful-statistics

Correlation coefficients r among body dimensions of US Air Force personnel ...older tables on soldiers: see the 1997 3rd edition of this book and Roebuck, Kroemer, Thomson (1975).

The easily measured stature is not a good predictor for most other body dimensions: An unpublished 1966 report (the authors may remain unnamed here) contained ratios between stature and many other body segments, even though most of them do not correlate well with stature. Since the ratios were given with three decimals, that misleading model seemed exact and has made ghostly re-appearances in the literature ever since its original misconception.

The weight distribution is non-normal: Statistical procedures are at hand for converting non-normal distributions into normal ones, especially the Box-Cox conversion: Freeman and Modarres (2006).

The validity of several assumptions regarding height-weight ratios was questioned: As discussed in detail by Andres (1984) and Speakman (1997).

A set of listed BMI ranges to assess whether the person is underweight...: https://www.cdc.gov/healthyweight/assessing/bmi/adult_bmi/index.html#Used.

Interpret the meaning of the BMI number: Prentice and Jebb (2001).

One in four will be over 65 years old: United Nations (2019).

Anthropometric differences among several groups of ethnic origins: Cheverud, Gordon, Walker et al. (1990).

Firefighters, police, guards are taller and heavier than Americans in all other occupations: Hsiao et al. (2002).

Truck drivers are heavier than the US general population: Guan, Hsiao, Bradtmiller et al. (2012).

Anthropometry of several ethnic, national, and geographic populations: older collections were listed in earlier editions of this book. Recent collections of measured and of estimated anthropometry data: Peebles and Norris (1998), Pheasant and Haslegrave (2006), Kroemer (2006, 2009, 2017).

International standards contain anthropometric information: ISO TR 7250; https://www.iso.org/standard/41249.html.

Simplified formulae may be used to establish minimal sample sizes: Gordon et al. (2014, pp 377–378) who refer to Zar (2010), Sokal and Rohlf (2011); ISO 15535:2012.

Stacking leg length plus torso length plus head height does not add up to stature: Robinette and McConville (1981).

Ratio scaling: Roebuck et al. (1975), Pheasant (1982), Roebuck (1995).

Estimating the relations among dimensions through regression equations: The estimation of body dimensions of American soldiers by Cheverud et al. (1990) is an example of this procedure. Roebuck (1995) discussed the implications in some detail, including the extension of this concept to develop multivariate regression equations, as well as to principal component analyses and boundary description analyses.

Consider the variability that exists naturally among "extra-ordinary" people of different body sizes, genders, ages, and abilities: Steenbekkers and Molenbroek (1990), Bradtmiller (2000, 2003), Froufe et al. (2002), Rebelo (2003), Kroemer (2006), Lueder and Rice (2007), Jones et al. (2018), Wiggermann et al. (2019) discuss how "extra-ordinary" populations can deviate in size, strength, or performance capabilities from the normative adult.

It is unlikely ever to encounter a person who displays mean values in several, many or even all dimensions: Kroemer et al. (2003), Kroemer (2006, 2009), Robinette and Hudson (2006).

Establishing size tariffs can become a rather complex procedure: Roebuck et al. (1975), McConville, Chap. 8 in NASA/Webb (1978), Robinette and McConville (1981), Roebuck (1995), Robinette and Veith (2019).

Summary

Traditional techniques for measuring the human body rely on sets of simple measuring scales, such as rods and tapes. New developments use computer-based three-dimensional electronic scanning techniques.

Many currently available data describe soldiers, traditionally males. Data describing large samples of civilians are still scarce.

Statistical variations in body dimensions result from differences among individuals, from secular body size development, aging, changes in health and fitness, and from changing local population composition such as due to immigration. Most changes influence the anthropometry of an entire population only subtly and slowly. However, in fitting equipment to a defined population subgroup, the engineering designer must consider the body dimensions that are specific to the group.

Body proportions are vastly different among individuals. Hence, neither generalized body types (somatotypes) nor single-percentile phantoms (such as the "average person") are suitable means to describe body dimensions or capabilities for engineering purposes. No single-percentile person exists; design for the average body fits nobody. Instead, ranges of body sizes (and of mobility, strengths, and other capabilities) establish suitable design criteria. Appropriate statistical methods may be applied to valid anthropometric data to generate useful data for input to design of tasks and equipment.

Glossary

Abduct To move away from the body or one of its parts; opposite of adduct.

Acromion The highest point on the lateral edge of the scapula above the shoulder joint. The acromion is often used as landmark in anthropometry: Acromial height is usually equated with shoulder height.

Adduct To move towards the body; opposite of abduct.

Anterior In front of the body; toward the front of the body; opposed to posterior.

Anterior In front of the body; toward the front of the body; opposed to posterior.

Anthropometry The systematic collection and interpretation of measurements of the human body.

Articulation Joint between bones.

Atlas The top cervical vertebra, supporting the skull.

Axis Center line of an object; midline about which rotation occurs

Bending See moment.

Breadth A straight-line, point-to-point horizontal measurement running across the body or a segment.

Canthus A corner or angle formed by the meeting of the eyelids.

Carpus The wrist bones, collectively.

Cervical Part of/pertaining to/the cervix (neck), especially the seven vertebrae at the top of the spinal column.

Circumference A closed measurement that follows a body contour; hence the measurement usually is not circular.

Compression The pressure (strain) generated in material caused by two opposing forces; opposite of tension.

Coronal plane Any vertical plane at right angles to the midsagittal plane (same as frontal plane).

Coccyx (Or tailbone) a triangular bone of fused vestigial vertebrae at the lower end of the spine.

Curvature A point-to-point measurement following a contour; this measurement is neither closed nor usually circular.

Density Mass of material per unit volume.

Depth A straight-line, point-to-point horizontal measurement running fore-aft the body.

Digit The thumb and the four fingers of the hand.

Distal Away from the center, peripheral; opposite of proximal.

Distance A straight-line, point-to-point measurement between landmarks on the body.

Dominant The hand or foot exclusively used for certain actions.

Dorsal Toward the back or spine; also pertaining to the top of hand or foot, opposite of palmar, plantar, and ventral.

Ear-Eye line An easily established reference line for the tilt angle of the head. It runs through the right meatus (ear hole) and the right external canthus (meeting corner of the eye lids). The EE line is angled about 11 degrees above the Frankfurt plane; see there.

Ergonomics The application of scientific principles, methods and data drawn from a variety of disciplines to the design of engineered systems in which people play significant roles.

Extend To move adjacent segments so that the angle between them is increased, as when the leg is straightened; opposite of flex.

External Away from the central long axis of the body; the outer portion of a body segment.

Flex To move a joint in such a direction as to bring together the two parts which it connects, as when the elbow is bent; opposite of extend.

Flexibility Term occasionally used instead of mobility.

Force In physics, a vector that can accelerate a mass. As per Newton's second law, the product of mass and acceleration; the proper unit is the Newton, with $1 \text{ N} = 1 \text{ kg m s}^{-2}$. On Earth, one kg applies a (weight) force of 9.81 N (1 lb exerts 4.44 N) to its support due to gravity. Muscular force often is described as tension (pressure or stress) multiplied by transmitting cross-sectional area.

Frankfurt Plane The former standard horizontal plane for orientation of the head. The plane is established by a line passing through the right tragion (approximate ear hole) and the lowest point of the right orbit (eye socket), with both eyes on the same level: see Ear-Eye line.

Frontal plane (Or coronal plane) any vertical plane at right angle to the mid-sagittal plane.

Height A straight-line, point-to-point vertical measurement.

Inferior Below, lower, in relation to another structure.

Internal Near the central long axis of the body; the inner portion of a body segment.

Lateral Lying near or toward the sides of the body; opposite of medial.

Lordosis Forward curvature of the spine; opposite of kyphosis.

Lumbar Part of/pertaining to/the five vertebrae atop the sacrum.

Medial Lying near or toward the midline of the body; opposite of lateral.

Medial plane The vertical plane which divides the body into right and left halves; same as mid-sagittal plane.

Mid-sagittal plane The vertical plane which divides the body into right and left halves; same as medial plane.

Mobility The ability to move segments of the body.

Moment The product of force and its lever arm when trying to rotate or bend an object; the stress in material generated by two opposing forces that try to bend the material about an axis perpendicular to its long axis; see also Torque.

Orbit The eye socket.

Palmar Pertaining to the palm (inside) of the hand; opposite of dorsal.

Patella The kneecap.

Pelvis The bones of the "pelvic girdle" consisting of illium, pubic arch and ischium which compose either half of the pelvis.

Phalanges The bones of the fingers and toes (singular, phalanx).

Physis The body as distinguished from the mind.

Plantar Pertaining to the sole of the foot.

Popliteal Pertaining to the ligament behind the knee or to the part of the leg behind the knee.

Posterior Pertaining to the back of the body; opposite of anterior.

Protuberance Protruding part of a bone.

Proximal The (section of a) body segment nearest the head (or the center of the body); opposite of distal.

Radius The bone of the forearm on its thumb side.

Reach A point-to-point measurement following the long axis of the arm or leg.

Sacrum A triangular bone of fused rudimentary vertebrae (S1-S5) near the lower end of the spine between the lumbar spine and coccyx.

Sagittal Pertaining to the medial (mid-sagittal) plane of the body, or to a parallel plane.

Scapula The shoulder blade.

Secular Referring to events that appear over long stretches of time.

Somatotyping Categorizing body builds into different types. (Greek *soma*, body)

Spine The column of vertebrae.

Spinal process of a vertebra The posterior prominence.

Stature Height of the standing human body.

Sub A prefix designating below or under.

Superior Above, in relation to another structure; higher.

Supra Prefix designating above or on.

Tailbone (or sacrum, coccyx) triangular bone of fused rudimentary vertebrae at the lower end of the spine.

Tension The strain in material generated by two opposing forces that try to stretch the material; opposite of compression.

Thoracic Part of/pertaining to/the thorax (chest), especially the twelve vertebrae in the middle of the spinal column.

Torsion The stress in material generated by two opposing forces that try to twist the material about its long axis; see also moment.

Torque A force applied at a lever arm that twists or rotates something about its central axis; see also Moment

Tragion Conical eminence of the auricle (pinna, external ear) in the front wall of the ear hole.

Transverse plane Horizontal plane through the body, orthogonal to the medial and frontal planes.

Ventral Pertaining to the anterior (abdominal) side of the trunk.

Vertebra A bone of the spine.

Vertex The top of the head.

Volar Pertaining to the sole of the foot or the palm of the hand.

References

ISO 7250-1:2008 Basic human body measurements for technological design—Part 1: Body measurement definitions and landmarks. International Organization for Standardization. ISO, Geneva

Andres R (1984) Mortality and obesity. the rationale for age-specific height-weight tables. In: Andres R, Bierman EL, Hazzard WR (eds) Principles of geriatric medicine. McGraw-Hill, New York, pp 311–318

Bradtmiller B (2000) Anthropometry for persons with disabilities: needs in the twenty-first century. Paper presented at RESNA 2000 annual conference and research symposium, Orlando, FL. Arlington, Rehabilitation Engineering and Assistive Technology Society of North America

Bradtmiller B (2003) Anthropometry of users of wheeled mobility aids: a critical review of recent work. www.ap.buffalo.edu, Accessed 11 Jan 2006

Cheverud J, Gordon CC, Walker RA, Jacquish C, Kohn L, Moore A, Yamashita N (1990) 1988 Anthropometric survey of US Army personnel. Technical reports 90/031 through 90/036. US Army Natick Research, Development, and Engineering Center, Natick

Eastman Kodak Company (ed) (1983) Ergonomic design for people at work. Van Nostrand Reinhold, New York

Fluegel B, Greil H, Sommer K (1986) Anthropologischer atlas. Tribuene, Berlin

Freeman J, Modarres R (2006) Inverse Box–Cox: the power-normal distribution. Stat Probab Lett 76(8):764–772

Froufe T, Ferreira F, Rebelo F (2002) Collection of anthropometric data from primary schoolchildren. In: 166–171 in Proceedings of CybErg 2002, the 3rd international cyberspace conference on ergonomics. London, International Ergonomics Association Press; and personal communications by F. Rebelo, February 2003

Garrett JW, Kennedy KW (1971) A collation of anthropometry. AMRL-TR-68-1. Aerospace Medical Research Laboratories, Wright-Patterson Air Force Base

Gordon CC, Churchill T, Clauser CE, Bradtmiller B, McConville JT, Tebbetts I, Walker RA (1989) 1988 Anthropometric survey of US Army personnel. Summary statistics interim report. Technical Report NATICK/TR-89/027. U.S. Army Natick Research, Development and Engineering Center, Natick

Gordon CC, Blackwell CL, Bradtmiller B, Parham JL, Barrientos P, Paquette SP, Corner BD, Carson JM, Venezia JC, Rockwell BM, Mucher M, Kristensen S (2014) 2012 Anthropometric survey of U.S. Army personnel, methods and summary statistics. Technical Report Natick/TR-15/007. U.S. Army Natick Soldier Research, Development and Engineering Center, Natick

Gould SJ (1981) The mismeasure of man. Norton, New York

Guan J, Hsiao H, Bradtmiller B, Kau TY, Reed MR, Jahns SK, Loczi J, Hardee HL, Piamonte DP (2012) U.S. truck driver anthropometric study and multivariate anthropometric models for cab designs. Hum Factors. 2012 Oct; 54(5):849–71

Han H, Nam Y, Choi K (2010) Comparative analysis of 3D body scan measurements and manual measurements of size Korea adult females. Int J Ind Ergon 40:530–540

Hertzberg HTE (1968) The conference on standardization of anthropometric techniques and terminology. Am J Phys Anthropol 28:1–16

Hinkelmann K, Kempthorne O (1994) Design and analysis of experiments. In: Introduction to experimental design, vol 1. Wiley, New York

Hinkelmann K, Kempthorne O (2005) Design and analysis of experiments. In: Advanced experimental design. Wiley-Interscience, New York

ILO International Labour Office (ed) (1986) Introduction to work study. 3rd ed. International Labour Office, Geneva

ISO 15535:2012 General requirements for establishing anthropometric databases. International Organization for Standardization. ISO, Geneva

ISO 20685:2010 3-D scanning methodologies for internationally compatible anthropometric databases, International Organization for Standardization. ISO, Geneva

ISO/TR 7250-2:2010 Basic human body measurements for technological design—Part 2: Statistical summaries of body measurements from national populations. International Organization for Standardization. ISO, Geneva

Jones MLH, Ebert SM, Reed MP, Klinich KD (2018) Development of a three-dimensional body shape model of young children for child restraint design. Comput Methods Biomech Biomed Eng 21(15):784–794. https://doi.org/10.1080/10255842.2018.1521960

Kroemer KHE (1989) Engineering anthropometry. Ergonomics 32:767–784

Kroemer KHE, Kroemer AD (2005) Office ergonomics. Korean ed. Kukje, Seoul

Kroemer KHE (2006) "Extra-ordinary" ergonomics: how to accommodate small and big persons, the disabled and elderly, expectant mothers and children. CRC Press, Boca Raton

Kroemer KHE (2009, 2017). Fitting the human. 6th (7th) ed. CRC Press, Boca Raton

Kroemer Elbert KE, Kroemer HB, Kroemer Hoffman AD (2018) Ergonomics: how to design for ease and efficiency, 3rd edn. London, Academic Press

Kroemer KHE, Kroemer HB, Kroemer-Elbert KE (2003) Ergonomics: how to design for ease and efficiency, 2nd edn. Prentice-Hall, Pearson Education, Upper Saddle River

Li P, Paquette S (2019) Predicting anthropometric measurements from 3D body scans: methods and evaluation. In: Karwowski W, Trzcielinski S, Mrugalska B (eds) Advances in manufacturing, production management and process control. AHFE 2019. Advances in intelligent systems and computing, vol 971. Springer, Cham. https://doi.org/10.1007/978-3-030-20216-3_52

Lohman TG, Roche AF, Martorell R (eds) (1988) Anthropometric standardization reference manual. Human Kinetics, Champaign

Lueder R, Rice VB (eds) (2007) Ergonomics for children. CRC Press, Boca Raton

Marras WS, Karwowski K (eds) (2006) The occupational ergonomics handbook. In: Fundamentals and assessment tools for occupational ergonomics, 2nd ed. CRC Press, Boca Raton

Martin R (1914) Lehrbuch der anthropologie. 1st ed. (In German). Fischer, Jena

McConville JT, Robinette KM, Churchill T (1981) An anthropometric data base for commercial design applications. Final Report NSF DAR-80 09 861. Anthropology Research Project, Yellow Springs

NASA/Webb (eds) (1978) Anthropometric sourcebook, vol 3. NASA Reference Publication 1024. LBJ Space Center, Houston

Peebles L, Norris B (1998) Adultdata. The handbook of adult anthropometric and strength measurements—data for design safety. DTI/Pub 2917/3 k/6/98/NP. Department of Trade and Industry, London

Pheasant S (1982) A technique for estimating anthropometric data from the parameters of the distribution of stature. Ergonomics 25:981–992

Pheasant S, Haslegrave CM (2006) Anthropometry, ergonomics and the design of work. Taylor & Francis, London

Prentice AM, Jebb SA (2001) Beyond body mass index. Obes Rev 2:141–147

Proctor RW, Van Zandt T (1994) Human factors in simple and complex systems. Allyn and Baco, Boston

Pujades S, Mohler BJ, Thaler A, Tesch J, Mahmood N, Hesse N, Buelthoff HH, Black MJ (2019) The virtual caliper: rapid creation of metrically accurate avatars from 3D measurements. IEEE Trans Vis Comput Grap PP(99):1. https://doi.org/10.1109/tvcg.2019.2898748

Robinette KM, Hudson JA (2006) Anthropometry. Chapter 12. In: Salvendy G Handbook of human factors and ergonomics. 3rd ed. Wiley, New York

Robinette KM, McConville JT (1981) An alternative to percentile models. SAE Technical Paper 810217. Society of Automotive Engineers, Warrendale

Robinette K, Veitch DE (2019) Use of anthropometry and fit databases to improve the bottom-line. In: Karwowski W, Trzcielinski S, Mrugalska B (eds) Advances in manufacturing, production management and process control. AHFE 2019. Advances in intelligent systems and computing, vol 971. Springer, Cham. https://doi.org/10.1007/978-3-319-96065-4_49

Roebuck JA, Kroemer KHE, Thomson WG (1975) Engineering anthropometry methods. Wiley, New York

Roebuck JA (1995) Anthropometric methods—designing to fit the human body. Human Factors and Ergonomics Society, Santa Monica

Roebuck JA, Kroemer KHE, Thomson WG (1975) Engineering anthropometry methods. Wiley, New York

Sokal RR, Rohlf FJ (2011) Biometry: the principles and practices of statistics in biological research, 4th edn. Freeman, New York

Speakman JL (1997) Doubly labelled water: theory and practice. Chapman & Hall, London

Steenbekkers LPA, Molenbroek JFM (1990) Anthropometric data of children for non-specialist users. Ergonomics 33:421–429

Strokina AN, Pakhomova BA (1999) Anthropo-ergonomic atlas. In Russian, ISBN 5-2111-04102-X. Moscow State University Publishing House, Moscow. Also: Arsen Purundjan, personal communication, 18 Sept 2002

Thaler A, Piryankova I, Stefanucci JK, Pujades S, de la Rosa S, Streuber S, Romero J, Black MJ, Mohler BJ (2018) Visual perception and evaluation of photo-realistic self-avatars from 3D body scans in males and females. Front. ICT. https://doi.org/10.3389/fict.2018.00018

United Nations, News Release (2019) New York, UN Department of Economic and Social Affairs

Wang MJJ, Wang EMY, Lin YC (2002) Anthropometric data book of the Chinese people in Taiwan. The Ergonomics Society of Taiwan, Hsinchu

Wiggermann N, Bradtmiller B, Bunnell S, Hildebrand C, Archibeque J, Ebert S, Reed MP, Jones MLH (2019) Anthropometric dimensions of individuals with high body mass index. Hum Factors 2019 Apr 29:18720819839809. https://doi.org/10.1177/0018720819839809. [Epub ahead of print]

Williges RC (2007) CADRE: computer-aided design reference for experiments, Electronic Book CD-ROM-07-01. Virginia Polytechnic Institute and State University, Blacksburg

Zar JH (2010) Biostatistical analysis, 5th edn. Prentice Hall, Englewood Cliffs

Index

© Springer Nature Switzerland AG 2020
K. H. E. Kroemer et al., *Engineering Physiology*,
https://doi.org/10.1007/978-3-030-40627-1

Printed in the United States
by Baker & Taylor Publisher Services